EMS Tracts in Mathematics 13

EMS Tracts in Mathematics

This series includes advanced texts and monographs covering all fields in pure and applied mathematics. *Tracts* will give a reliable introduction and reference to special fields of current research. The books in the series will in most cases be authored monographs, although edited volumes may be published if appropriate. They are addressed to graduate students seeking access to research topics as well as to the experts in the field working at the frontier of research.

1 Panagiota Daskalopoulos and Carlos E. Kenig, *Degenerate Diffusions*
2 Karl H. Hofmann and Sidney A. Morris, *The Lie Theory of Connected Pro-Lie Groups*
3 Ralf Meyer, *Local and Analytic Cyclic Homology*
4 Gohar Harutyunyan and B.-Wolfgang Schulze, *Elliptic Mixed, Transmission and Singular Crack Problems*
5 Gennadiy Feldman, *Functional Equations and Characterization Problems on Locally Compact Abelian Groups*
6 Erich Novak and Henryk Woźniakowski, *Tractability of Multivariate Problems. Volume I: Linear Information*
7 Hans Triebel, *Function Spaces and Wavelets on Domains*
8 Sergio Albeverio et al., *The Statistical Mechanics of Quantum Lattice Systems*
9 Gebhard Böckle and Richard Pink, *Cohomological Theory of Crystals over Function Fields*
10 Vladimir Turaev, *Homotopy Quantum Field Theory*
11 Hans Triebel, *Bases in Function Spaces, Sampling, Discrepancy, Numerical Integration*
12 Erich Novak and Henryk Woźniakowski, *Tractability of Multivariate Problems. Volume II: Standard Information for Functionals*

Laurent Bessières
Gérard Besson
Sylvain Maillot
Michel Boileau
Joan Porti

Geometrisation of 3-Manifolds

European Mathematical Society

Authors:

Laurent Bessières and
Gérard Besson
Université Joseph Fourier, Grenoble
Institut Fourier – UMR 5582 CNRS-UJF
B.P. 74
38402 Saint Martin d'Hères Cedex
France

E-mail: lbessier;g.besson@ujf-grenoble.fr

Michel Boileau
Université Paul Sabatier
Institut de Mathématiques de Toulouse –
UMR 5219 CNRS-UPS
118, route de Narbonne
31062 Toulouse Cedex 9
France

E-mail: boileau@picard.ups-tlse.fr

Sylvain Maillot
Université Montpellier II
Institut de Mathématiques et de Modélisation
de Montpellier, UMR 5149 CNRS
Case Courrier 051, Place Eugène Bataillon
34095 Montpellier Cédex 5
France

E-mail: smaillot@math.univ-montp2.fr

Joan Porti
Departament de Matemàtiques
Universitat Autònoma de Barcelona
08193 Bellaterra
Spain

E-mail: porti@mat.uab.cat

2010 Mathematical Subject Classification: 57-02; 57M50, 53C44

Key words: Ricci flow, geometrisation, hyperbolisation, three-dimensional topology, geometric structure, Haken manifolds, JSJ decomposition

ISBN 978-3-03719-082-1

The Swiss National Library lists this publication in The Swiss Book, the Swiss national bibliography, and the detailed bibliographic data are available on the Internet at http://www.helveticat.ch.

Contact address:

European Mathematical Society Publishing House
Seminar for Applied Mathematics
ETH-Zentrum FLI C4
CH-8092 Zürich
Switzerland

Phone: +41 (0)44 632 34 36
Email: info@ems-ph.org
Homepage: www.ems-ph.org

Typeset using the authors' TEX files: I. Zimmermann, Freiburg
Printing and binding: Druckhaus Thomas Müntzer GmbH, Bad Langensalza, Germany
∞ Printed on acid free paper
9 8 7 6 5 4 3 2 1

Preface

The aim of this book is to give a proof of Thurston's Geometrisation Conjecture, solved by G. Perelman in 2003. Perelman's work completes a program initiated by R. Hamilton, using a geometric evolution equation called *Ricci flow*. Perelman presented his ideas in three very concise manuscripts [Per02], [Per03a], [Per03b]. Important work has since been done to fill in the details. The first set of notes on Perelman's papers was posted on the web in June 2003 by B. Kleiner and J. Lott. These notes have progressively grown to the point where they cover the two papers [Per02], [Per03b]. The final version has been published as [KL08]. A proof of the Poincaré Conjecture, following G. Perelman, is given in the book [MT07] by J. Morgan and G. Tian. Another text covering the Geometrisation Conjecture following Perelman's ideas is the article [CZ06a] by H.-D. Cao and X.-P. Zhu. Alternative approaches to some of Perelman's arguments were given by T. Colding and W. Minicozzi [CM07], T. Shioya and T. Yamaguchi [SY05], the authors of the present book [BBB$^+$07], [BBB$^+$10], J. Morgan and G. Tian [MT08], J. Cao and J. Ge [CG09], B. Kleiner and J. Lott [KL10].

One goal of this book is to present a proof more attractive to topologists. For this purpose, we have endeavoured to reduce its analytical aspects to blackboxes and refer to some well-known and well-written sources (e.g. [MT07],[CK04]). At various points, we have favoured topological and geometric arguments over analytic ones.

The bulk of Perelman's proof is the construction and study of a kind of generalised solution of the Ricci flow equation called *Ricci flow with surgery*. In our treatment, this part has been simplified by replacing Ricci flow with surgery by a variant, which we call *Ricci flow with bubbling-off*. This is a sort of discontinuous, piecewise smooth Ricci flow on a fixed manifold. For the last part of the proof, we provide a completely different argument, based on the preprint [BBB$^+$07], which relies on topological arguments and Thurston's hyperbolisation theorem for Haken manifolds. We use a technique borrowed from the proof of the orbifold theorem ([BLP05]).

We have tried to make the various parts of the proof as independent as possible, in order to clarify its overall structure. The book has four parts. The first two are devoted to the construction of Ricci flow with bubbling-off. These two parts, combined with [CM07] or [Per03a], are sufficient to prove the Poincaré Conjecture, as we explain in Section 1.2.3. Part III is concerned with the long-time behaviour of Ricci flow with bubbling-off. Part IV completes the proof of the Geometrisation Conjecture and can be read independently of the rest. The material found here is of topological and geometric nature.

The idea of writing this book originated during the party that followed the conference held in honour of Larry Siebenmann in December 2005. It came after several workshops held in Barcelona, München and Grenoble devoted to the reading of Perelman's papers and the Kleiner–Lott notes. A first version of this book was handed out as lecture notes during the trimestre on the Ricci curvature held in I.H.P. (Paris) in May 2008.

The results of this monograph have been announced in the survey article [Mai08]. We warn the reader that there are a few discrepancies between that article and the present book, due to changes in terminology and minor adjustments in statements of theorems. For an introduction to the Poincaré Conjecture that follows the approach of this book, see [Mai09].

Many expository texts about Perelman's work have been written. We recommend, among others, [And04], [Bes06b], [Mor05], [Lot07], [Mor07], [Tao06], [Yau06], [Bes05], [Bes06a], [BBB06], [Bes07].

The authors wish to thank the Agence Nationale de la Recherche for its support under the programs FOG (ANR-07-BLAN-0251-01) and GROUPES (ANR-07-BLAN-0141), as well as the Feder/ Micinn grant MTM2006-04546. We have also been supported by the Clay Mathematics Institute, the Fondation des Sciences Mathématiques de Paris, the Centre Emile Borel, and the Centre de Recerca Matemàtica. The third author thanks the Institut Universitaire de France for his support. The fourth author thanks the Institut de Recherche Mathématique Avancée (UMR CNRS 7501 and Université de Strasbourg) where he held a position during most of the writing of this book. The last author received the prize "ICREA Acadèmia", funded by the Generalitat de Catalunya.

We warmly thank B. Kleiner, J. Lott, J. Morgan and G. Tian for numerous fruitful exchanges. Part of this work originated in workshops organized at Barcelona, Grenoble and München. We thank all the participants of these activities, in particular J. Dinkelbach, V. Bayle, Y. Colin de Verdière, S. Gallot, B. Leeb, L. Rozoy, T. Schick and H. Weiss. The trimestre on the Ricci curvature organized at I.H.P. in the spring 2008 has played an essential role and we thank the participants and particularly E. Aubry, R. Bamler, R. Conlon, Z. Djadli and D. Semmes.

We warmly thank C. Vernicos, for inviting us to give lectures at NUI Maynooth, and for the numerous exchanges held there. The fourth author would like to thank Burkhard Wilking, Christoph Böhm, and the differential geometry team at Münster for fruitful exchanges. Finally, the authors wish to thank O. Biquard, T. Delzant, L. Guillou, P. Pansu and M. Paun for discussions and insights.

Contents

Chapter 1
The Geometrisation Conjecture

1.1 Introduction

At the international conference of mathematics devoted to *Some Questions of Geometry and Topology*, organized in 1935 by the University of Geneva, W. Threlfall started his lecture with the following words:

> *"Nous savons tous que le problème d'homéomorphie des variétés à n dimensions, posé par Poincaré, est un des plus intéressants et des plus importants de la Géométrie. Il est intéressant en soi car il a donné naissance à la Topologie combinatoire ou algébrique, théorie comparable par son importance à la Théorie des fonctions classique. Il est important par ses applications à la Cosmologie, où il s'agit de déterminer l'aspect de l'espace de notre intuition et de la physique. Dès l'instant où le physicien a envisagé la possibilité de considérer l'espace de notre intuition, espace où nous vivons, comme clos, la tâche du mathématicien est de lui proposer un choix d'espaces clos, et même de les énumérer tous, comme il le ferait pour les polyèdres réguliers. ...Malheureusement le problème n'est complètement résolu que pour deux dimensions."*

The question discussed by Threlfall is the *homeomorphism problem* for compact 3-dimensional manifolds, which asks for an algorithm to decide whether two such manifolds M, N are homeomorphic. During the last 70 years this problem was central in 3-dimensional topology. It came down to W. Thurston's Geometrisation Conjecture at the end of the 1970s. Finally this conjecture was solved in 2003 by G. Perelman.

A classical result states that every compact surface admits a Riemannian metric with constant curvature. This result contains the topological classification of surfaces mentioned by Threlfall. Moreover, some important properties of surfaces, e.g. linearity of the fundamental group, can be deduced from this fact. Geometric structures on surfaces are also central in studying their mapping class groups.

Recall that a Riemannian manifold X is called *homogeneous* if its isometry group $\mathrm{Isom}(X)$ acts transitively on X. We call X *unimodular* if it has a quotient of finite volume. A *geometry* is a homogeneous, simply-connected, unimodular Riemannian manifold. A manifold M is *geometric* if M is diffeomorphic to the quotient of a geometry X by a discrete subgroup of $\mathrm{Isom}(X)$ acting freely on X. We also say that *M admits a geometric structure modelled on X*. By extension, a manifold with boundary is *geometric* if its interior is geometric.

In dimension 2 the situation is rather special, since a geometry is always of constant curvature. This is no longer true in dimension 3. Thurston observed that there are, up to a suitable equivalence relation, eight *maximal*[1] 3-dimensional geometries: those of constant curvature $\mathbb{S}^3$, $\mathbb{E}^3$, and $\mathbb{H}^3$; the product geometries $\mathbb{S}^2 \times \mathbb{E}^1$ and $\mathbb{H}^2 \times \mathbb{E}^1$; the twisted product geometries Nil and $\widetilde{\mathrm{SL}(2,\mathbb{R})}$, and finally Sol, which is the only simply-connected, unimodular 3-dimensional Lie group which is solvable but not nilpotent.

A 3-manifold is *spherical* (resp. *Euclidean*, resp. *hyperbolic*) if it admits a geometric structure modelled on $\mathbb{S}^3$ (resp. $\mathbb{E}^3$, resp. $\mathbb{H}^3$). It follows from standard arguments in Riemannian geometry that a 3-manifold is spherical (resp. Euclidean, resp. hyperbolic) if and only if it admits a complete Riemannian metric of constant sectional curvature equal to $+1$ (resp. 0, resp. -1). A metric of constant positive sectional curvature is often called a *round metric*.

The classification of spherical 3-manifolds was completed by H. Seifert and W. Threlfall in 1932 [TS31], [TS33]. A key observation for this classification is that every finite subgroup of SO(4) acting freely and orthogonally on $\mathbb{S}^3$ commutes with the action of some SO(2)-subgroup of SO(4). This SO(2)-action induces a circle foliation on the manifold where each circle has a saturated tubular neighbourhood. This phenomenon led Seifert [Sei33] to classify all 3-manifolds carrying such a foliation by circles, nowadays called *Seifert fibred* manifolds. It turns out that a compact 3-manifold is Seifert fibred if and only if it admits a geometric structure modelled on one of the six following geometries: $\mathbb{S}^3$, $\mathbb{E}^3$, $\mathbb{S}^2 \times \mathbb{E}^1$, Nil, $\mathbb{H}^2 \times \mathbb{E}^1$, $\widetilde{\mathrm{SL}(2,\mathbb{R})}$.

Poincaré's Fundamental Polyhedron Theorem gives a general method for building a hyperbolic 3-manifold by identifying the faces of a convex polyhedron of $\mathbb{H}^3$. However, this construction does not provide a criterion to decide whether a compact 3-manifold admits a hyperbolic structure. It was only in 1977 that such a criterion was conjectured by W. Thurston, and proved for a large class of compact 3-manifolds, the so-called *Haken* 3-manifolds. The fact that this criterion holds true is part of a wider conjecture proposed by Thurston, the Geometrisation Conjecture, which gives a general picture of all compact 3-manifolds.

The content of the Geometrisation Conjecture is that every compact 3-manifold splits along a finite collection of embedded surfaces into canonical geometric pieces. In order to be more precise, we need a definition: a closed connected surface in a compact orientable 3-manifold M is *essential* if it is π_1-injective and if it does not bound a 3-ball nor cobound a product with a connected component of ∂M.

Conjecture 1.1.1 (Thurston's Geometrisation Conjecture). *The interior of any compact, orientable* 3*-manifold can be split along a finite collection of essential, pairwise disjoint, embedded* 2*-spheres and* 2*-tori into a canonical collection of geometric* 3*-manifolds after capping off all boundary spheres by* 3*-balls.*

Such a decomposition will be called a *geometric decomposition*. A special case of the Geometrisation Conjecture is the so-called *Elliptisation Conjecture*, which asserts that every closed orientable 3-manifold with finite fundamental group is spherical. The

[1] In this context, *maximal* means that there is no $\mathrm{Isom}(X)$-invariant Riemannian metric on X whose isometry group is strictly larger than $\mathrm{Isom}(X)$.

famous Poincaré Conjecture is itself the special case of the Elliptisation Conjecture where the fundamental group is trivial:

Conjecture 1.1.2 (Poincaré Conjecture). *If M is a closed, simply-connected 3-manifold, then M is diffeomorphic to the 3-sphere.*

The Geometrisation Conjecture has far-reaching consequences for 3-manifold topology. For instance, it implies that every closed, orientable, aspherical[2] 3-manifold is determined, up to homeomorphism, by its fundamental group. This is a special case of the so-called Borel Conjecture.

The existence of the topological decomposition required for Thurston's Geometrisation Conjecture follows from two central results in 3-manifold topology: the Kneser decomposition theorem and the JSJ-splitting theorem. An orientable 3-manifold M is *irreducible* if any embedding of the 2-sphere into M extends to an embedding of the 3-ball into M. The *connected sum* of two orientable 3-manifolds is the orientable 3-manifold obtained by pulling out the interior of a 3-ball in each manifold and gluing the remaining parts together along the boundary spheres.

The first stage of the decomposition is due to H. Kneser for the existence [Kne29], and to J. Milnor for the uniqueness [Mil62]:

Theorem 1.1.3 (Kneser's decomposition). *Every compact, orientable 3-manifold is a connected sum of 3-manifolds that are either homeomorphic to $\mathbb{S}^1 \times \mathbb{S}^2$ or irreducible. Moreover, the connected summands are unique up to ordering and orientation-preserving homeomorphism.*

This result reduces the Geometrisation Conjecture to the case of irreducible manifolds. The second stage of the decomposition is more subtle. An orientable surface of genus at least 1, embedded in a 3-manifold, is *incompressible* if it is π_1-injective. An embedded torus in a compact, orientable, irreducible 3-manifold M is *canonical* if it is essential and can be isotoped off any incompressible embedded torus (see [NS97]). A compact orientable 3-manifold M is *atoroidal* if it contains no essential embedded torus and is not homeomorphic to a product $T^2 \times [0,1]$ or a twisted product $K^2 \tilde{\times} [0,1]$.

Theorem 1.1.4 (JSJ-splitting). *A maximal (possibly empty) collection of disjoint, non-parallel, canonical tori is finite and unique up to isotopy. It cuts M into 3-submanifolds that are atoroidal or Seifert fibred.*

This splitting is called the *JSJ-splitting* because it was constructed by W. Jaco and P. Shalen [JS79], and by K. Johannson [Joh79].

A compact, orientable, irreducible 3-manifold is *Haken* if its boundary is not empty, or if it contains a closed essential surface. In the mid 1970s W. Thurston proved the Geometrisation Conjecture for Haken 3-manifolds (see Otal [Ota01], [Ota98]):

Theorem 1.1.5 (Thurston's hyperbolisation theorem). *The Geometrisation Conjecture is true for Haken 3-manifolds.*

[2]A manifold is *aspherical* if all its higher homotopy groups vanish.

In particular, any atoroidal Haken 3-manifold M is hyperbolic or Seifert fibred.

The main goal of this book is to explain the proof of the remaining and most difficult case, dealing with closed, atoroidal manifolds:

Theorem 1.1.6 (G. Perelman). *Let M be a closed, orientable, irreducible, atoroidal 3-manifold. Then:*

(i) *If $\pi_1(M)$ is finite, then M is spherical.*

(ii) *If $\pi_1(M)$ is infinite, then M is hyperbolic or Seifert fibred.*

A standard argument using Kneser's theorem and van Kampen's theorem shows that part (i) of Theorem 1.1.6 implies the Elliptisation Conjecture.[3] More generally, Theorems 1.1.5 and 1.1.6 together imply:

Corollary 1.1.7 (Geometrisation theorem). *The Geometrisation Conjecture is true for all compact orientable 3-manifolds.*

Since the fundamental group of an irreducible Seifert manifold is either finite or contains $\mathbb{Z}^2$ (see e.g. [BMP03]), one also obtains as a corollary:

Corollary 1.1.8 (Hyperbolisation theorem). *A closed, orientable, irreducible 3-manifold is hyperbolic if and only if $\pi_1(M)$ is infinite and does not contain a subgroup isomorphic to $\mathbb{Z}^2$.*

In fact, Perelman's proof deals with all cases and allows us to recover the geometric splitting of the manifold along spheres and tori. However, in this book we focus on Theorem 1.1.6.

The main ingredient of the proof is the *Ricci flow,* which is an evolution equation introduced by R. Hamilton. In Section 1.2, we review general facts about this equation, and define an object called *Ricci flow with bubbling-off,* which is a variation on Perelman's *Ricci flow with surgery* (cf. also [Ham97]). We then state an existence result and deduce the Elliptisation Conjecture from this. In Section 1.3 we tackle the long-time behaviour of Ricci flow with bubbling-off and the case where $\pi_1(M)$ is infinite.

Remark 1.1.9. For more historical background on the Geometrisation Conjecture, see e.g. [BMP03] and the references therein.

---

[3] Indeed, let M be a closed, orientable 3-manifold with finite fundamental group. By Theorem 1.1.3, it is a connected sum of closed, orientable 3-manifolds M_i which are irreducible or homeomorphic to $S^1 \times S^2$. By the van Kampen theorem, $\pi_1(M)$ is a free product of the fundamental groups of the M_i's, so by standard group theory, $\pi_1(M_i)$ is finite for each i, and trivial except for at most one value of i. Since $\pi_1(S^1 \times S^2)$ is infinite, this implies that each M_i is irreducible. It is also atoroidal, since $\mathbb{Z}^2$ does not embed into a finite group. Hence we can apply Theorem 1.1.6 (i) to each M_i. It follows that all M_i's are spherical, and all but at most one are homeomorphic to S^3. Therefore M is spherical.

1.2 Ricci flow and elliptisation

1.2.1 Ricci flow

Notation. If g is a Riemannian metric, we denote by $R_{\min}(g)$ the minimum of its scalar curvature, by Ric_g its Ricci tensor, and by $\mathrm{vol}(g)$ its volume.

Let M be a closed, orientable, irreducible 3-manifold. In the Ricci flow approach to geometrisation, one studies solutions of the evolution equation

$$\frac{dg}{dt} = -2\,\mathrm{Ric}_{g(t)}, \tag{1.1}$$

called the *Ricci flow equation*, which was introduced by R. Hamilton. A solution is an *evolving metric* $\{g(t)\}_{t\in I}$, i.e., a 1-parameter family of Riemannian metrics on M defined on an interval $I \subset \mathbb{R}$. In [Ham82], Hamilton proved that for any metric g_0 on M, there exists $\varepsilon > 0$ such that equation (1.1) has a unique solution defined on $[0, \varepsilon)$ with initial condition $g(0) = g_0$. Thus there exists $T \in (0, +\infty]$ such that $[0, T)$ is the maximal interval where the solution to (1.1) with initial condition g_0 is defined. When T is finite, one says that the Ricci flow has a *singularity* at time T. Ideally, one would like to see the geometry of M appear by looking at the metric $g(t)$ when t tends to T (whether T be finite or infinite). To understand how this works, we first consider some (very) simple examples, where the initial metric is locally homogeneous.

Example 1. If g_0 has constant sectional curvature K, then the solution is given by $g(t) = (1 - 4Kt)g_0$. Thus in the spherical case, where $K > 0$, we have $T < \infty$, and as t goes to T, the manifold shrinks to a point while remaining of constant positive curvature.

By contrast, in the hyperbolic case, where $K < 0$, we have $T = \infty$ and $g(t)$ expands indefinitely, while remaining of constant negative curvature. In this case, the rescaled solution $\tilde{g}(t) := (4t)^{-1}g(t)$ converges to the metric of constant sectional curvature -1.

Example 2. If M is the product of a circle with a surface F of genus at least 2, and g_0 is a product metric whose second factor has constant (negative) curvature, then $T = \infty$; moreover, the solution is a product of a constant metric on S^1 with an expanding metric on F.

In this case, $g(\cdot)$ does not have any convergent rescaling. However, one can observe that the rescaled solution $\tilde{g}(\cdot)$ defined above *collapses with bounded sectional curvature*, i.e., has bounded sectional curvature and injectivity radius going to 0 everywhere as t goes to $+\infty$.

Several important results were obtained by Hamilton [Ham82], [Ham86], [Ham93], [Ham95b], [Ham99]. The following two can be viewed as partial results on the Geometrisation Conjecture:[4]

[4]Note that we have reformulated Hamilton's results, whose original statements concerned *normalised* Ricci flow.

Theorem 1.2.1 ([Ham82]). *Let M be a closed, orientable 3-manifold and g_0 be a Riemannian metric of positive Ricci curvature on M. Then $T < \infty$, and the volume-rescaled Ricci flow $\operatorname{vol}(g(t))^{-2/3} g(t)$ converges (modulo diffeomorphisms) to a metric of positive constant sectional curvature as $t \to T$. In particular, M is spherical.*

Theorem 1.2.2 ([Ham99]). *Let M be a closed, orientable 3-manifold and $g(\cdot)$ be a Ricci flow on M defined on $[0, +\infty)$. If the sectional curvature of the rescaled flow $\tilde{g}(\cdot)$ is bounded independently of t, then one of the following assertions holds:*

(i) *As t tends to infinity, $\tilde{g}(\cdot)$ converges modulo diffeomorphisms to a hyperbolic metric.*

(ii) *As t tends to infinity, $\tilde{g}(\cdot)$ collapses with bounded sectional curvature.*

(iii) *M contains an incompressible torus.*

The general case, however, is more difficult, because it sometimes happens that $T < \infty$ while the behavior of $g(t)$ as t tends to T does not allow us to determine the topology of M. One possibility is the so-called *neck pinch*, where part of M looks like a thinner and thinner cylindrical neck as one approaches the singularity. This can happen even if M is irreducible (see [AK04] for an example where $M = S^3$); thus neck pinches may not give any useful information on the topology of M.

1.2.2 Ricci flow with bubbling-off

Throughout the book, we suppose that M is closed, orientable and irreducible. In order to simplify the presentation, we also assume that M is *$\mathbb{R}P^2$-free*, i.e., does not contain any submanifold diffeomorphic to $\mathbb{R}P^2$. This is not much of a restriction because the only closed, irreducible 3-manifold that does contain an embedded copy of $\mathbb{R}P^2$ is $\mathbb{R}P^3$, which is a spherical manifold.[5]

As we already explained, one of the main difficulties in the Ricci flow approach to geometrisation is that singularities unrelated to the topology of M may appear. Using maximum principle arguments, one shows that singularities in a 3-dimensional Ricci flow can occur only when the scalar curvature tends to $+\infty$ somewhere (see e.g. [Ham95b]). One of Perelman's major breakthroughs was to give a precise local description of the geometry at points of large scalar curvature: every such point has a so-called *canonical neighbourhood* U. For instance U may be an *ε-neck* (i.e., almost homothetic to the product of the round 2-sphere of unit radius with an interval of length $2\varepsilon^{-1}$) or an *ε-cap* (i.e., a 3-ball such that a collar neighbourhood of ∂U is an ε-neck) as pictured below.

See Chapter 3 for the precise definitions and Chapter 4 for more information on ε-necks and ε-caps.

[5] To see this, observe that if M is an orientable 3-manifold containing an embedded projective plane P, then the tubular neighbourhood U of P is a twisted I-bundle over $\mathbb{R}P^2$. Its boundary is an embedded 2-sphere S in M. If M is irreducible, then S bounds a 3-ball B, so $M = U \cup B$.

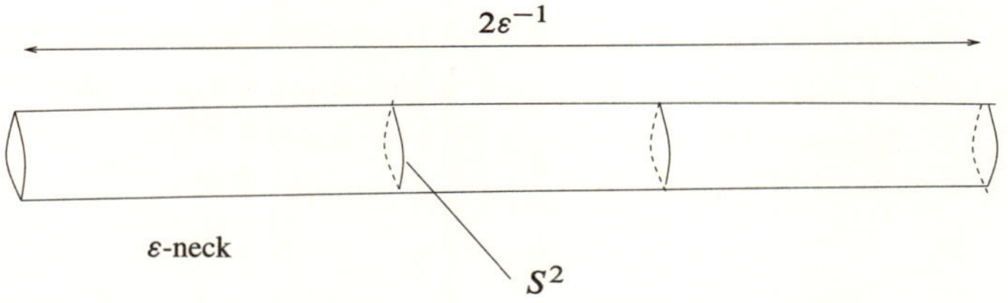

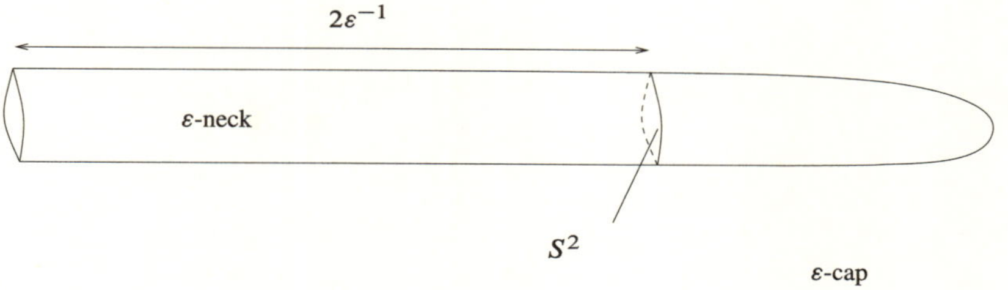

To avoid singularities, we fix a large number Θ, which plays the role of a curvature threshold. As long as the maximum of the scalar curvature is less than Θ, Ricci flow is defined. If it reaches Θ at some time t_0, then there are two possibilities: if the minimum of the scalar curvature of the time-t_0 metric is large enough, then every point has a canonical neighbourhood. As we shall see, this implies that M is spherical (cf. Corollary 3.2.5). Thus the construction stops in this case.[6]

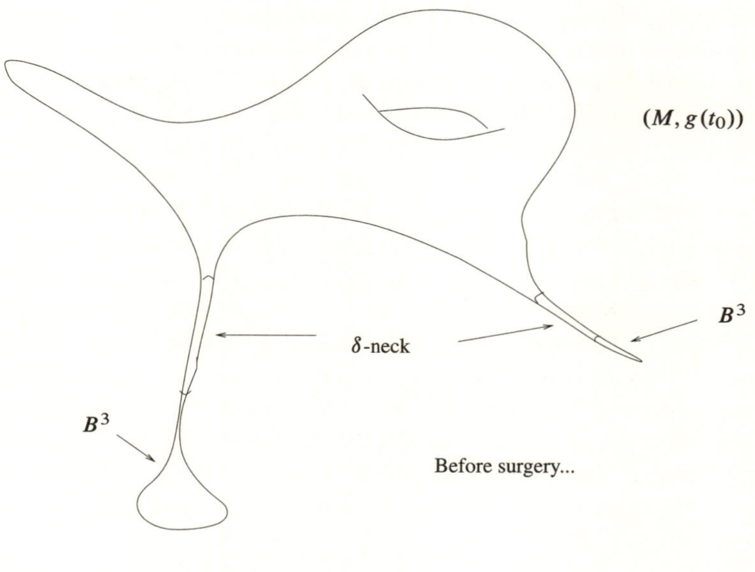

[6]This situation is analogous to Perelman's *extinction* case.

Otherwise, we modify $g(t_0)$ so that the maximum of the scalar curvature of the new metric, denoted by $g_+(t_0)$, is at most $\Theta/2$. This modification is called *metric surgery*.

It consists in replacing the metric in some 3-balls containing regions of high curvature by a special type of ε-caps called *almost standard caps*.

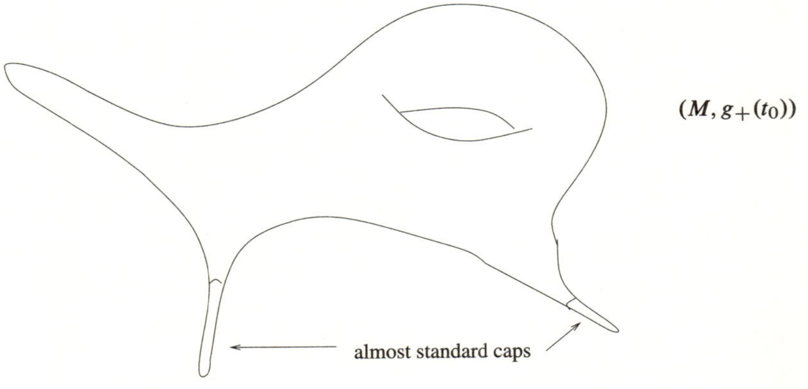

....after surgery

Then we start the Ricci flow again, using $g_+(t_0)$ as a new initial metric. This procedure is repeated as many times as necessary. It follows from estimates on the Ricci flow that it takes a certain amount of time for the maximum of the scalar curvature to double its value. Hence once Θ is fixed, surgery times cannot accumulate.

The main difficulty is to choose Θ and do metric surgery in such a way that the construction can indeed be iterated. In practice, we will fix three parameters $r, \delta, \kappa > 0$, where r is the canonical neighbourhoods' scale, i.e., is such that all points of scalar curvature at least r^{-2} have canonical neighbourhoods, δ is a small number describing the precision of surgery, and κ prevents the metric from locally collapsing. The threshold Θ is then determined by the numbers r, δ (cf. Theorem 5.2.4 and Definition 5.2.5).

In this way, we construct an evolving metric $\{g(t)\}_t$ which is piecewise $\mathcal{C}^1$-smooth with respect to t, and such that for each *singular time* t_0 (i.e., value of t such that $g(\cdot)$ is not $\mathcal{C}^1$ in a neighbourhood of t_0) the map $t \mapsto g(t)$ is left-continuous and has a right-limit $g_+(t_0)$. The result of this construction is called *Ricci flow with* (r, δ, κ)*-bubbling-off.*[7] Since its precise definition is quite intricate, we begin with the simpler, more general definition of a *Ricci flow with bubbling-off*, which retains the essential features that are needed in order to deduce the Elliptisation Conjecture.

We say that an evolving metric $\{g(t)\}_t$ as above is a *Ricci flow with bubbling-off* if equation (1.1) is satisfied at all nonsingular times, and for every singular time t_0 one has

[7]Note that the parameters r, δ, κ, and therefore Θ, are constant on each compact time interval, but will change as t goes to infinity.

(i) $R_{\min}(g_+(t_0)) \geqslant R_{\min}(g(t_0))$, and

(ii) $g_+(t_0) \leqslant g(t_0)$.

The following theorem is one of the main results of this monograph; its proof occupies Parts I and II:

Theorem 1.2.3 (Finite-time existence of Ricci flow with bubbling-off). *Let M be a closed, orientable, irreducible 3-manifold. Then*

(i) *M is spherical, or*

(ii) *for every $T > 0$ and every Riemannian metric g_0 on M, there exists a Ricci flow with bubbling-off $g(\cdot)$ on M, defined on $[0, T]$, with initial condition $g(0) = g_0$.*

By iteration, one immediately obtains existence on $[0, +\infty)$ of Ricci flow with bubbling-off on closed, orientable, irreducible 3-manifolds which are *not spherical*. However, a lot of work remains to be done in order to obtain the necessary estimates for hyperbolisation (see Section 1.3); this occupies Part III. Let us begin with the application of Ricci flow with bubbling-off to elliptisation.

Remark 1.2.4. Ricci flow with bubbling-off with a given initial condition is of course absolutely not unique. Even with the restricted definition of (r, δ)-bubbling-off there is still no uniqueness result, due to possible freedom in the choice of the place where surgery is performed. The question of finding a *canonical* surgery procedure for Ricci flow in dimension 3 is still open.

Remark 1.2.5. There are a few differences between Ricci flow with bubbling-off and Perelman's Ricci flow with δ-cutoff. An obvious one is that a Ricci flow with bubbling-off is an evolving metric on a fixed manifold rather than an evolving manifold. This simplification is made possible by the extra topological assumptions on M. Another, perhaps more significant difference is that surgery occurs *before* the Ricci flow becomes singular, rather than *at* the singular time. Our construction is in this respect closer in spirit to the surgery process envisioned by Hamilton [Ham97].

1.2.3 Application to elliptisation

To deduce the first part of Theorem 1.1.6 from Theorem 1.2.3, all we need is the following result (cf. [Per03a], [CM05]:)

Theorem 1.2.6 (Finite-time extinction). *Let M be a closed, orientable, irreducible 3-manifold with finite fundamental group. For each Riemannian metric g_0 on M there is a number $T(g_0) > 0$ such that if $g(\cdot)$ is a Ricci flow with bubbling-off on M defined on some interval $[0, T]$ and with initial condition $g(0) = g_0$, then $T < T(g_0)$.*

To see why Theorems 1.2.3 and 1.2.6 together imply Theorem 1.1.6 (i), let M be a closed, orientable, irreducible 3-manifold with finite fundamental group. Let g_0

be an arbitrary Riemannian metric on M. Let $T(g_0)$ be a positive number given by Theorem 1.2.6. If M were not spherical, then we would be in the second case of Theorem 1.2.3. Applying this with $T = T(g_0)$ gives a contradiction.

We now give a proof of Theorem 1.2.6 using a result of Colding and Minicozzi [CM05], [CM07]. See [Mai09] for a somewhat more detailed discussion.

Let M be a 3-manifold satisfying the hypotheses and g_0 be a Riemannian metric on M. Let Ω be the space of smooth maps $f: S^2 \times [0,1] \to M$ such that $f(S^2 \times \{0\})$ and $f(S^2 \times \{1\})$ are points. Since $\pi_1(M)$ is finite, we have $\pi_3(M) \neq 0$ (see, e.g., [Mai08], Lemma 2.2). It follows that there exists $f_0 \in \Omega$ which is not homotopic to a constant map [MM88], Lemma 3. Let ξ be the homotopy class of f_0. We set

$$W(g) := \inf_{f \in \xi} \max_{s \in [0,1]} E(f(\cdot, s)),$$

where E denotes the energy

$$E(f(\cdot, s)) = \frac{1}{2} \int_{S^2} |\nabla_x f(x,s)|_g^2 \, dx.$$

Let $\{g(t)\}_{t \in [0,T]}$ be a Ricci flow with bubbling-off such that $g(0) = g_0$. First consider the function $t \mapsto R_{\min}(t)$. At regular times it is continuous and satisfies (see Corollary C.2.2 in the Appendix)

$$\frac{d^+}{dt} R_{\min} \geqslant \frac{2}{3} R_{\min}^2(t), \tag{1.2}$$

where we use $\frac{d^+}{dt}$ to denote the limsup of forward difference quotients.

If t_0 is a singular time, then it follows from Definition 2.2.1 (i) that the left limit of $R_{\min}$ at t_0 is no greater than its right limit.

As a consequence, $R_{\min}(t)$ satisfies the following *a priori* lower bound:

$$R_{\min}(t) \geqslant \frac{R_{\min}(0)}{1 - 2tR_{\min}(0)/3}. \tag{1.3}$$

At regular times, the function $t \mapsto W(g(t))$ is continuous, and since ξ is nontrivial, by [CM05], [CM07] it satisfies

$$\frac{d^+}{dt} W(g(t)) \leqslant -4\pi - \frac{1}{2} R_{\min}(t) W(g(t)) \leqslant -4\pi + \frac{R_{\min}(0)}{\frac{4tR_{\min}(0)}{3} - 2} W(g(t)). \tag{1.4}$$

Now if t_0 is a singular time, it follows from Definition 2.2.1 (ii) that the left limit of W at t_0 is no smaller than its right limit. Since the right-hand side of (1.4) is not integrable and $W(g(t))$ cannot become negative, this easily implies an upper bound on T depending only on $R_{\min}(0)$ and $W(g_0)$. Thus the proof of Theorem 1.2.6 is complete.

For future reference, we note the following byproduct of Theorem 1.2.3 and (1.3):

Theorem 1.2.7. *Let M be a closed, orientable, irreducible 3-manifold. If M has a metric of positive scalar curvature, then M is spherical.*

Remark 1.2.8. Perelman's argument for finite-time extinction uses an invariant $A(\alpha(g))$ in place of the width $W(g)$. This invariant is defined using the second homotopy group of the space of loops in M.

1.3 3-manifolds with infinite fundamental group

Let M be a closed, orientable, irreducible 3-manifold with infinite fundamental group. Our next goal is to prove that M is hyperbolic, or Seifert fibred, or contains an incompressible torus, which gives part (ii) of Theorem 1.1.6. First we need to refine the existence theorem for Ricci flow with bubbling-off in order to include more geometric information.

Note that since M has infinite fundamental group, it is not spherical, hence, by Theorem 1.2.7, does not carry any metric with positive scalar curvature.

We adopt the convention that *throughout the book, all hyperbolic 3-manifolds are complete and of finite volume.*

1.3.1 Long-time behaviour of the Ricci flow with bubbling-off

We consider, for any metric g on M the two following scale invariant quantities:

$$\widehat{R}(g) := R_{\min}(g) \cdot \mathrm{vol}(g)^{2/3} \leqslant 0,$$

and

$$\widehat{V}(g) = \left(\frac{R_{\min}(g)}{-6}\right)^{3/2} \mathrm{vol}(g) = \left(\frac{\widehat{R}(g)}{-6}\right)^{3/2} \geqslant 0.$$

In particular, when H is a hyperbolic manifold, we let $\widehat{V}(H) = \mathrm{vol}(H)$ denote $\widehat{V}(g_{\mathrm{hyp}})$, and $\widehat{R}(H)$ denote $\widehat{R}(g_{\mathrm{hyp}})$, where g_{hyp} is the hyperbolic metric. On the other hand, note that $\widehat{V}(g) = 0$ if $R_{\min}(g) = 0$.

Those quantities are monotonic along the Ricci flow on a closed manifold, as long as $R_{\min}$ remain nonpositive (see Corollary C.2.2). This is also true in presence of bubbling-off by conditions (i) and (ii) of Definition 2.2.1. As a consequence, we have:

Proposition 1.3.1. *Let $\{g(t)\}$ be a Ricci flow with bubbling-off on M. Then*

(i) *the function $t \mapsto \widehat{R}(g(t))$ is nondecreasing;*
(ii) *the function $t \mapsto \widehat{V}(g(t))$ is nonincreasing.*

In order to state the main theorem of this subsection, we need two definitions. The first is standard and goes back to Gromov, the second is specific to our situation and is essentially due to Perelman.

Definition 1.3.2. Let $k > 0$ be an integer. Let (M_n, g_n, x_n) be a sequence of pointed Riemannian manifolds, and let $(M_\infty, g_\infty, x_\infty)$ be a pointed Riemannian manifold. We shall say that (M_n, g_n, x_n) *converges to* $(M_\infty, g_\infty, x_\infty)$ in the $\mathcal{C}^k$-sense if there exists a sequence of numbers $\varepsilon_n > 0$ tending to zero, and a sequence of $\mathcal{C}^{k+1}$-diffeomorphisms ψ_n from the metric ball $B(x_\infty, \varepsilon_n^{-1}) \subset M_\infty$ to the metric ball $B(x_n, \varepsilon_n^{-1}) \subset M_n$ such that $\psi_n^*(g_n) - g_\infty$ has $\mathcal{C}^k$-norm less than ε_n everywhere. We say that the sequence *subconverges* if it has a convergent subsequence.

Definition 1.3.3. Let (X, g) be a Riemannian 3-manifold. We say that a point $x \in X$ is *ε-thin with respect to* g if there exists a radius $\rho \in (0, 1]$ such that the ball $B(x, \rho)$ has the following two properties: all sectional curvatures on this ball are bounded below by $-\rho^{-2}$, and the volume of this ball is less than $\varepsilon\rho^3$. Otherwise, x is *ε-thick with respect to g*.

We are now in position to state the main result of this subsection (cf. [Per03b], Sections 6 and 7). Recall that we denote by $\tilde{g}(t)$ the rescaled evolving metric $(4t)^{-1}g(t)$, the rescaling being motivated by the example of hyperbolic manifolds.

Theorem 1.3.4 (Thin-thick decomposition). *Let M be a closed, orientable, irreducible, non-spherical* 3*-manifold. For every Riemannian metric g_0 on M, there exists a Ricci flow with bubbling-off $g(\cdot)$ defined on* $[0, +\infty)$ *with the following properties:*

(i) $g(0) = g_0$.

(ii) *The volume of the rescaled metric $\tilde{g}(t)$ is bounded independently of t.*

(iii) *For every $\varepsilon > 0$ and every sequence $(x_n, t_n) \in M \times [0, +\infty)$, if t_n tends to $+\infty$ and x_n is ε-thick with respect to $\tilde{g}(t_n)$ for each n, then there exists a hyperbolic* 3*-manifold H and a basepoint $* \in H$ such that the sequence $(M, \tilde{g}(t_n), x_n)$ subconverges to $(H, g_{\text{hyp}}, *)$ in the $\mathcal{C}^2$-sense.*

(iv) *For every sequence $t_n \to \infty$, the sequence $\tilde{g}(t_n)$ has controlled curvature in the sense of Perelman.*

The proof of this theorem occupies Part III.

Remark 1.3.5. "Controlled curvature in the sense of Perelman" is a technical property, which is weaker than a global two-sided curvature bound, but suffices for some limiting arguments. For the precise definition, see 13.1.1.

Remark 1.3.6. The third conclusion may be vacuous, i.e., there need not exist a sequence of uniformly thick basepoints in spacetime with time going to infinity. The simplest example of this is the case where the manifold M is Euclidean and g_0 is a flat metric: then $g(t) = g_0$ for all t, and $\operatorname{diam}(\tilde{g}(t))$ goes to zero as t goes to $+\infty$. One says that $\tilde{g}(\cdot)$ "collapses to a point" in this case. For a more detailed discussion of collapsing, see Chapter 13.

Next is a fairly straightforward corollary of Theorem 1.3.4:

Corollary 1.3.7. *Let M be a closed, orientable, irreducible, non-spherical 3-manifold. For every Riemannian metric g_0 on M, there exists an infinite sequence of Riemannian metrics $g_1, \dots, g_n, \dots$ with the following properties:*

(i) *The sequence $(\mathrm{vol}(g_n))_{n \geqslant 0}$ is bounded.*

(ii) *For every $\varepsilon > 0$ and every sequence $x_n \in M$, if x_n is ε-thick with respect to g_n for each n, then the sequence (M, g_n, x_n) subconverges in the pointed $\mathcal{C}^2$ topology to a pointed hyperbolic 3-manifold $(H, g_{\mathrm{hyp}}, *)$, where $\mathrm{vol}(H) = \hat{V}(H) \leqslant \hat{V}(g_0)$.*

(iii) *The sequence g_n has controlled curvature in the sense of Perelman.*

Proof. Applying Theorem 1.3.4, we obtain a Ricci flow with bubbling-off $\{g(t)\}_{t \in [0,\infty)}$ with initial condition g_0. Pick any sequence $t_n \to +\infty$ and set $g_n := \bar{g}(t_n)$ for $n \geqslant 1$. Assertions (i), (iii) and the first part of (ii) follow directly from the conclusions of Theorem 1.3.4. To see that $\hat{V}(H) \leqslant \hat{V}(g_0)$, remark that $(\hat{V}(g_n))_{n \geqslant 0}$ is nonincreasing. Indeed, it follows from Proposition 1.3.1 and the scale invariance of $\hat{V}$. On the other hand, $R_{\min}(g_{\mathrm{hyp}}) \geqslant \liminf R_{\min}(g_n)$ and $\mathrm{vol}(H) \leqslant \liminf \mathrm{vol}(g_n)$, hence $\hat{V}(H) \leqslant \liminf \hat{V}(g_n) \leqslant \hat{V}(g_0)$. □

1.3.2 Hyperbolisation

We now explain the proof of the second part of Theorem 1.1.6. We shall define two topological invariants $V_0(M)$ and $\bar{V}(M)$. Let M be a closed 3-manifold. For us, a *link* in M is a (possibly empty, possibly disconnected) closed 1-submanifold of M. A link is *hyperbolic* if its complement is a hyperbolic 3-manifold. We shall study the volume of these complements endowed with the (unique) hyperbolic metric.

Definition 1.3.8. We define

$$V_0(M) := \inf \{\mathrm{vol}(M \setminus L) \mid L \text{ hyperbolic link } \subset M\}.$$

Notice that $V_0(M)$ is finite because any closed 3-manifold contains a hyperbolic link [Mye82]. Since the set of volumes of hyperbolic 3-manifolds is well-ordered by Thurston–Jørgensen theory (see e.g. [Gro81a], [Mai10] and the references therein), this infimum is always realised by some hyperbolic 3-manifold H_0; in particular, it is positive. We note that M is hyperbolic if and only if $H_0 = M$ (see e.g. [Bes00]).

Definition 1.3.9. For any closed manifold M we define

$$\bar{V}(M) := \inf \{\mathrm{vol}(g) \mid R_{\min}(g) \geqslant -6\}.$$

We then have the following proposition proved in Appendix E (Proposition E.1.7). It can be thought of as a criterion for hyperbolicity.

Proposition 1.3.10. *Let M be any closed, orientable 3-manifold. Suppose that the inequality $\bar{V}(M) \geqslant V_0(M)$ holds. Then equality holds, M is hyperbolic, and the hyperbolic metric realises $\bar{V}(M)$ and $V_0(M)$.*

Proof. It is a result of the following fact. Let H_0 be a hyperbolic manifold homeomorphic to the complement of a link L_0 in M and whose volume realises $V_0(M)$. It is sufficient to show that L_0 is empty. If this is not true we show that M carries a metric g_ε such that $\mathrm{vol}(g_\varepsilon) < V_0(M)$ and $R_{\min}(g_\varepsilon) \geqslant -6$, hence proving that $\bar{V}(M) < V_0(M)$; a contradiction. This can be done by a direct construction as in Anderson [And02], p. 21–23. We give an alternative proof following Salgueiro [Sal09] in Appendix E. (Cf. also Section 6 in [BBB+07].) □

An immediate corollary is the following:

Corollary 1.3.11. *For every closed, orientable 3-manifold M one has $\bar{V}(M) \leqslant V_0(M)$.*

Let us now again assume that M is a closed, orientable, irreducible 3-manifold whose fundamental group is infinite. These topological hypotheses imply that M does not carry any metric of positive scalar curvature. We then easily check that

$$\bar{V}(M) = \inf\{\hat{V}(g) \mid g \text{ metric on } M\}. \tag{1.5}$$

In the (rather special) case where M has a metric of *nonnegative* scalar curvature, then $\bar{V}(M) = 0$.

At last, we can state a second result which, together with Proposition 1.3.10, implies the second part of Theorem 1.1.6, i.e., the 'hyperbolisation' part of the Geometrisation Conjecture.

Theorem 1.3.12. *Let M be a closed, orientable, irreducible 3-manifold with infinite fundamental group. If $\bar{V}(M) < V_0(M)$, then M is a Seifert fibred manifold or contains an incompressible torus.*

Notice that in contrast to Proposition 1.3.10 the proof of Theorem 1.3.12 does use the Ricco flow.

Theorem 1.3.12 relies on the following statement which we shall prove in Part IV and which is independent of the results of Parts I–III:

Theorem 1.3.13 (Weak collapsing). *Let M be a non-simply connected, closed, orientable, irreducible 3-manifold. Suppose that there exists a sequence g_n of Riemannian metrics on M satisfying the following properties:*

(1) *The sequence $\mathrm{vol}(g_n)$ is bounded.*
(2) *For every $\varepsilon > 0$ and every sequence $x_n \in M$, if x_n is ε-thick with respect to g_n for each n, then the sequence (M, g_n, x_n) subconverges in the pointed $\mathcal{C}^2$-topology to a pointed hyperbolic 3-manifold $(H, g_{\mathrm{hyp}}, *)$, where $\mathrm{vol}(H) < V_0(M)$.*
(3) *The sequence g_n has locally controlled curvature in the sense of Perelman.*

Then M is a Seifert fibred manifold or contains an incompressible torus.

Proof of Theorem 1.3.12 *assuming Theorem* 1.3.13. Take a metric g_0 on M such that $\widehat{V}(g_0) < V_0(M)$, which exists by definition of $\overline{V}(M)$. Applying Corollary 1.3.7 with initial condition g_0, we find a sequence g_n satisfying exactly the assumptions of Theorem 1.3.13. Indeed, if H is a hyperbolic limit appearing in (2), we have by Corollary 1.3.7 (ii) that $\mathrm{vol}(H) = \widehat{V}(H) \leqslant \widehat{V}(g_0) < V_0(M)$. Hence the conclusion follows from Theorem 1.3.13. □

Remark 1.3.14. Let us explain again why Proposition 1.3.10 and Theorem 1.3.12 implies the Hyperbolisation Theorem as stated in 1.1.8: the fundamental group of a Seifert manifold is either finite or contains $\mathbb{Z}^2$ (see e.g. [BMP03]); similarly, if M contains an incompressible torus, then its fundamental group contains $\mathbb{Z}^2$.

Remark 1.3.15. Notice that Theorem 1.3.12 and Proposition 1.3.10 have an important consequence. It allows to show that if M is hyperbolic then, among the Riemannian metric with $R_{\min} \geqslant -6$, the hyperbolic metric is the unique one which realises $\overline{V}(M)$. An even stronger version can be proved (see [And05], p. 133, and [KL08], Section 93).

1.4 Some consequences of geometrisation

1.4.1 The homeomorphism problem

The main consequence of the Geometrisation Conjecture is the solution of the homeomorphism problem for 3-manifolds. The homeomorphism problem asks for an algorithm to determine whether or not two compact 3-manifolds are homeomorphic. Since the set of homeomorphism classes of 3-manifolds is countable, this leads to a classification of 3-manifolds, that is to say a complete list of 3-manifolds without duplication.

We discuss here only the case of *closed* orientable 3-manifolds and do not try to indicate the simplest algorithm. The general case needs a more sophisticated version of the Jaco–Shalen–Johannson decomposition, in order to deal with boundary issues. For a detailed account of the proof in the general case, we refer to the lectures on the homeomorphism problem for 3-manifolds given by W. Jaco at the Peking Summer School in July 2005 and available at the following address:

http://cauchy.math.okstate.edu/~jaco/pekinglectures.htm

Theorem 1.4.1. *There is an algorithm to decide if two closed, orientable* 3*-manifolds are homeomorphic.*

The proof outlined below follows Sela's solution of the homeomorphism problem for closed orientable 3-manifolds admitting a geometric decomposition [Sel95], Section 10. It is based on a succession of works culminating with Perelman's proof of the Geometrisation Conjecture.

Step 1. *Given a closed, orientable, triangulated* 3*-manifold M, there is an algorithm to either construct a prime decomposition of M or conclude that M is homeomorphic to* S^3.

This result follows the Rubinstein–Thompson algorithm to recognize S^3 [Rub97], [Rub95], Thompson and the work of W. Jaco and J. Tollefson [JT95]. See also [Hak61], [JR89] and [JLR02].

Step 2. *Given a closed, orientable, irreducible, triangulated* 3*-manifold M, there is an algorithm to decide if M is Haken.*

Such an algorithm is given in [JO84].

Step 3. *The homeomorphism problem is algorithmically solvable for closed, orientable Haken* 3*-manifolds.*

This result follows from the works of W. Haken [Hak62], G. Hemion [Hem79] and S. Matveev [Mat97], [Mat03].

Steps 1 to 3 reduce the homeomorphism problem to the case of non-Haken 3-manifolds. By Theorem 1.1.6, these manifolds fall into two classes: they are either Seifert fibred or hyperbolic.

The next two steps deal with irreducible, non-Haken 3-manifolds having a Seifert fibred structure.

Step 4. *There is an algorithm which decides if a given closed, orientable, irreducible, triangulated* 3*-manifold M which is not Haken is a Seifert fibred space.*

This follows from work of T. Li [Li06] and J. H. Rubinstein [Rub95] and [Rub04]. Notice that since the manifold M is known to be hyperbolic or Seifert fibred, Z. Sela gives an algorithm at the level of the fundamental group to decide whether M is a non-Haken Seifert fibred space or a hyperbolic manifold, by running two algorithms at the same time: one to detect the word hyperbolicity of $\pi_1(M)$ and the second to look for a central element in $\pi_1(M)$ using the fact that the word problem is solvable in $\pi_1(M)$, since it is residually finite (see [Sel95], Section 10).

Step 5. *The homeomorphism problem is algorithmically solvable for non-Haken, Seifert fibred* 3*-manifolds.*

These Seifert fibred 3-manifolds are either spherical with finite fundamental group or their fundamental group is a central extension of a Euclidean or hyperbolic triangle group by an infinite cyclic group. One can algorithmically determine if such a manifold M is spherical, because its first homology group $H_1(M;\mathbb{Z})$ is finite and there are finitely many spherical 3-manifolds with given homology, see [Orl72]. Therefore, using the solution of the word problem, one can check if $\pi_1(M)$ is isomorphic to a finite group in the list. Then the homeomorphism problem for spherical 3-manifolds is algorithmically solvable, since, except for lens spaces, spherical manifolds are determined by their fundamental group, which is finite. Moreover, one can also decide when two lens spaces are homeomorphic.

By [OVZ67], non-Haken, Seifert fibred 3-manifolds with infinite fundamental group are determined, up to homeomorphism, by their fundamental group. Z. Sela has shown how to solve algorithmically the isomorphism problem for these fundamental groups, using the uniqueness of the central extension, see [Sel95], Section 10.

Another more geometrical approach is to use the fact that these manifolds have Heegaard genus 2 and that the homeomorphism problem is algorithmically solvable for 3-manifolds having Heegaard genus 2 (see [Rub97], [Rub04]).

Hence we are left with the case of manifolds having a hyperbolic structure:

Step 6. *Given two closed, orientable, hyperbolic* 3*-manifolds, there is an algorithm to decide whether or not they are homeomorphic.*

The Mostow rigidity theorem states that homotopy equivalent, closed, hyperbolic 3-manifolds are homeomorphic. Then the solution of the homeomorphism problem for closed hyperbolic 3-manifolds follows from Sela's algorithm to solve the isomorphism problem for word hyperbolic groups, since we know the constant δ of hyperbolicity [Sel95], Section 7.

We remark that, according to Manning's algorithm to recognize hyperbolicity of 3-manifolds [Man02], given a closed orientable, triangulated hyperbolic 3-manifold, there is an algorithm to construct the hyperbolic structure.

In contrast, the homeomorphism problem is not solvable for closed triangulated manifolds of dimension $n \geqslant 4$. Any finitely presented group can be realized as the fundamental group of a closed 4-manifold. This fact shows the impossibility of classifying 4-manifolds since it is impossible to classify finitely presented groups. Here is a more precise statement, see [Mar60] and also [CL06].

Theorem 1.4.2 (Markov). *For every natural number $n \geqslant 4$, one can build a combinatorial n-dimensional manifold M^n such that there is no algorithm to decide whether a PL n-manifold is PL-homeomorphic to M^n.*

In fact, according to [CL06], all manifolds of dimension $n \geqslant 5$ and all manifolds of the type $M^4 \# 14(S^2 \times S^2)$ satisfy the conclusion of the theorem.

1.4.2 Fundamental group

In general, closed, orientable 3-manifolds are not determined, up to homeomorphism, by their fundamental group. However, the solution of the Geometrisation Conjecture implies that it is the case for closed, orientable, prime 3-manifolds, unless they are lens spaces. This result shows the importance of the fundamental group in the study of 3-manifolds and solves the Borel Conjecture for closed, aspherical 3-manifolds.

Theorem 1.4.3. *Let M, N be two closed, orientable, prime 3-manifolds. If M is not a lens space, then M is homeomorphic to N if and only if $\pi_1(M)$ is isomorphic to $\pi_1(N)$.*

By Perelman's proof of the Geometrisation Conjecture, a closed prime 3-manifold is either homeomorphic to $S^2 \times S^1$ or it is irreducible and either Haken, hyperbolic or Seifert fibred. Furthermore, the Geometrisation Conjecture implies that two manifolds with isomorphic fundamental group belong to the same class. Then the theorem is due to F. Waldhausen [Wal68a] when they are Haken, to Mostow rigidity when they are hyperbolic and to P. Scott [Sco83] when they are Seifert fibred with infinite fundamental group. In the remaining case where the two manifolds are spherical and have finite fundamental groups, the theorem follows from the work of Seifert and Threlfall (see for example [Orl72]).

As said above, any finitely presented group can be realized as the fundamental group of a closed 4-manifold. In contrast, closed 3-manifolds admit topological and geometrical properties that put strong constraints on their fundamental groups. For example the existence of a Heegaard decomposition shows that a closed 3-manifold (fundamental) group admits a finite *balanced* presentation, i.e., a presentation with as many relators as generators.

The existence of a geometric decomposition implies that any 3-manifold group is the fundamental group of a *graph of groups* whose vertices are discrete subgroups of isometries of the 3-dimensional geometries, and the edge groups are trivial or isomorphic to $\mathbb{Z}^2$. This graph of group structure reduces in principle any problem on 3-manifold groups to combination theorems for fundamental groups of geometric manifolds, which are linear groups.

It is still not known whether 3-manifold groups are linear, however they share some fundamental properties with finitely generated linear groups. For example they are *residually finite* and they satisfy *Tits alternative*: a 3-manifold group contains a non-abelian free group unless it is virtually solvable.

A consequence of this algebraic structure is that some classical decision problems formulated by Dehn in 1910 are solvable in the class of closed 3-manifold groups.

The *word problem* (Dehn 1910) asks for an algorithm to decide whether or not a word in the generators represents the trivial element. Its solution for a 3-manifold group relies on early work by Waldhausen [Wal68b] and more recent work on automatic groups, cf. [ECH$^+$92]. The solution of the word problem is also a consequence of the property that 3-manifold groups are residually finite.

The *conjugacy problem* (Dehn 1910) asks for an algorithm to decide whether or not a pair of words in the generators are conjugate. Given an algorithm to solve the word problem, the solution for 3-manifold groups is due to J.-P. Préaux [Pré06], [Pré05].

The solutions to these problems use their topological versions, i.e., given a 3-manifold with some finite topological data from which the group presentation is deduced. It turns out that the existence of solutions for these problems does not depend on the finite presentation involved, but only on the isomorphism class of G.

The *isomorphism problem* (Tietze 1908, Dehn 1910) asks for an algorithm to decide whether or not two finite presentations of groups in a given class present isomorphic groups. Given two compact 3-manifolds M and N, the existence of an algorithm to decide if $\pi_1(M)$ and $\pi_1(M)$ are isomorphic follows from the solution of the homeomorphism problem.

Theorem 1.4.4. *The word problem, the conjugacy problem and the isomorphism problem are solvable in the class of fundamental groups of closed, orientable 3-manifolds.*

The proof of this theorem does not require only that the groups are known to be isomorphic to fundamental groups of 3-manifolds, but that the topological types of these manifolds are known. It follows from Theorem 1.4.3 that only finitely many closed, orientable 3-manifolds can have a fundamental group isomorphic to a given group G, and that this number is bounded by the order of the torsion part of the abelianization of G. Hence, given a presentation of a closed, orientable 3-manifold group, one can use Tietze transformations to find algorithmically the 3-manifolds realizing this fundamental group in the enumerable list of all closed orientable 3-manifolds.

By work of Stallings [Sta62], the property for a *finitely presented* group to belong to the class of the fundamental groups of compact 3-manifolds is *markovian*, and thus it is algorithmically undecidable. More precisely, given any non-empty class of connected 3-manifolds, there is no algorithm for deciding whether or not a finite presentation of a group defines a group isomorphic to the fundamental group of an element of this class.

Stallings' result raises the question of characterizing the fundamental groups of closed, orientable, irreducible 3-manifolds. There are actually two main conjectures related to this question. Due to the solution of the Geometrisation Conjecture, it turns out that for the class of word hyperbolic groups these two conjectures are equivalent.

A group G is an orientable *n-dimensional Poincaré duality group* or a PD(n)*-group* if for every $\mathbb{Z}G$-module A there is an isomorphism between $H^i(G; A)$ and $H_{n-i}(G; A)$. Equivalently, G is a PD(n)-group if it acts freely, properly discontinuously and cocompactly on a contractible cell complex X with the same cohomology with compact supports as $\mathbb{R}^n$: $H_c^*(X, \mathbb{Z}) \cong H_c^*(\mathbb{R}^n, \mathbb{Z})$. Fundamental groups of closed, orientable, irreducible 3-manifolds are PD(3)-groups. A good reference for this topic is [Bro82].

The first conjecture, due to C. T. C. Wall in the 1960s, asks whether this cohomological duality is sufficient to characterize the fundamental group of closed irreducible, orientable 3-manifolds, cf. C. T. C. Wall's survey [Wal04].

Conjecture 1.4.5 (Wall). *Every finitely presented* PD(3)*-group is isomorphic to the fundamental group of some closed, orientable, irreducible* 3*-manifold.*

B. Eckmann, P. Linnell and H. Müller have shown that PD(2)-groups are surface groups, see [Eck87]. For $n \geqslant 4$, M. Davis [Dav98] gave examples of PD(n)-groups which are not finitely presentable, and thus cannot be fundamental groups of closed n-dimensional manifolds.

B. Bowditch [Bow04] has shown that every PD(3)-group with a non-trivial infinite cyclic normal subgroup is isomorphic to the fundamental group of a Seifert fibred manifold. Together with results by J. Hillman [Hil87] and C.B. Thomas [Tho84], this implies that every PD(3)-group which contains a non-trivial, finitely generated, normal subgroup of infinite index is isomorphic to the fundamental group of some 3-manifold which is either Seifert fibred or a surface bundle.

The solution of the Geometrisation Conjecture implies that a PD(3)-group is isomorphic to a 3-manifold group if and only if it contains a subgroup isomorphic to the fundamental group of a closed, orientable, irreducible 3-manifold, see [Hil08].

M. Dunwoody and E. Swenson [DS00] have proved a *torus theorem* for PD(3)-groups. Every PD(3)-group has one of the following properties:

(i) it is the fundamental group of a Seifert 3-manifold,
(ii) it splits over a subgroup isomorphic to $\mathbb{Z} \oplus \mathbb{Z}$,
(iii) it is *atoroidal*, i.e., contains no $\mathbb{Z} \oplus \mathbb{Z}$.

If the fundamental group of a closed, orientable, irreducible 3-manifold is atoroidal, Perelman's hyperbolization theorem shows that it is a discrete cocompact subgroup of $\mathrm{PSL}(2, \mathbb{C})$. Gehring and Martin showed that some fundamental properties of these groups depend only on a topological property of the dynamics of their action on the sphere at infinity S^2 of the 3-dimensional hyperbolic space $\mathbb{H}^3$. They call this property the *convergence property*, and the groups of homeomorphisms of S^2 satisfying this condition have been subject of much attention.

Let Γ be a group acting by homeomorphisms on S^2. Denote by ρ the associated homomorphism from Γ to $\mathrm{Homeo}(S^2)$. Then Γ is a called a *convergence group* if for each sequence $\{g_n\}$ of elements of Γ such that the $\rho(g_n)$'s are pairwise distinct, there exist points $a, b \in S^2$ and a subsequence $\{g_{n_k}\}$ such that $\lim \rho(g_{n_k})(x) = a$ uniformly on compact subsets of S^2 that do not contain b.

Denote by $\Theta(S^2)$ the set of triples $(x, y, z) \in S^2 \times S^2 \times S^2$ such that x, y, z are pairwise distinct, topologized as a subset of $S^2 \times S^2 \times S^2$. Then Γ is a convergence group if and only if the induced action of Γ on $\Theta(S^2)$ is proper. The convergence group Γ is called *uniform* if the induced action on $\Theta(S^2)$ is cocompact.

The second conjecture, due to J. Cannon, states that the convergence property is sufficient to characterize discrete cocompact subgroups of $\mathrm{PSL}(2, \mathbb{C})$:

Conjecture 1.4.6 (Cannon). *If Γ is a uniform convergence group acting on S^2, then Γ has a finite normal subgroup K such that Γ/K is a cocompact subgroup of* $\mathrm{PSL}(2, \mathbb{C})$.

B. Bowditch has shown that a uniform convergence group on S^2 is word hyperbolic with boundary homeomorphic to the sphere S^2. By M. Bestvina and G. Mess [BM91], a word hyperbolic group is a PD(3)-group if and only if its boundary is homeomorphic to S^2, that is to say if and only if it is a uniform convergence group on S^2. It follows then from the solution of the Geometrisation Conjecture that for word hyperbolic groups, Cannon's conjecture is equivalent to Wall's conjecture.

1.5 Final remarks

1.5.1 Comparison with Perelman's original arguments

One important difference between Perelman's approach and ours is that he does not use Thurston's hyperbolisation theorem for Haken manifolds. By studying the long-time behaviour of Ricci flow with surgery, Perelman gets a thin–thick decomposition theorem (cf. Theorem 1.3.4). He proves that the thick part is hyperbolic and that the thin part is a graph manifold. This last result uses a collapsing theorem for manifolds with toric boundary. In order to prove that the boundary components of the thick part (if any) are incompressible in M, he suggests two arguments. The first one, borrowed from Hamilton [Ham99], has two steps: first prove that the hyperbolic pieces *persist* and then argue by contradiction using the Meeks–Yau geometric loop theorem and the evolution of the area of a minimal compressing disk under Ricci flow. The second one, which inspired our argument in Section 1.3.2, uses earlier ideas of Anderson [And02] together with the monotonicity along the Ricci flow of the first eigenvalue of some Schrödinger-type operator.

A version of Perelman's collapsing theorem for closed 3-manifolds is given in the appendix of a paper by Shioya and Yamaguchi [SY05]. They use deep results of Alexandrov space theory, including Perelman's stability theorem [Per91] (see also the paper by V. Kapovitch [Kap07]) and a fibration theorem for Alexandrov spaces, proved by Yamaguchi [Yam96]. Other proofs of Perelman's collapsing theorem can be found in papers by Morgan–Tian [MT08], Cao–Ge [CG09], and Kleiner–Lott [KL10]. For more details see Section 13.4.

1.5.2 Beyond geometrisation

Hamilton–Perelman technology now has applications which go beyond the geometrisation of closed 3-manifolds. Dinkelbach and Leeb [DL09] use an equivariant version of Perelman's Ricci flow with surgery to finish off Thurston's geometrisation program for orbifolds, including the nonorientable case. In [BBM09] the construction of Parts I and II of this monograph is generalised to complete 3-manifolds with bounded geometry. This result has an application to the classification of complete Riemannian 3-manifolds with bounded geometry and uniformly positive scalar curvature. Codá Marques [CM09] uses a refinement of Perelman's construction to prove the arcwise connectedness of the moduli space of metrics with positive scalar curvature on a closed, orientable 3-manifold.

In [AST07], Agol, Storm and Thurston use the monotonicity of scalar curvature along the Ricci flow and the existence of solutions for $\mathcal{C}^0$ initial conditions, to prove the following. Let (M, g) be a hyperbolic 3-manifold with minimal surface boundary, then $\operatorname{vol}(M, g) \geqslant \frac{1}{2} v_3 \| DM \|$, where DM denotes the double of M across its boundary, $\| DM \|$, the simplicial volume, and v_3 denotes the volume of the regular ideal hyperbolic

3-simplex (see Section 13.2). This result is in turn used by D. Gabai, R. Meyerhoff and P. Milley to determine the closed, orientable hyperbolic 3-manifold of smallest volume (see e.g. [Mai10] and the references therein).

Part I

Ricci flow with bubbling-off: definitions and statements

Chapter 2
Basic definitions

In order to prove the finite-time existence theorem for Ricci flow with bubbling-off (Theorem 1.2.3), we need to introduce a more precise notion, called Ricci flow with (r, δ, κ)-bubbling-off. As we already explained in Section 1.2.2, the parameters r, δ control the scale of the canonical neighbourhoods, and the precision of the metric surgery. The parameter κ refers to the important notion of κ-noncollapsing, introduced by Perelman in [Per02], which will be defined in Chapter 4. In Chapter 5, we will state an existence theorem for Ricci flow with (r, δ, κ)-bubbling-off (Theorem 5.3.1). This result implies Theorem 1.2.3. In order to make the proof of Theorem 5.3.1 more digestible, and in particular to clarify the interdependence of the various parameters, we will organise this proof in three independent propositions, called Propositions A, B, C. These propositions will be proved in Part II.

In this section, we fix some conventions regarding Riemannian geometry, such as the definitions of curvature and closeness of metrics. We also formalise the fundamental notions of *evolving metric* and *Ricci flow with bubbling-off.*

Throughout this section, we let M be an n-dimensional manifold, where $n \geqslant 2$, and U be an open subset of M.

2.1 Riemannian geometry conventions

Let g be a Riemannian metric on M. Then

$$\mathrm{Riem}(X, Y, Z, T) = g(D_Y D_X Z - D_X D_Y Z + D_{[X,Y]} Z, T)$$

is the *Riemann curvature tensor* and for any $x \in M$, we denote by $\mathrm{Rm}(x) \colon \bigwedge^2 T_x M \to \bigwedge^2 T_x M$ the *curvature operator* defined by

$$\langle \mathrm{Rm}(X \wedge Y), Z \wedge T \rangle = \mathrm{Riem}(X, Y, Z, T),$$

where $\wedge$ and $\langle \cdot, \cdot \rangle$ are normalised so that $\{e_i \wedge e_j \mid i < j\}$ is an orthonormal basis if $\{e_i\}$ is.

We denote by $|\mathrm{Rm}(x)|$ the operator norm of Rm, which is also the maximum of the absolute values of the sectional curvatures at x. We let $R(x)$ denote the scalar curvature of x. The infimum (resp. supremum) of the scalar curvature of g on M is denoted by $R_{\min}(g)$ (resp. $R_{\max}(g)$).

We write $d \colon M \times M \to [0, \infty)$ for the distance function associated to g. For $r > 0$ we denote by $B(x, r)$ the open ball of radius r around x.

Finally, if x, y are points of M, we denote by $[xy]$ a geodesic segment, that is a minimizing curve, connecting x to y. This is a (common) abuse of notation, since such a segment is not unique in general.

Closeness of metrics

Definition 2.1.1. Let T be a tensor on U, and N be a nonnegative integer. We write

$$\|T\|_{N,U,g}^2 := \sup_{x\in U} \sum_{k=0}^{N} |\nabla_g^k T(x)|_g^2.$$

Here the pointwise norm is the Euclidean one. More precisely, if S is a p-linear map on T_xU then,

$$|S|_g^2 = \sum S(e_{i_1},\dots,e_{i_p})^2,$$

where $\{e_i\}$ is a g-orthonormal basis at x. Let ε be a positive number and g_0 be a Riemannian metric on U. One says that g is *ε-close* to g_0 on U if

$$\|g - g_0\|_{[\varepsilon^{-1}],U,g_0} < \varepsilon.$$

One says that g is *ε-homothetic* to g_0 on U if there exists $\lambda > 0$ such that λg is ε-close to g_0 on U.

A pointed Riemannian manifold (U, g, x) is called *ε-close* to a model $(U_0, g_0, *)$ if there exists a $\mathcal{C}^{[\varepsilon^{-1}]+1}$-diffeomorphism ψ from U_0 to U sending $*$ to x and such that the pull-back metric $\psi^*(g)$ is ε-close to g_0 on U_0. We say that (U, g, x) is *ε-homothetic* to $(U_0, g_0, *)$ if there is a positive number λ such that $(U, \lambda g, x)$ is ε-close to $(U_0, g_0, *)$. The map ψ is called an *ε-isometry* or an *ε-homothety*.

A Riemannian manifold (U, g) is *ε-close* (resp. *ε-homothetic*) to (U_0, g_0) if there exist points $x \in U$ and $* \in U_0$ such that the pointed manifold (U, g, x) is ε-close (resp. ε-homothetic) to $(U_0, g_0, *)$. If there is no ambiguity on the relevant metrics, we may simply say that U is ε-close (resp. *ε-homothetic*) to U_0.

2.2 Evolving metrics and Ricci flow with bubbling-off

Let M be an n-manifold. An *evolving metric* on M defined on an interval $[a, b]$ is a map $t \mapsto g(t)$ from $[a, b]$ to the space of $\mathcal{C}^\infty$ Riemannian metrics on M. In the sequel, this space is endowed with the $\mathcal{C}^2$ topology. We say that an evolving metric $g(\cdot)$ is *piecewise* $\mathcal{C}^1$ if there is a finite subdivision $a = t_0 < t_1 < \dots < t_p = b$ of the interval of definition with the following property: for each $i \in \{0, \dots, p-1\}$, there exists a metric $g_+(t_i)$ on M such that the map defined on $[t_i, t_{i+1}]$ by sending t_i to $g_+(t_i)$ and every $t \in (t_i, t_{i+1}]$ to $g(t)$ is $\mathcal{C}^1$-smooth.

For $t \in [a,b]$, we say that t is *regular* if $g(\cdot)$ is $\mathcal{C}^1$-smooth in a neighbourhood of t. Otherwise t is called *singular*. By definition, the set of singular times is finite. If $t \in (a,b)$ is a singular time, then it follows from the definition that the map $g(\cdot)$ is continuous from the left at t, and has a limit from the right, denoted by $g_+(t)$.

There are similar definitions where the domain of definition $[a,b]$ is replaced by an open or a half-open interval I. When I has infinite length, the set of singular times may be infinite, but must be discrete as a subset of $\mathbb{R}$.

We often view an evolving metric $g(\cdot)$ on a manifold M as a 1-parameter family of metrics indexed by the interval I; thus we use the notation $\{g(t)\}_{t\in I}$. For each pair $(x,t) \in M \times I$, we have an inner product on T_xM; whenever necessary, we use the notation $g(x,t)$ for this inner product.

Definition 2.2.1. Let $I \subset \mathbb{R}$ be an interval. A piecewise $\mathcal{C}^1$ evolving metric $t \mapsto g(t)$ on M defined on I is said to be a *Ricci flow with bubbling-off* if

(i) the Ricci flow equation $\frac{\partial g}{\partial t} = -2\,\mathrm{Ric}$ is satisfied at all regular times;

(ii) for every singular time $t \in I$ we have

- (a) $R_{\min}(g_+(t)) \geqslant R_{\min}(g(t))$, and
- (b) $g_+(t) \leqslant g(t)$.

Let $\{g(t)\}_{t\in I}$ be a Ricci flow with bubbling-off on M. For $t \in I$, the *time-t singular locus* of this solution, denoted by Σ_t, is the closure of the set $\{x \in M \mid g(x,t) \neq g_+(x,t)\}$. If $\Sigma_t \neq \emptyset$ then t is singular. The converse will be true for Ricci flows with bubbling-off that will be considered later in this book: if t is singular then $g(t) \neq g_+(t)$ at some point.

Definition 2.2.2. A product subset $X \times [a,b] \subset M \times I$ is *scathed* if $X \times [a,b)$ contains a singular point. Otherwise, we say that X is *unscathed*. Similarly, a curve $\gamma\colon [a,b] \to M$ is *scathed* if there exists $t \in [a,b)$ such that $\gamma(t) \in \Sigma_t$. Otherwise it is *unscathed*.

Note that a subset $X \times [a,b]$ may be unscathed even if there is a singular point in $X \times \{b\}$. This definition is consistent with our choice of making the map $t \mapsto g(t)$ continuous from the left.

A Ricci flow with bubbling-off without singular point is of course a true solution of the Ricci flow equation. For brevity, we sometimes call such a solution a *Ricci flow*.

It will occasionally be useful to consider partially defined Ricci flows, where the domain of definition of $t \mapsto g(x,t)$ is allowed to depend on $x \in X$:

Definition 2.2.3. Let $(a,b]$ a time interval. A *partial Ricci flow* on $U \times (a,b]$ is a pair $(\mathcal{P}, g(\cdot,\cdot))$, where $\mathcal{P} \subset U \times (a,b]$ is an open subset which contains $U \times \{b\}$ and $(x,t) \mapsto g(x,t)$ is a smooth map defined on $\mathcal{P}$ such that the restriction of g to any subset $V \times I \subset \mathcal{P}$ is a Ricci flow on $V \times I$.

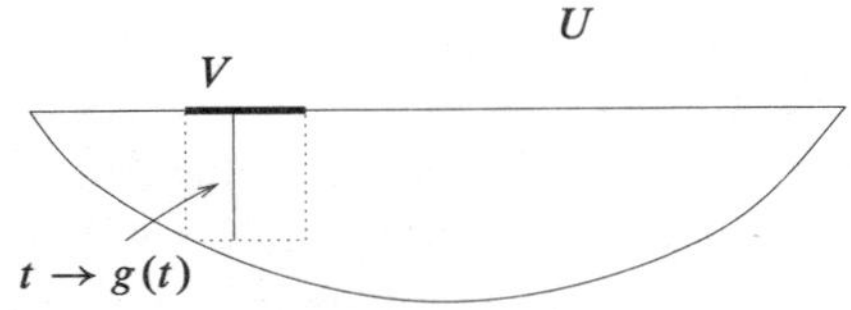

Notation. If $\{g(t)\}_{t\in I}$ is an evolving metric on M, and $(x,t) \in M \times I$, then we use the notation $\mathrm{Rm}(x,t)$, $R(x,t)$ to denote the curvature operator and the scalar curvature respectively. For brevity we set $R_{\min}(t) := R_{\min}(g(t))$ and $R_{\max}(t) := R_{\max}(g(t))$.

We use $d_t(\cdot,\cdot)$ for the distance function associated to $g(t)$. The ball of radius r around x for $g(t)$ is denoted by $B(x,t,r)$.

Definition 2.2.4. If $T(\cdot)$ is an evolving tensor on $U \times [a,b]$, $g(\cdot)$ an evolving metric, and N a nonnegative integer, we write

$$\|T(\cdot)\|_{N,U\times[a,b],g(\cdot)} = \sup_{t\in[a,b]} \|T(t)\|_{N,U,g(t)}.$$

Given two evolving metrics $g_0(\cdot)$, $g(\cdot)$ on $U \times [a,b]$, we say that $g(\cdot)$ is ε-close to $g_0(\cdot)$ on $U \times [a,b]$, if

$$\|g(\cdot) - g_0(\cdot)\|_{[\varepsilon^{-1}],U\times[a,b],g_0(\cdot)} < \varepsilon.$$

Equivalently,

$$\sup_{t\in[a,b]} \|g(t) - g_0(t)\|_{[\varepsilon^{-1}],U,g_0(t)} < \varepsilon.$$

Closeness of evolving metrics. Given two pointed evolving Riemannian manifolds $(U_0, g_0(\cdot), *)$ and $(U, g(\cdot), x)$, where $g_0(\cdot)$ and $g(\cdot)$ are defined on $[a,b]$, we say that $(U, g(\cdot), x)$ is *ε-close* to $(U_0, g_0(\cdot), *)$ if there exists a $\mathcal{C}^{[\varepsilon^{-1}]+1}$-diffeomorphism ψ from U_0 to U sending $*$ to x, and such that $\psi^* g(\cdot)$ is ε-close to $g_0(\cdot)$ on $U_0 \times [a,b]$. We say that $(U, g(\cdot))$ is *ε-close* to $(U_0, g_0(\cdot))$ if there exist points $x \in U$ and $x_0 \in U_0$ such that this property holds.

Pointed convergence of evolving metrics. We say that a sequence of pointed evolving metrics $(M_k, \{g_k(t)\}_{t\in I}, (x_k, t_0))$ *converges smoothly* to a pointed evolving metric $(M_\infty, \{g_\infty(t)\}_{t\in I}, (x_\infty, t_0))$ if there exists an exhaustion of M by open sets U_k, such that $x \in U_k$ for all k, and smooth maps $\psi_k \colon U_k \to M_k$, diffeomorphic onto their images, sending x to x_k, such that $\psi_k^* g_k(\cdot) - g(\cdot)$ and all its derivatives converge to zero uniformly on compact subsets of $M \times I$.

Definition 2.2.5. Let $(\{g(t)\}_{t\in I})$ be an evolving metric on M, $t_0 \in I$ and $Q > 0$. The *parabolic rescaling* with factor Q at time t_0 is the evolving metric

$$t \mapsto Q\, g\Big(t_0 + \frac{t}{Q}\Big).$$

Remark 2.2.6. Any parabolic rescaling of a Ricci flow is a Ricci flow.

Distance comparison estimates under Ricci flow. We gather below some usual distance comparison estimates under Ricci flow.

Lemma 2.2.7 (Multiplicative distance-distortion). *Let $g(\cdot)$ be a Ricci flow on U, defined for $t \in [t_1, t_2]$. Suppose that $|\mathrm{Rm}| \leqslant \Lambda$ on $U \times [t_1, t_2]$. Then*

$$e^{-2(n-1)\Lambda(t_2-t_1)} \leqslant \frac{g(t_2)}{g(t_1)} \leqslant e^{2(n-1)\Lambda(t_2-t_1)}. \tag{2.1}$$

Proof. If $|\mathrm{Rm}| \leqslant \Lambda$ then $|\mathrm{Ric}| \leqslant (n-1)\Lambda$. Let $x \in U$ and $u \in T_xU$ be a nonzero vector. Then

$$\left|\frac{d}{dt} g_{(x,t)}(u,u)\right| = |2\,\mathrm{Ric}_{(x,t)}(u,u)| \leqslant 2(n-1)\Lambda g_{(x,t)}(u,u),$$

hence $|\frac{d}{dt} \ln g_{(x,t)}(u,u)| \leqslant 2(n-1)\Lambda$. Integrating between t_1 and t_2 gives the result. □

Remark 2.2.8. In particular, if $t_2 - t_1 \leqslant (2\Lambda(n-1))^{-1}$, we have

$$4^{-1} \leqslant \frac{g(t_2)}{g(t_1)} \leqslant 4,$$

and

$$B\left(x, t_2, \frac{r}{2}\right) \subset B(x, t_1, r) \subset B(x, t_2, 2r)$$

if $B(x, t_2, 2r)$ is compactly contained in U. This will be used repeatedly.

Remark 2.2.9. The upper bound (resp. lower bound) in (2.1) is true under the assumption $\mathrm{Ric} \geqslant -(n-1)\Lambda$ (resp. $\mathrm{Ric} \leqslant (n-1)\Lambda$).

Lemma 2.2.10 (Additive distance-distortion, [Ham95b], Section 17). *Let $g(\cdot)$ be a Ricci flow on M, and K be a constant. Suppose that $d_t(x_0, x_1) \geqslant 2r_0$ and* $\mathrm{Ric} \leqslant (n-1)K$ *for all $x \in B(x_0, t_0, r_0) \cup B(x_1, t_0, r_0)$. Then*

$$\frac{d}{dt} d_t(x_0, x_1) \geqslant -2(n-1)\left(\frac{2}{3}Kr_0 + r_0^{-1}\right), \tag{2.2}$$

at time $t = t_0$.

Here is another distance-distortion lemma.

Lemma 2.2.11 ([Ham95b], Theorem 17.12). *There exists a constant $C(n) > 0$ such that if $g(\cdot)$ is a Ricci flow on M such that* $\mathrm{Ric} \leqslant (n-1)K$ *for some constant $K > 0$, then for all x_0, x_1 in M,*

$$\frac{d}{dt} d_t(x_0, x_1) \geqslant -C(n)K^{1/2}. \tag{2.3}$$

Notes. The convention we have chosen for the curvature operator is the same as Morgan–Tian's. Note that other conventions are to be found in the Ricci flow literature, in particular in Hamilton's papers.

The word *unscathed* was introduced by Kleiner and Lott in [KL08]. Our definition is slightly different from theirs.

The definition of Ricci flow with bubbling-off is new. As noted in the introduction (see Remark 1.2.5), there are two main differences from Perelman's Ricci flow with surgery: first, the manifold does not change with time; second, more importantly surgery is done *before* curvature blow-up (compare with [Ham97]). This spares us the need to analyse the structure of the manifold at blow-up time as in [Per03b], Section 3. However, the proof of existence uses contradiction arguments which are quite close to Perelman's (cf. [Per03b], Sections 4, 5).

Chapter 3
Piecing together necks and caps

As explained in the introduction, one of Perelman's main achievements is to show that in a 3-dimensional Ricci flow, points of sufficiently high scalar curvature have so-called *canonical neighbourhoods*. The precise definition, which is rather technical, will be given in Chapter 4. In this chapter, we study weaker, simpler notions called *ε-necks* and *ε-caps*, and prove some topological results about them.

3.1 Necks, caps and tubes

3.1.1 Necks

The *standard ε-neck* is the Riemannian product $N_\varepsilon := S^2 \times (-\varepsilon^{-1}, \varepsilon^{-1})$, where the S^2 factor is round of scalar curvature 1. Its metric is denoted by g_{cyl}. We fix a basepoint $*$ in $S^2 \times \{0\}$.

Definition 3.1.1. Let (M, g) be a Riemannian 3-manifold and $x \in M$. A neighbourhood $N \subset M$ of x is called an *ε-neck centred at x* if (N, g, x) is ε-homothetic to $(N_\varepsilon, g_{\mathrm{cyl}}, *)$.

If N is an ε-neck and $\psi\colon N_\varepsilon \to N$ is a *parametrisation*, i.e., a diffeomorphism such that some rescaling of $\psi^*(g)$ is close to g_{cyl}, then the sphere $S = \psi(S^2 \times \{0\})$ is called the *middle sphere* of N. This is slightly abusive, since the parametrisation is not unique.

3.1.2 Caps and tubes

Definition 3.1.2. Let (M, g) be a Riemannian 3-manifold and x be a point of M. We say that U is an *ε-cap centred at x* if U is the union of two sets V, W such that $x \in \operatorname{Int} V$, V is a closed topological 3-ball, $\bar{W} \cap V = \partial V$, and W is an ε-neck. A subset V as above is called a *core* of U.

Definition 3.1.3. An *ε-tube* is an open subset $U \subset M$ which is equal to a finite union of ε-necks, and whose closure in M is diffeomorphic to $S^2 \times I$.

3.2 Gluing results

There is a function $\varepsilon \mapsto f_3(\varepsilon)$ tending to zero as ε tends to zero, such that if N is an ε-neck with scaling factor λ and $x, y \in N$, then we have

$$|\lambda R(x) - 1| \leqslant f_3(\varepsilon), \quad \left|\frac{R(x)}{R(y)} - 1\right| \leqslant f_3(\varepsilon).$$

Definition 3.2.1. We fix a constant $\varepsilon_0 \in (0, 10^{-3})$ sufficiently small so that $f_3(\varepsilon)$, $f_3(10\varepsilon) \leqslant 10^{-2}$ if $\varepsilon \in (0, 2\varepsilon_0]$.

Lemma 3.2.2. *Let $\varepsilon \in (0, 2\varepsilon_0]$. Let (M, g) be a Riemannian 3-manifold. Let y_1, y_2 be points of M. Let $U_1 \subset M$ be an ε-neck centred at y_1 with parametrisation $\psi_1 \colon S^2 \times (-\varepsilon^{-1}, \varepsilon^{-1}) \to U_1$ and middle sphere S_1. Let $U_2 \subset M$ be a 10ε-neck centred at y_2 with middle sphere S_2. Call $\pi : U_1 \to (-\varepsilon^{-1}, \varepsilon^{-1})$ the composition of ψ_1^{-1} with the projection of $S^2 \times (-\varepsilon^{-1}, \varepsilon^{-1})$ onto its second factor.*

Assume that $y_2 \in U_1$ and $|\pi(y_2)| \leqslant (2\varepsilon)^{-1}$. Then the following conclusions hold:

(i) *U_2 is contained in U_1.*

(ii) *The boundary components of $\partial\overline{U_2}$ can be denoted by S_+, S_- in such a way that*

$$\pi(S_-) \subset [\pi(y_2) - (10\varepsilon)^{-1} - 10, \pi(y_2) - (10\varepsilon)^{-1} + 10]$$

and

$$\pi(S_+) \subset [\pi(y_2) + (10\varepsilon)^{-1} - 10, \pi(y_2) + (10\varepsilon)^{-1} + 10].$$

(iii) *The spheres S_1, S_2 are isotopic in U_1.*

Proof. Let $\psi_2 \colon S^2 \times (-(10\varepsilon)^{-1}, (10\varepsilon)^{-1}) \to U_2$ be a parametrisation, and call λ_i the scaling factor of ψ_i for $i = 1, 2$. By choice of ε, the quotient λ_2/λ_1 is close to 1, so the map $\psi_1^{-1} \circ \psi_2$ is almost an isometry. Then (i) and (ii) follow from straightforward distance computations.

Let us show assertion (iii). By Corollary A.2.2, it suffices to prove that S_2 is not null-homologous in U_1. Assume that on the contrary, S_2 is null-homologous in U_1. Topologically, U_2 is a tubular neighbourhood of S_2, so the inclusion $S_2 \to U_2$ is a homotopy equivalence. Hence the induced homomorphism $H_2(U_2) \to H_2(U_1)$ has zero image.

Set $S := \psi_1(S^2 \times \{\pi(y_2)\})$. By (ii), the diameter of S is less than the distance between y_2 and $\partial\overline{U_2}$. It follows that $S \subset U_2$. Now the inclusion $S \to U_1$ is a homotopy equivalence, so the induced homomorphism $H_2(S) \to H_2(U_1)$ is an isomorphism. This is impossible. □

We now treat the case of closed manifolds.

Proposition 3.2.3. *Let $\varepsilon \in (0, 2\varepsilon_0]$. Let (M, g) be a closed, orientable Riemannian 3-manifold. Let X be a compact, connected, nonempty subset of M such that every point of X is the centre of an ε-neck or an ε-cap. Then there exists an open subset $U \subset M$ containing X such that either*

(a) *U is equal to M and diffeomorphic to S^3 or $S^2 \times S^1$, or*

(b) *U is a 10ε-cap, or*

(c) *U is a 10ε-tube.*

In addition, if M is irreducible and every point of X is the centre of an ε-neck, then Case (c) *holds.*

Proof. First we deal with the case where X is covered by ε-necks.

Lemma 3.2.4. *If every point of X is the centre of an ε-neck, then there exists an open set U containing X such that $U = M \cong S^2 \times S^1$, or U is a 10ε-tube.*

Proof. Let x_0 be a point of X and N_0 be a 10ε-neck centred at x, contained in an ε-neck U_0, also centred at x. If $X \subset N_0$ we are done. Otherwise, since X is connected, we can pick a point $x_1 \in X \cap N_0$ and a 10ε-neck N_1 centred at x_1, with x_1 arbitrarily near the boundary of N_0. By Lemma 3.2.2, an appropriate choice of x_1 ensures that $N_1 \subset U_0$ and the middle spheres of N_0 and N_1 are isotopic. In particular, the closure of $N_0 \cup N_1$ is diffeomorphic to $S^2 \times I$, so $N_0 \cup N_1$ is a 10ε-tube.

If $X \subset N_0 \cup N_1$, then we can stop. Otherwise, we go on, producing a sequence $x_0, x_1, x_2, \ldots$ of points of X together with a sequence $N_0, N_1, N_2 \ldots$ of 10ε-necks centred at those points, such that $N_0 \cup \cdots \cup N_k$ is a 10ε-tube. Since X is compact, the scalar curvature on X is bounded. Hence definite volume is added at each step. This implies that the construction must stop. Then either we have found a 10ε-tube containing X, or $N_0 \cap N_k \neq \emptyset$. In the latter case, we can adjust the choice of x_k and N_k so that the necks N_i cover M, and can be cyclically ordered in such a way that two consecutive necks form a tube. Since M is orientable, it follows that M is diffeomorphic to $S^2 \times S^1$. □

To complete the proof of Proposition 3.2.3, we need to deal with the case where there is a point $x_0 \in X$ which is the centre of an ε-cap C_0. By definition of a cap, a collar neighbourhood of the boundary of C_0 is an ε-neck U_0. If $X \not\subset C_0$, pick a point x_1 close to the boundary of C_0. If x_1 is the centre of a 10ε-neck N_1, then we apply Lemma 3.2.2 again to find that $C_1 := C_0 \cup N_1$ is an ε-cap. If x_1 is the centre of an ε-cap C_1, then by an appropriate choice of x_1 we can make sure that $\partial C_0 \cap \partial C_1$ is empty. Thus either $C_0 \subset C_1$, or $M = C_0 \cup C_1$.

Again we iterate the construction, producing an increasing sequence of 10ε-caps $C_0 \subset C_1 \subset \cdots$. For the same reason as before, the process stops, either when $X \subset C_n$, or when we have expressed M as the union of two caps. In the latter case, it follows from Alexander's theorem that M is diffeomorphic to S^3.

Finally, the addendum follows from Lemma 3.2.4, since $S^2 \times S^1$ is reducible. □

Putting $X = M$, we obtain the following corollary:

Corollary 3.2.5. *Let $\varepsilon \in (0, 2\varepsilon_0]$. Let (M, g) be a closed, orientable Riemannian 3-manifold. If every point of M is the centre of an ε-neck or an ε-cap, then M is diffeomorphic to S^3 or $S^2 \times S^1$. In particular, if M is irreducible, then M is diffeomorphic to S^3.*

Here is another consequence of Proposition 3.2.3:

Corollary 3.2.6. *Let $\varepsilon \in (0, 2\varepsilon_0]$. Let (M, g) be a closed, orientable Riemannian 3-manifold. Let X be a compact submanifold of M such that every point of X is the centre of an ε-neck or an ε-cap. Then one of the following conclusions holds:*

(a) *M is diffeomorphic to S^3 or $S^2 \times S^1$, or*

(b) *there exists a finite collection $N_1, \dots, N_p$ of 10ε-caps and 10ε-tubes with disjoint closures such that $X \subset \bigcup_i N_i$.*

Proof. First apply Proposition 3.2.3 to each component of X. If Case (a) does not occur, then we have found a finite collection $N'_1, \dots, N'_q$ of 10ε-caps and 10ε-tubes which cover X. If the closures of those caps and tubes are disjoint, then we get conclusion (b).

Otherwise, we pick two members N'_i, N'_j is the collection whose closures are not disjoint. By adding one or two necks if necessary, we can merge them into a larger subset $U \supset N'_i \cup N'_j$. Observe that

- merging two tubes produces a tube or shows that $M \cong S^2 \times S^1$,
- merging two caps shows that $M \cong S^3$,
- merging a cap with a tube yields a cap.

Hence after finitely many merging operations, we are done. □

We now describe the situation for open manifolds.

Definition 3.2.7. Let (X, g) be a (not necessarily complete) Riemannian 3-manifold and ε be a positive number. An *ε-horn* is an open subset $U \subset X$ which has the following properties:

(i) U is a union of ε-necks;

(ii) U is diffeomorphic to $S^2 \times \mathbb{R}$;

(iii) the scalar curvature is bounded on one end of U and tends to $+\infty$ on the other end.

Lemma 3.2.8. *Let $\varepsilon \in (0, 2\varepsilon_0]$, (X, g) be a Riemannian 3-manifold, and $\gamma : [0, 1) \to X$ be a continuous map such that every point of the image of γ is the centre of an ε-neck, and $R(\gamma(s)) \to +\infty$ as $s \to 1$. Then there exists a 10ε-horn $U \subset X$ containing the image of γ.*

Proof. Set $x_0 := \gamma(0)$. Using the construction described in the proof of Lemma 3.2.4, we produce an infinite sequence of points $x_0, x_1, x_2, \dots$ in the image of γ, and a sequence of necks $N_0, N_1, N_2, \dots$ centred at those points, such that for every k, $N_0 \cup \dots \cup N_k$ is a tube, obtained from $N_0 \cup \dots \cup N_{k-1}$ by gluing a thickened sphere at one end. The union U of all the N_k's is thus diffeomorphic to $S^2 \times \mathbb{R}$.

To see that it contains the whole image of γ, suppose it does not. Choose $s \in [0, 1)$ such that $\gamma(s) \notin U$. Then all the x_k's are images of numbers in the compact interval $[0, s]$. Hence we have an upper bound for their scalar curvatures, which gives a lower bound for the distance $d(x_k, x_{k+1})$. This in turn implies that $d(x_0, x_k)$ goes to infinity, which is a contradiction. □

Proposition 3.2.9. *Let (M, g) be an open Riemannian 3-manifold. If every point of M is the centre of a $2\varepsilon_0$-neck, then M is diffeomorphic to $S^2 \times \mathbb{R}$.*

Proof. The proof is similar to that of Lemma 3.2.8, except that after the choice of x_0 and N_0, we pick two points x_{-1}, x_1, one near each boundary component of N_0. In this way we construct biinfinite sequences $\dots, x_{-2}, x_{-1}, x_0, x_1, x_2, \dots$ and $\dots, N_{-2}, N_{-1}, N_0, N_1, N_2, \dots$ such that for each k, N_k is a neck centred at x_k, and each finite union $N_{-k}, \dots, N_k$ is a tube, obtained from the previous one by gluing a thickened sphere at each end..

The union of these tubes is diffeomorphic to $S^2 \times \mathbb{R}$. To see that it is the whole of M, we argue as in the proof of Lemma 3.2.8: if some point $y \in M$ is not in the union of the N_k's, then we connect it to x_0 by a continuous path and use the compactness of this path to obtain a contradiction.

Therefore, M is diffeomorphic to $S^2 \times \mathbb{R}$. □

3.3 More results on ε-necks

Definition 3.3.1. Let (M, g) be a Riemannian 3-manifold and N be a neck in M. Let $[xy]$ be a geodesic segment such that $x, y \notin N$. We say that $[xy]$ *traverses* N if the geometric intersection number of $[xy]$ and the middle sphere of N is odd.

Lemma 3.3.2. *Let $\varepsilon \in (0, 10^{-1}]$. Let (M, g) be a Riemannian 3-manifold, $N \subset M$ be an ε-neck, and S be a middle sphere of N. Let $[xy]$ be a geodesic segment such that $x, y \in M \setminus N$ and $[xy] \cap S \neq \emptyset$. Then $[xy]$ traverses N. In particular, if S is separating in M, then x, y lie in different components of $M \setminus S$.*

Proof. Let $[x'y']$ be a subsegment of $[xy]$ such that $x', y' \in \partial\bar{N}$ and $[x'y'] \cap S \neq \emptyset$. Since $[x'y']$ is minimising, we have $d(x', y') \geqslant 2d(S, \partial\bar{N})$. By choice of ε, this is greater than the diameter of each component of $\partial\bar{N}$. Hence x', y' must belong to two distinct such components.

This proves that there is exactly one subsegment $[x'y']$ such that $x', y' \in \partial\bar{N}$, $[x'y'] \cap S \neq \emptyset$ and $(x'y') \subset N$. After perturbation, this segment must intersect S transversally in an odd number of points. □

Corollary 3.3.3. *Let $\varepsilon \in (0, 10^{-1}]$. Let (M, g) be a Riemannian 3-manifold, $U \subset M$ be an ε-cap and V be a core of U. Let x, y be points of $M \setminus U$ and $[xy]$ a geodesic segment connecting x to y. Then $[xy] \cap V = \emptyset$.*

Proof. By definition, $U = N \cup V$ where N is an ε-neck of M. Let S be the middle sphere of N. It is isotopic to ∂V, and separates M in two connected components, one of them containing x, y, the other V. If $[xy] \cap V \neq \emptyset$ then $[xy]$ intersects S. The last assertion of Lemma 3.3.2 gives the contradiction. □

Notes. Results similar to those in this chapter can be found in Chapter 19 of Morgan and Tian's book [MT07] and the paper by Dinkelbach and Leeb [DL09]. The versions we give are sometimes weaker, but sufficient for the applications in this book.

Chapter 4
κ-noncollapsing, canonical geometry and pinching

The purpose of this chapter is to introduce some notions that are fundamental for understanding the behaviour of Ricci flow, in particular on 3-manifolds. We begin with the κ-noncollapsing property, which was defined and used by Perelman in the first breakthrough result of [Per02], to obtain an estimate on the local injectivity radius. This property allows us to use compactness theorems to study the singularities of Ricci flow. This leads to the notion of κ-solutions, which are models for the singularities, and are described in Section 4.2. The classification of these models then brings up the concept of canonical neighbourhoods, whose precise definition is given in Section 4.2.2. In Section 4.3 we introduce the standard solution, which is the model for the almost standard caps which are added during the surgery process. We conclude the chapter with some curvature pinching properties of Ricci flow, which were discovered by Hamilton and Ivey (Section 4.4).

4.1 κ-noncollapsing

Let M be an n-manifold. Let $\{g(t)\}_{t\in I}$ be an evolving metric on M. A (backward) *parabolic neighbourhood* of a point $(x,t)\in M\times I$ is a set of the form

$$P(x,t,r,-\Delta t) = \{(x',t')\in M\times I \mid x'\in B(x,t,r), t'\in[t-\Delta t,t]\}.$$

In particular, the set $P(x,t,r,-r^2)$ is called a *parabolic ball* of radius r. Notice that after parabolic rescaling with factor r^{-2} at time t, it becomes a parabolic ball $P(x,0,1,-1)$ of radius 1.

Definition 4.1.1. Fix $\kappa, r>0$. We say that $g(\cdot)$ is *κ-collapsed* at (x,t) on the scale r if for all $(x',t')\in P(x,t,r,-r^2)$ one has $|\mathrm{Rm}(x',t')|\leqslant r^{-2}$, and

$$\operatorname{vol} B(x,t,r) < \kappa r^n.$$

Otherwise, $g(\cdot)$ is *κ-noncollapsed* at (x,t) on the scale r. We say that the evolving metric $g(\cdot)$ is *κ-noncollapsed* on the scale r if it is κ-noncollapsed on this scale at every point of $M\times I$.

Remark 4.1.2. $g(\cdot)$ is κ-noncollapsed at (x,t) on the scale r if

$$|\mathrm{Rm}|\leqslant r^{-2} \text{ on } P(x,t,r,-r^2) \quad\text{implies}\quad \operatorname{vol} B(x,t,r)\geqslant \kappa r^n.$$

Remark 4.1.3. If $|\mathrm{Rm}| \leqslant r^{-2}$ on $P(x,t,r,-r^2)$, a κ-noncollapsing property at (x,t) on the scale r gives a positive lower bound for the injectivity radius of $g(t)$ at x which depends only on n, κ and r (cf. Theorem B.1.2).

Lemma 4.1.4 (κ-noncollapsing property closed in time). *In the case when $n = 3$, let $g(\cdot)$ be a Ricci flow with bubbling-off defined on an interval $(a,b]$, $x \in M$ and $r, \kappa > 0$. If for all $t < b$, $g(\cdot)$ is κ-noncollapsed at (x,t) on all scales less than or equal to r, then it is κ-noncollapsed at (x,b) on the scale r.*

Proof. Assume that $|\mathrm{Rm}| \leqslant r^{-2}$ on $P(x,b,r,-r^2)$. Fix $s < r$. There exists $t_s < b$ such that for all $t \in [t_s, b)$ we have $P(x,t,s,-s^2) \subset P(x,b,r,-r^2)$. Without loss of generality, we assume that there is no singular time in $[t_s,b)$. Now on $P(x,t,s,-s^2)$ we have $|\mathrm{Rm}| \leqslant r^{-2} \leqslant s^{-2}$. Applying the noncollapsing hypothesis at (x,t) on the scale s, we deduce that $\operatorname{vol} B(x,t,s) \geqslant \kappa s^3$. The result follows by letting first t tend to b, then s tend to r. □

4.2 κ-solutions

4.2.1 Definition and main results

Definition 4.2.1. Let $\kappa > 0$. A Ricci flow $(M, g(\cdot))$ is called a *κ-solution* if the following holds:

(i) It is *ancient*, i.e., defined on $(-\infty, 0]$.

(ii) For each $t \in (-\infty, 0]$, the metric $g(t)$ is complete, and $\mathrm{Rm}(\cdot, t)$ is nonnegative and bounded.

(iii) There exists $(x,t) \in M \times (-\infty, 0]$ such that $\mathrm{Rm}(x,t) \neq 0$.

(iv) For every $r > 0$, the evolving metric $g(\cdot)$ is κ-noncollapsed on the scale r.

A κ-solution is *round* if for each t, the metric $g(t)$ has constant positive sectional curvature.

The *asymptotic volume* of a complete Riemannian manifold (M^n, g) of nonnegative curvature is defined by

$$\mathcal{V}(g) := \lim_{r\to\infty} \frac{\operatorname{vol} B(x,r)}{r^n}.$$

One easily sees, using Bishop–Gromov inequality B.1.1, that the limit is well defined and does not depend on $x \in M$. We state here fundamental results on κ-solutions.

Theorem 4.2.2 (Vanishing asymptotic volume, [Per02], Section 11.4, [KL08], Proposition 41.13). *For any κ-solution $(M, g(\cdot))$, one has $\mathcal{V}(g(t)) = 0$ for all t.*

Theorem 4.2.3 (Compactness theorem, [Per02], Section 11.7, [KL08], Theorem 46.1). *For all $\kappa > 0$, the space of pointed 3-dimensional κ-solutions $(M, g(\cdot), (x,0))$ such that $R(x,0) = 1$ is compact for the pointed convergence of flows.*

Theorem 4.2.4 ([Per02], Corollary 11.3, [KL08], Corollary 40.1 and Section 43). *Every 2-dimensional κ-solution is round.*

Theorem 4.2.5 (Universal κ, [Per03b], Section 1.5, [KL08], Proposition 50.1). *There exists a universal constant $\kappa_{\mathrm{sol}} > 0$ such that any 3-dimensional κ-solution is round or a κ_{sol}-solution.*

4.2.2 Canonical neighbourhoods

The *cylindrical flow* is $S^2 \times \mathbb{R}$ together with the product Ricci flow on $(-\infty, 0]$, where the first factor is round, normalised so that the scalar curvature at time 0 is identically 1. We denote this evolving metric by $g_{\mathrm{cyl}}(t)$.

Definition 4.2.6. Let $\varepsilon > 0$. Let M be a 3-manifold, $\{g(t)\}_{t\in I}$ be an evolving metric on M, and $(x_0, t_0) \in M \times I$. An open subset $N \subset M$ is called a *strong ε-neck* centred at (x_0, t_0) if there is a number $Q > 0$ such that $(N, \{g(t)\}_{t\in[t_0-Q^{-1},t_0]}, x_0)$ is unscathed and, after parabolic rescaling with factor Q at time t_0, ε-close to $(S^2 \times (-\varepsilon^{-1}, \varepsilon^{-1}), \{g_{\mathrm{cyl}}(t)\}_{t\in[-1,0]}, *)$.

Remarks 4.2.7. (i) Fix $Q > 0$, and consider the flow $(S^2 \times \mathbb{R}, Q g_{\mathrm{cyl}}(tQ^{-1}))$ restricted to $(-Q, 0]$. Then for every $x \in S^2 \times \mathbb{R}$, the point $(x, 0)$ is the centre of a strong ε-neck for all $\varepsilon > 0$.

(ii) If (x_0, t_0) is the centre of a strong ε-neck, then there is a neighbourhood Ω of x_0 such that for all $x \in \Omega$, (x, t_0) is the centre of a strong ε-neck: one can use the same set N and factor Q, but change the parametrisation so that the basepoint $*$ is sent to x rather than x_0. Choosing Ω smaller, we also have that (x, t_0) is the centre of a strong ε_0-neck for all evolving metric $\tilde{g}(\cdot)$ sufficiently close to $g(\cdot)$ is the $\mathcal{C}^{[\varepsilon^{-1}]}$ topology.

(iii) By abuse of language, we shall sometimes call strong ε-neck the parabolic set $N \times [t_0 - Q^{-1}, t_0] \subset M \times I$, rather than the set $N \subset M$.

Definition 4.2.8. Let $\varepsilon, C > 0$ and $\{g(t)\}_{t\in I}$ be an evolving metric on M. We say that an open subset $U \subset M$ is an *(ε, C)-cap centred at* (x_0, t_0) if U is an ε-cap which satisfies furthermore the following estimates for the metric $g(t_0)$: $R(x_0) > 0$ and there exists $r \in (C^{-1}R(x_0)^{-\frac{1}{2}}, CR(x_0)^{-\frac{1}{2}})$ such that

(i) $\overline{B(x_0, r)} \subset U \subset B(x_0, 2r)$,
(ii) the scalar curvature function restricted to U has values in a compact subinterval of $(C^{-1}R(x_0), CR(x_0))$,
(iii) $\operatorname{vol} U > C^{-1}R(x_0)^{-3/2}$ and if $B(y, s) \subset U$ satisfies $|\mathrm{Rm}| \leqslant s^{-2}$ on $B(y, s)$ then
$$C^{-1} < \frac{\operatorname{vol} B(y, s)}{s^3}. \tag{4.1}$$

Moreover, the following inequalities hold true on U:

(iv)
$$|\nabla R(x)| < CR(x)^{\frac{3}{2}}, \tag{4.2}$$

(v)
$$|\Delta R(x) + 2|\mathrm{Ric}(x)|^2| < CR(x)^2, \tag{4.3}$$

(vi)
$$|\nabla \mathrm{Rm}(x)| < C|\mathrm{Rm}(x)|^{\frac{3}{2}}, \tag{4.4}$$

Remarks 4.2.9. (i) The diameter of U is at most $4r$, which in turn is bounded above by a function of C and $R(x_0)$.

(ii) The estimate (4.3) implies the following scale-invariant bound on the time-derivative of R:
$$\left|\frac{\partial R}{\partial t}(x,t)\right| < CR(x,t)^2. \tag{4.5}$$

(iii) For suitable $C = C(\varepsilon)$, any strong ε-neck also satisfies (i)–(vi).

(iv) The property of being the centre of an (ε, C)-canonical neighbourhood is open with respect to changing the point, and the metric in the $\mathcal{C}^{[\varepsilon^{-1}]}$ topology.

(v) If U is an (ε, C)-cap, then the set of all centres of U is open.

Definition 4.2.10. Let ε, C be positive numbers and $\{g(t)\}_{t\in I})$ be an evolving metric on M. We say that a point (x,t) admits (or is the centre of) an (ε, C)-*canonical neighbourhood* if x is the centre of an (ε, C)-cap for $g(t)$ or if x is the centre of a strong ε-neck which satisfies (i)–(vi) of Definition 4.2.8.

In the next theorem, we say that M is $\mathbb{R}P^2$-free if it does not contain any embedded $\mathbb{R}P^2$.

Theorem 4.2.11 ([Per02], Section 1.5, [KL08], Lemma 59.7). *For all $\varepsilon > 0$ there exists $C_{\mathrm{sol}} = C_{\mathrm{sol}}(\varepsilon)$ such that, if $(M, \{g(t)\}_{t\in(-\infty,0]})$ is a 3-dimensional $\mathbb{R}P^2$-free κ-solution, then M is a spherical manifold, or every $(x,t) \in M \times (-\infty, 0]$ is the centre of an $(\varepsilon, C_{\mathrm{sol}})$-canonical neighbourhood.*

4.3 The standard solution I

4.3.1 Definition and main results

We will consider a Riemannian manifold $\mathcal{S}_0$ satisfying a number of properties, such as: $\mathcal{S}_0$ is diffeomorphic to $\mathbb{R}^3$, it is complete, the sectional curvature is nonnegative and bounded, it is rotationally symmetric, a neighbourhood of infinity is isometric to the product $S^2 \times [0, +\infty)$.

This manifold will be called the *standard initial metric* and its centre of symmetry will be denoted by p_0. For technical reasons, one needs to be more precise in the choice of $\mathcal{S}_0$, so the definition will be given in Chapter 7. In this section, we record some properties of $\mathcal{S}_0$ for future reference. For brevity, a Ricci flow is said to be *complete* (resp. to have *bounded sectional curvature*) if each of its time slices has this property.

Theorem 4.3.1 ([Per03b], Section 2, [KL08], Sections 61–64, and [CZ06b], [LT05] for the uniqueness). *The Ricci flow with initial condition $\mathcal{S}_0$ has a maximal solution defined on $[0, 1)$, which is unique among complete Ricci flows of bounded sectional curvature.*

Definition 4.3.2. This solution is called the *standard solution.*

Proposition 4.3.3 ([KL08], Lemma 60.3). *There exists $\kappa_{\text{st}} > 0$ such that the standard solution is κ_{st}-noncollapsed on all scales.*

Proposition 4.3.4 ([Per03b], Claim 5, [KL08], Lemma 63.1). *For every $\varepsilon > 0$ there exists $C_{\text{st}}(\varepsilon) > 0$ such that for all $(x, t) \in \mathcal{S}_0 \times [0, 1)$, if $t > 3/4$ or $x \in B(p_0, 0, \varepsilon^{-1})$, then (x, t) has an $(\varepsilon, C_{\text{st}})$-canonical neighbourhood. Otherwise, there is an ε-neck $B(x, t, \varepsilon^{-1} r)$ such that $P(x, t, \varepsilon^{-1} r, -t)$ is ε-close to the corresponding subset of the cylindrical flow.*

Moreover, we have the estimate $R_{\min}(t) \geqslant \text{const}_{\text{st}}(1 - t)^{-1}$ for some constant $\text{const}_{\text{st}} > 0$.

4.3.2 Neck strengthening

The following lemma is a variant of [MT07], Proposition 15.2. Let K_{st} be the supremum of the sectional curvatures of the standard solution on $[0, 4/5]$.

Lemma 4.3.5 (Neck strengthening). *For all $\varepsilon \in (0, 10^{-4})$ there exists $\beta = \beta(\varepsilon) \in (0, 10^{-3})$ such that the following holds. Let a, b be real numbers satisfying $a < b < 0$ and $|b| \leqslant 3/4$, let $(M, g(\cdot))$ be a 3-dimensional Ricci flow with bubbling-off defined on $(a, 0]$, and $x \in M$ be a point such that*

(i) *$R(x, b) = 1$;*

(ii) *(x, b) is the centre of a strong $\beta\varepsilon$-neck;*

(iii) *$P(x, b, (\beta\varepsilon)^{-1}, |b|)$ is unscathed and satisfies $|\text{Rm}| \leqslant 2K_{\text{st}}$.*

Then $(x, 0)$ is the centre of a strong ε-neck.

Proof. We argue by contradiction, assuming that there exists a number ε, a sequence $\beta_k \to 0$, sequences a_k, b_k with $a_k < b_k$, $b_k \in [-3/4, 0]$, and for each k a Ricci flow with bubbling-off $(M_k, \{g_k(t)\}_{a_k, 0]})$ with a point $x_k \in M_k$ such that

(i) $R(x_k, b_k) = 1$;
(ii) (x_k, b_k) is the centre of a strong $\beta_k\varepsilon$-neck N_k of scaling factor Q_k;
(iii) $P(x_k, b_k, (\beta_k\varepsilon)^{-1}, |b_k|)$ is unscathed and satisfies $|\mathrm{Rm}| \leqslant 2K_{\mathrm{st}}$, but
(iv) $(x_k, 0)$ is not the centre of a strong ε-neck.

Here is the picture:

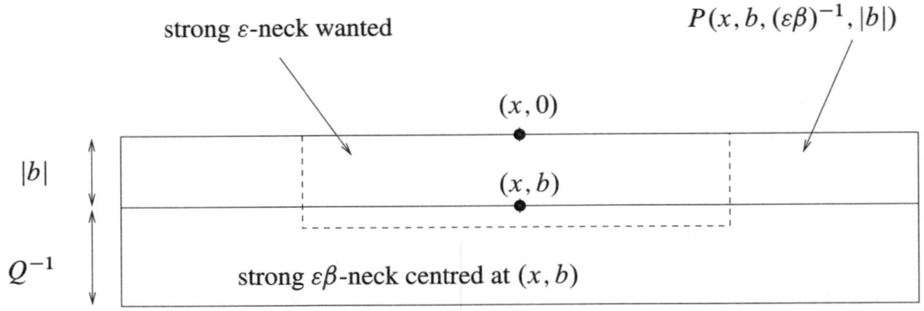

From assumptions (i), (ii) we first deduce that $|\mathrm{Rm}| \leqslant 2$ on $M_k \times [b_k - Q_k^{-1}, b_k]$ for all k. Using the curvature bound $|\mathrm{Rm}| \leqslant 2K_{\mathrm{st}}$ on $P(x_k, b_k, (\beta_k\varepsilon)^{-1}, |b_k|)$, the Distance-Distortion Lemma 2.2.7 gives a constant c such that $B(x_k, 0, c(\beta_k\varepsilon)^{-1}) \subset B(x_k, b_k, (\beta_k\varepsilon)^{-1})$. Hence we have $|\mathrm{Rm}| \leqslant \max\{2K_{\mathrm{st}}, 2\}$ on $P(x_k, 0, c(\beta_k\varepsilon)^{-1}, b_k - Q_k^{-1})$.

Using the injectivity radius estimate (Theorem B.1.2), we deduce from the closeness with the cylinder at (x_k, b_k), and from the bounded curvature, a positive lower bound on the injectivity radius at $(x_k, 0)$.

Note that the normalisation $R(x_k, b_k) = 1$ implies that $Q_k \to 1$. The Compactness Theorem C.3.1 gives a subsequence of $(M_k, \{g_k(t)\}_{t \in [b_k - Q_k^{-1}, 0]}, (x_k, 0))$ which converges to a complete pointed Ricci flow $(M_\infty, g_\infty(\cdot), (x_\infty, 0))$ of bounded curvature, defined on $(b-1, 0]$, where $b := \lim b_k \in [-3/4, 0]$. In particular, the sequence of pointed Riemannian manifolds $(M_k, g_k(b_k - (2Q_k)^{-1}), x_k)$ converges smoothly to $(M_\infty, g_\infty(b - 1/2), x_\infty)$.

From the hypothesis that N_k is a strong $\beta_k\varepsilon$-neck of scaling factor Q_k centred at (x_k, b_k), we deduce that $(M_k, Q_k g_k(b_k - (2Q_k)^{-1}), x_k)$ converges to $(S^2 \times \mathbb{R}, g_{\mathrm{cyl}}(-1/2), *)$. Hence $(M_\infty, g_\infty(b - 1/2), x_\infty)$ is isometric to this cylinder. By the Chen–Zhu uniqueness theorem [CZ06c], we deduce that the limit flow $(M_\infty, \{g_\infty(t)\}_{t \in [b-1/2, 0]})$ is the cylindrical flow up to parabolic scaling. The same conclusion holds on $(b-1, 0]$ since N_k is a strong $\beta_k\varepsilon$-neck for each k.

We have $R(x_\infty, b) = 1$, hence $R(x_\infty, 0) = (1+b)^{-1} \geqslant 1$. By Remark 4.2.7 (i) we are done. □

4.4 Curvature pinched toward positive

In this section, we discuss the so-called Hamilton–Ivey pinching property of the Ricci flow.

Let (M, g) be a 3-manifold and $x \in M$ be a point. We denote by $\lambda(x) \geqslant \mu(x) \geqslant \nu(x)$ the eigenvalues of the curvature operator $\mathrm{Rm}(x)$. According to our conventions, all sectional curvatures lie in the interval $[\nu(x), \lambda(x)]$. Moreover, $\lambda(x)$ (resp. $\nu(x)$) is the maximal (resp. minimal) sectional curvature at x. If C is a real number, we sometimes write $\mathrm{Rm}(x) \geqslant C$ instead of $\nu(x) \geqslant C$. Likewise, $\mathrm{Rm}(x) \leqslant C$ means $\lambda(x) \leqslant C$.

It follows that the eigenvalues of the Ricci tensor are equal to $\lambda + \mu$, $\lambda + \nu$, and $\mu + \nu$; as a consequence, the scalar curvature $R(x)$ is equal to $2(\lambda(x) + \mu(x) + \nu(x))$.

For evolving metrics, we use the notation $\lambda(x,t)$, $\mu(x,t)$, and $\nu(x,t)$, and correspondingly write $\mathrm{Rm}(x,t) \geqslant C$ for $\nu(x,t) \geqslant C$, and $\mathrm{Rm}(x,t) \leqslant C$ for $\lambda(x,t) \leqslant C$.

Definition 4.4.1. Let ϕ be a nonnegative function. A metric g on M has *ϕ-almost nonnegative curvature* if R takes its values in the domain of ϕ and if $\mathrm{Rm} \geqslant -\phi(R)$ on M.

Remark 4.4.2. In this case, we also have a bound above of the curvature operator by a function of the scalar curvature, namely

$$\frac{R}{2} + 2\phi(R) \geqslant \frac{R}{2} - \mu - \nu = \lambda \geqslant \mathrm{Rm} \geqslant \nu \geqslant -\phi(R).$$

Now we consider a family of positive functions $(\phi_t)_{t \geqslant 0}$ defined as follows. Let $s_t = \frac{e^2}{1+t}$ and define

$$\phi_t \colon [-2s_t, +\infty) \to [s_t, +\infty)$$

as the reciprocal of the increasing function

$$s \mapsto 2s(\ln(s) + \ln(1+t) - 3).$$

As in [MT07], we use the following definition.

Definition 4.4.3. Let $I \subset [0, \infty)$ be an interval and $\{g(t)\}_{t \in I}$ be an evolving metric on M. We say that $g(\cdot)$ has *curvature pinched toward positive at time t* if for all $x \in M$ we have

$$R(x,t) \geqslant -\frac{6}{4t+1}, \tag{4.6}$$

$$\mathrm{Rm}(x,t) \geqslant -\phi_t(R(x,t)). \tag{4.7}$$

We say that $g(\cdot)$ has *curvature pinched toward positive* if it has curvature pinched toward positive at each $t \in I$.

Remark 4.4.4. If we set $X := \max\{0, |\nu(x,t)|\}$, equation (4.7) is equivalent to[1]

$$R(x,t) \geqslant 2X\,(\ln X + \ln(1+t) - 3)\,. \tag{4.8}$$

The content of the Hamilton–Ivey theorem, which is proved using a time-dependent version of the maximum principle, is that curvature pinching toward positive propagates forward under Ricci flow:

Proposition 4.4.5 ([Ham99], [Ive93]). *Let a, b be two real numbers such that $0 \leqslant a < b$. Let $\{g(t)\}_{t\in[a,b]}$ be a Ricci flow solution on M such that $g(\cdot)$ has curvature pinched toward positive at time a. Then $g(\cdot)$ has curvature pinched toward positive.*

The following lemmas will be useful.

Lemma 4.4.6. (i) $\phi_t(s) = \frac{\phi_0((1+t)s)}{1+t}$.

(ii) *$\frac{\phi_t(s)}{s}$ decreases to 0 as s tends to $+\infty$.*

(iii) *$\frac{\phi_0(s)}{s} = \frac{1}{4}$ if $s = 4e^5$.*

Proof. (i) is left to the reader. For (ii), we have

$$0 < \frac{\phi_0(s)}{s} = u \Leftrightarrow \phi_0(s) = su \Leftrightarrow s = 2(su)[\ln(su) - 3] \Leftrightarrow \psi(u) = s = \frac{1}{u}e^{3+\frac{1}{2}u}.$$

Now, since $\psi'(u) < 0$, ψ and $s \to \phi(s)/s$ are decreasing. Finally $\psi(1/4) = 4e^5$ proves (iii). □

Property (ii) will ensure that limits of suitably rescaled evolving metrics with curvature pinched toward positive will have nonnegative curvature operator (see Proposition 6.1.6). Set $\bar{s} := 4e^5$.

Lemma 4.4.7 (Pinching lemma). *Assume $g(\cdot)$ has curvature pinched toward positive and let $t \geqslant 0$, $r > 0$ be such that $(1+t)r^{-2} \geqslant \bar{s}$. If $R(x,t) \leqslant r^{-2}$ then $|\mathrm{Rm}(x,t)| \leqslant r^{-2}$.*

Proof. By assumption, we have at (x,t),

$$\begin{aligned}\mathrm{Rm} \geqslant -\phi_t(R) = -\frac{\phi_0((1+t)R)}{1+t} &\geqslant -\frac{\phi_0((1+t)r^{-2})}{1+t}\\ &= -\frac{\phi_0((1+t)r^{-2})}{(1+t)r^{-2}}r^{-2} \geqslant -\frac{1}{4}r^{-2}\end{aligned}$$

by Lemma 4.4.6. On the other hand

$$\mathrm{Rm} \leqslant \lambda = \frac{R}{2} - \nu - \mu \leqslant \frac{r^{-2}}{2} + 2\left(\frac{r^{-2}}{4}\right) = r^{-2}.$$ □

[1] The factor 2 comes from our choices or normalisation of the curvature operator Rm (cf. [Ham99]).

Remarks 4.4.8. This lemma will be used in two kinds of settings:

(i) $r^{-2} \geqslant \bar{s}$. For example, if $R(x,t) \leqslant Q = r^{-2}$ where, say, $Q \geqslant 10^6$, then $|\mathrm{Rm}| \leqslant Q$. Hence if the scalar curvature is large, it bounds the curvature operator.

(ii) $tr^{-2} \geqslant \bar{s}$, which is equivalent to $r \in (0, \bar{r}\sqrt{t}]$ if one sets $\bar{r} = \bar{s}^{-1/2}$. This will be used in the study of the long-time behaviour of the Ricci flow with bubbling-off.

Notes. Except for the Hamilton–Ivey pinching ([Ham99], [Ive93]), most of the material of this chapter comes from Perelman [Per02], [Per03b]. There are slightly different definitions of canonical neighbourhoods in the literature. Ours is closest to the one in Morgan–Tian's book [MT07]. Our definition is more restrictive because we exclude spherical manifolds and some ε-caps diffeomorphic to $\mathbb{R}P^3 \setminus \{\mathrm{pt}\}$. This is consistent with the topological hypotheses put on the manifold throughout the book. It allows us to slightly simplify the exposition. In the proof of the strengthening lemma (Lemma 4.3.5) we use the Chen–Zhu uniqueness theorem for simplicity only.

Chapter 5
Ricci flow with (r, δ, κ)-bubbling-off

In this chapter we assume that M is a closed, orientable, irreducible 3-manifold. We also assume that M is not spherical. (cf. discussion at the beginning of Section 1.2.2).

In particular, M is $\mathbb{R}P^2$-free, and not diffeomorphic to S^3 or $S^2 \times S^1$. Moreover, any manifold obtained from a sequence of Riemannian metrics on M by taking a pointed limit automatically has the same properties (except compactness and irreducibility, of course).

We introduce a version with parameters of Ricci flow with bubbling-off, which we call Ricci flow with (r, δ, κ)-bubbling-off. The bulk of the definition consists of a canonical neighbourhood property at a scale r, and a metric surgery governed by *surgery parameters* r, δ and other ones computed with them, e.g. the curvature threshold used to launch the surgery process. Metric surgery and the *associated cutoff parameters* are precisely defined in Section 5.2. With this in hand, we state the main finite-time existence theorem of the book (Theorem 5.3.1), which asserts the existence, given any normalised initial data and finite interval $[0, T]$, of a Ricci flow with (r, δ, κ)-bubbling-off on $[0, T]$ with such initial data. We reduce the proof of this result to three independent propositions, called A, B and C, whose demonstrations occupy all Part II of the book. Finally, we give in Section 5.4 a long-time existence theorem, obtained more or less by iteration of the first one.

5.1 Let the constants be fixed

Recall that ε_0 has been fixed in Chapter 3 (cf. Definition 3.2.1).

Definition 5.1.1. Let $\beta_0 := \beta(\varepsilon_0)$ be the parameter given by Lemma 4.3.5. Finally, set

$$C_0 := \max\{100{\varepsilon_0}^{-1}, 2C_{\text{sol}}(\varepsilon_0/2), 2C_{\text{st}}(\beta_0\varepsilon_0/2)\},$$

where C_{sol} is defined by Theorem 4.2.11, and C_{st} by Proposition 4.3.4.

Definition 5.1.2. Let $r > 0$. An evolving metric $\{g(t)\}_{t\in I}$ on M has property $(\text{CN})_r$ if for all $(x,t) \in M \times I$, if $R(x,t) \geqslant r^{-2}$, then (x,t) admits an (ε_0, C_0)-canonical neighbourhood.

Definition 5.1.3. Let $\kappa > 0$. An evolving metric $\{g(t)\}_{t\in I}$ on M has property $(\text{NC})_\kappa$ if $\{g(t)\}$ is κ-noncollapsed on all scales less than or equal to 1.

Remark 5.1.4. If $(\text{NC})_\kappa$ is satisfied on some time interval (a,b), it is also satisfied on $(a,b]$. This follows immediately from Lemma 4.1.4.

Definition 5.1.5. A metric g on M is *normalised* if $\operatorname{tr}\operatorname{Rm}^2 \leqslant 1$ and each ball of radius 1 has volume at least half of the volume of the unit ball in Euclidean 3-space.

Remark 5.1.6. If $g(0)$ is normalised then $|\operatorname{Rm}| \leqslant 1$. This in turn implies that $g(\cdot)$ has curvature pinched toward positive at time 0. Indeed, $R(x,0) \geqslant -6$ and $\operatorname{Rm}(x,0) \geqslant -1 \geqslant -\phi_0(R(x,0))$, for all $x \in M$.

Remark 5.1.7. Since M is closed, any metric on M can be normalised by scaling.

Proposition 5.1.8. *There exists a constant $\kappa_{\text{norm}} > 0$ such that if g_0 is a normalised metric on M, then the maximal Ricci flow solution $g(\cdot)$ with initial condition g_0 is defined at least on $[0, 1/16]$, and on this interval it satisfies*

(i) $|\operatorname{Rm}| \leqslant 2$,

(ii) $(\mathrm{NC})_{\kappa_{\text{norm}}}$, *and*

(iii) $g(\cdot)$ *has curvature pinched toward positive.*

Proof. Inequality (i) follows from $(\operatorname{tr}\operatorname{Rm}^2)^{1/2} \leqslant 2$, which is the doubling time estimate given in Lemma 6.1 of [CLN06]. The property (ii) follows from the fact that on the time interval $[0, 1/16]$ the metrics are uniformly bi-Lipschitz because of the previous curvature bound. An explicit expression for the constant κ_{norm} may be given. Finally, (iii) is a consequence of Proposition 4.4.5 and the fact that if $g(0)$ is normalised then it has curvature pinched toward positive at time 0. □

Remark 5.1.9. Conclusion (i) implies $R \leqslant 12$, hence $(\mathrm{CN})_r$ is vacuously true for $r < (2\sqrt{3})^{-1}$.

Definition 5.1.10. We set $\kappa_0 := \min(\kappa_{\text{norm}}, \kappa_{\text{sol}}/2, \kappa_{\text{st}}/2)$.

5.2 Metric surgery and cutoff parameters

Recall from Section 4.3 that we denote by $\mathcal{S}_0$ the standard initial metric and p_0 its centre of symmetry (see Section 7.1 for the precise definition of $\mathcal{S}_0$).

Definition 5.2.1. Let δ, δ' be positive numbers. Let g be a Riemannian metric on M. Let (U, V, p, y) be a 4-tuple such that U is an open subset of M, V is a compact subset of U, $p \in \operatorname{Int} V$, $y \in \partial V$. Then (U, V, p, y) is called a *marked (δ, δ')-almost standard cap* if there exists a δ'-isometry $\psi \colon B(p_0, 5 + \delta^{-1}) \to (U, R(y)g)$, sending $B(p_0, 5)$ to $\operatorname{Int} V$ and p_0 to p. One calls V the *core* and p the *tip* (see Figure 5.1).

Theorem 5.2.2 (Metric surgery). *There exist $\delta_0 > 0$ and a function $\delta' \colon (0, \delta_0] \ni \delta \mapsto \delta'(\delta) \in (0, \varepsilon_0/10]$ tending to zero as $\delta \to 0$, with the following property:*

Let ϕ be a nondecreasing, nonnegative function, let $\delta \leqslant \delta_0$, let g be a Riemannian metric on M with ϕ-almost nonnegative curvature, and $N \subset M$ be a δ-neck whose middle sphere bounds a 3-ball $B \subset M$.

Then there exists a Riemannian metric g_+ on M such that the following holds:

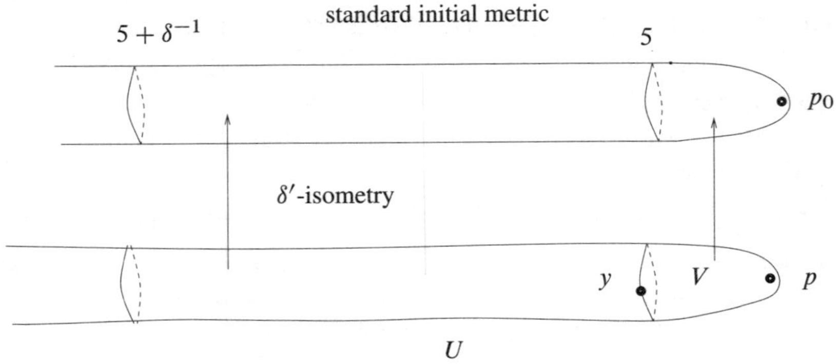

Figure 5.1. An almost standard cap.

(i) $g_+ = g$ *on* $M \setminus \operatorname{Int} B$ *and* $g_+ < g$ *on* $\operatorname{Int} B$.
(ii) *There exist* $p \in \operatorname{Int} B$ *and* $y \in \partial B$ *such that* $(N \cup B, B, p, y)$ *is a marked* $(\delta, \delta'(\delta))$*-almost standard cap with respect to* g_+.
(iii) g_+ *has* ϕ*-almost nonnegative curvature.*

This is proved in Chapter 7. Figure 5.2 summarizes the construction.

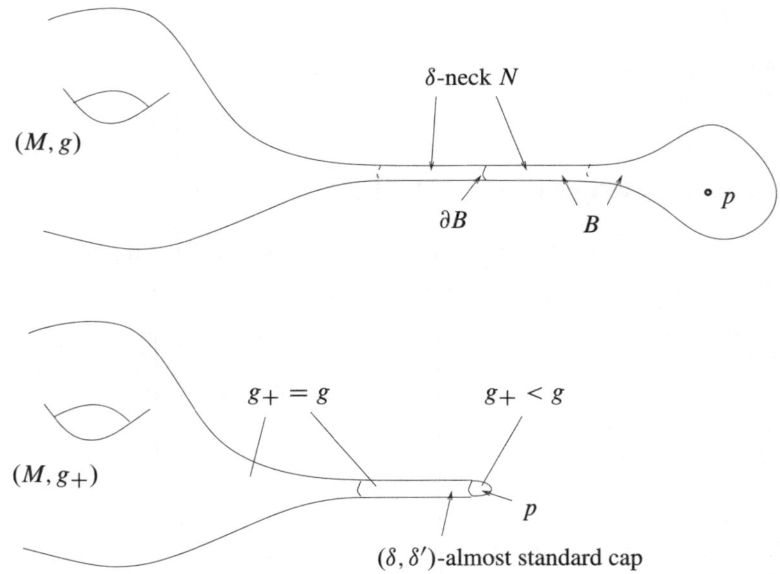

Figure 5.2. The metric surgery.

Definition 5.2.3. In the sequel we fix a function $\delta' \colon (0, \delta_0] \to (0, \varepsilon_0/10]$ with the above properties. A marked $(\delta, \delta'(\delta))$-almost standard cap will be simply called a *δ-almost standard cap*. An open subset U of M is called a *δ-almost standard cap* if there exist V, p and y such that (U, V, p, y) is a δ-almost standard cap.

Theorem 5.2.4 (Cutoff parameters). *For all $r, \delta > 0$, there exist $h \in (0, \delta r)$ and $D > 10$ such that if $g(\cdot)$ is a Ricci flow with bubbling-off on M defined on an interval $[a, b]$, with curvature pinched toward positive and satisfying* (CN)$_r$, *then the following holds:*

Suppose that $x, y, z \in M$ and $t \in [a, b]$ are such that $R(x, t) \leqslant 2r^{-2}$, $R(y, t) = h^{-2}$, $R(z, t) \geqslant Dh^{-2}$, and y lies on a $g(t)$-geodesic segment connecting x to z. Then (y, t) is the centre of a strong δ-neck.

This will be proved in Chapter 6. In the sequel we fix functions $(r, \delta) \mapsto h(r, \delta)$ and $(r, \delta) \mapsto D(r, \delta)$ with the above property. We set $\Theta(r, \delta) := 2D(r, \delta)h(r, \delta)^{-2}$. This number will be used as a curvature threshold for the surgery process, as sketched in Section 1.2.

Definition 5.2.5. We say that two real numbers r, δ are *surgery parameters* if $0 < r < 10^{-3}$ and $0 < \delta < \min(\varepsilon_0, \delta_0)$. The *associated cutoff parameters* are $h := h(r, \delta)$ and $\Theta := 2Dh^{-2}$.

Remark 5.2.6. The convention $r < 10^{-3}$ implies that $g(\cdot)$ satisfies (CN)$_r$ vacuously on $[0, 1/16]$ if $g(0)$ is normalised. See Remark 5.1.9.

Definition 5.2.7. Fix surgery parameters r, δ and let h, Θ be the associated cutoff parameters. Let $\{g(t)\}_{t \in I}$ be an evolving metric on M. Let g_+ be a Riemannian metric on M and $t_0 \in I$. We say that g_+ *is obtained from $g(\cdot)$ by (r, δ)-surgery at time t_0* if the following conditions are satisfied:

(i) For all $x \in M$, if $g_+(x) \neq g(x, t_0)$, then there exists a δ-almost standard cap (U, V, p, y) with respect to g_+ such that

  (a) $x \in \operatorname{Int} V$;

  (b) $R(y, t_0) = h^{-2}$;

  (c) (y, t_0) is the centre of a strong δ-neck;

  (d) $g_+ < g(t_0)$ on $\operatorname{Int} V$.

(ii) $R_{\max}(g(t_0)) = \Theta$ and $R_{\max}(g_+) \leqslant \Theta/2$.

We will refer to h as the *scale* of the δ-almost standard cap.

Definition 5.2.8. Fix surgery parameters r, δ and let h, D, Θ be the associated cutoff parameters. Let $I \subset [0, \infty)$ be an interval and $\{g(t)\}_{t \in I}$ be a Ricci flow with bubbling-off on M. We say that $\{g(t)\}_{t \in I}$ is a *Ricci flow with (r, δ)-bubbling-off* if it has the following properties:

(i) $g(\cdot)$ has curvature pinched toward positive and satisfies $R(x, t) \leqslant \Theta$ for all $(x, t) \in M \times I$.

(ii) For every singular time $t_0 \in I$, the metric $g_+(t_0)$ is obtained from $g(\cdot)$ by (r, δ)-surgery at time t_0.

(iii) Condition $(\mathrm{CN})_r$ holds.

Let $\kappa > 0$. A Ricci flow with (r, δ)-bubbling-off which in addition satisfies Condition $(\mathrm{NC})_\kappa$ will be called an *Ricci flow with (r, δ, κ)-bubbling-off.*

The following lemma is easy, but extremely important. It gives an estimate that bounds from below the elapsed time between two singular times.

Lemma 5.2.9 (Elapsed time between surgeries). *Let r, δ be surgery parameters. Let $\{g(t)\}_{t \in I}$ be a Ricci flow with (r, δ)-bubbling-off on M. Let $t_1 < t_2$ be two singular times. Then $t_2 - t_1 \geqslant (C_0 \Theta)^{-1}$.*

Proof. One can suppose that $g(\cdot)$ is smooth on $(t_1, t_2]$. Let $x \in M$ such that $R(x, t_2) = R_{\max}(t_2) = \Theta$. As $R_{\max}(g_+(t_1)) \leqslant \Theta/2$, there exists $t_+ \in [t_1, t_2]$ maximal such that $\lim_{t \to t_+, t > t_+} R(x, t) = \Theta/2$. In particular, (x, t) is the centre of an (ε_0, C_0)-canonical neighbourhood for all $t \in (t_+, t_2]$. Integrating inequality (4.5) of p. 40 on $(t_+, t_2]$ gives

$$-\frac{1}{R(x, t_2)} + \lim_{t \to t_+, t > t_+} \frac{1}{R(x, t)} \leqslant C_0(t_2 - t_+) \leqslant C_0(t_2 - t_1).$$

Hence $t_2 - t_1 \geqslant \frac{1}{C_0}\left(-\frac{1}{\Theta} + \frac{2}{\Theta}\right) = (C_0 \Theta)^{-1}$. □

5.3 Finite-time existence theorem for Ricci flow with (r, δ, κ)-bubbling-off

5.3.1 The statements

Recall that we are assuming that M is a closed, orientable, irreducible 3-manifold, and that ***M is not spherical***.

In this case, Theorem 1.2.3 asserts that for any $T > 0$ and any metric g_0 on M, there exists a Ricci flow with bubbling-off $g(\cdot)$ on M, defined on $[0, T]$, with initial condition $g(0) = g_0$. This follows immediately from the following technical result:

Theorem 5.3.1. *For every $T > 0$, there exist $r, \delta, \kappa > 0$ such that for any normalised metric g_0 on M, there exists a Ricci flow with (r, δ, κ)-bubbling-off defined on $[0, T]$ with initial condition g_0.*

Since any metric on M can be normalised by scaling, Theorem 5.3.1 implies Theorem 1.2.3.

Our next aim is to reduce Theorem 5.3.1 to three results, called Propositions A, B, C, which are independent of one another.

Proposition A below essentially says that if in a Ricci flow with bubbling-off, $R_{\max}$ reaches the threshold value Θ, then one can perform the (r, δ)-surgery and get a new metric g_+, which in particular satisfies $R_{\max}(g_+) \leqslant \Theta/2$. Its proof consists in putting together the cutoff parameters theorem and the metric surgery theorem, as well as some elementary topological arguments which are needed in order to find the collection of cutoff δ-necks.

Proposition A. *There exists a universal constant $\bar{\delta}_A \in (0, \delta_0)$ having the following property: let r, δ be surgery parameters, a, b be positive numbers with $a < b$, and $\{g(t)\}_{t\in(a,b]}$ be a Ricci flow with (r, δ)-bubbling-off on M. Suppose that $\delta \leqslant \bar{\delta}_A$, and $R_{\max}(b) = \Theta$.*

Then M admits a metric g_+ which is obtained from $g(\cdot)$ by (r, δ)-surgery at time b, and in addition satisfies

(i) *g_+ has ϕ_b-almost nonnegative curvature;*

(ii) *$R_{\min}(g_+) \geqslant R_{\min}(g(b))$.*

In the next proposition, we show that the canonical neighbourhood property is closed in time, for a suitable choice of parameters.

Proposition B. *For all $\kappa > 0$ there exist $r(\kappa) \in (0, 10^{-3})$ and $\bar{\delta}_B = \bar{\delta}_B(\kappa) \in (0, \delta_0)$ with the following property: let $\delta \leqslant \bar{\delta}_B$, $b > 0$ and $g(\cdot)$ be a Ricci flow with bubbling-off defined on $[0, b]$ with normalised initial condition. Assume that the restriction of $g(\cdot)$ to the half-open interval $[0, b)$ is a Ricci flow with (r, δ, κ)-bubbling-off. Then it satisfies* $(\mathrm{CN})_r$ *on all of* $[0, b]$.

The last proposition is a version of Perelman's noncollapsing theorem for Ricci flow with bubbling-off.

Proposition C. *For all $T > 0$ there exists $\kappa = \kappa(T) \in (0, \kappa_0)$ such that for all $r \in (0, 10^{-3})$ there exists $\bar{\delta}_C = \bar{\delta}_C(T, r) \in (0, \delta_0)$ with the following property.*

For all $0 < \delta \leqslant \bar{\delta}_C$ and $0 < b \leqslant T$, every Ricci flow with (r, δ)-bubbling-off defined on $[0, b)$ and having normalised initial condition satisfies $(\mathrm{NC})_\kappa$.

5.3.2 Proof of the finite-time existence theorem, assuming Propositions A, B, and C

We now prove Theorem 5.3.1, assuming the three propositions A, B and C.

Let $T > 0$. Proposition C gives a constant $\kappa = \kappa(T)$. Proposition B gives constants r, $\bar{\delta}_B$ depending on κ. Then apply Proposition C again to get a constant $\bar{\delta}_C$. Set $\delta := \min(\bar{\delta}_A, \bar{\delta}_B, \bar{\delta}_C, C_0^{-3/2}/10)$.

From r, δ we get the associated cutoff parameters h, Θ.

Let g_0 be a normalised metric on M. We denote by $\mathcal{X}$ the set of ordered pairs $(T', \{g(t)\}_{t\in[0,T')})$ consisting of a number $T' \in (0, T]$ and a Ricci flow with (r, δ, κ)-bubbling-off $\{g(t)\}_{t\in[0,T')}$ such that $g(0) = g_0$. By Proposition 5.1.8, $\mathcal{X}$ is nonempty.

It has a partial ordering, defined by $(T'_1,\{g_1(t)\}) \leqslant (T'_2,\{g_2(t)\})$ if $T'_1 \leqslant T'_2$ and $g_2(\cdot)$ is an extension of $g_1(\cdot)$.

We want to use Zorn's lemma to prove existence of a maximal element in $\mathcal{X}$. In order to do this, we consider an infinite chain, that is an infinite sequence of numbers $0 < T_1 < T_2 < \cdots T_n < \cdots < T$ and of Ricci flows with (r,δ,κ)-bubbling-off defined on the intervals $[0,T_n)$, and which extend one another. In this way we get an evolving metric $g(\cdot)$ defined on $[0,T_\infty)$, where T_∞ is the supremum of the T_n's. By Lemma 5.2.9, the set of singular times is a discrete subset of $\mathbb{R}$, so $g(\cdot)$ is a Ricci flow with (r,δ,κ)-bubbling-off.

Hence we can apply Zorn's lemma. Let $(T',g(\cdot)) \in \mathcal{X}$ be a maximal element. Its scalar curvature lies between -6 and Θ, hence it is bounded. Its curvature is pinched toward positive so the sectional curvature is also bounded. Hence by Proposition C.1.2 we may extend $g(\cdot)$ to a Ricci flow with bubbling-off on $[0,T']$, with $R_{\max}(T') \leqslant \Theta$. By Proposition B, Property $(\mathrm{CN})_r$ is satisfied on $[0,T']$. By Lemma 4.1.4 the same holds true for $(\mathrm{NC})_\kappa$ (cf. Remark 5.1.4). Hence we have obtained a Ricci flow with (r,δ,κ)-bubbling-off on the closed interval $[0,T']$.

To conclude, we prove by contradiction that $T' = T$. Assume that $T' < T$ and consider the following two cases.

Case 1. $R_{\max}(T') < \Theta$. Applying the short-time existence theorem for Ricci flow (Theorem C.1.1) with initial metric $g(T')$, we extend $g(\cdot)$ to a Ricci flow with bubbling-off defined on an interval $[0,T'+\alpha)$ for some $\alpha > 0$. We choose α sufficiently small so that we still have $R_{\max}(t) < \Theta$ on $[0,T'+\alpha)$. There are no singular times in $[T',T'+\alpha)$, and by Proposition 4.4.5 the extension has curvature pinched toward positive.

Lemma 5.3.2. *There exists an $\alpha' \in (0,\alpha]$ such that Condition $(\mathrm{CN})_r$ holds for $\{g(t)\}_{t\in[0,T'+\alpha')}$.*

Proof. Arguing by contradiction, we assume that there exist sequences $x_k \in M$ and $t_k \to T'$ such that $R(x_k,t_k) \geqslant r^{-2}$, but (x_k,t_k) does not have an (ε_0,C_0)-canonical neighbourhood. Since M is compact, we may assume that the sequence x_k converges and call x its limit.

We have $R(x,T') \geqslant r^{-2}$, so (x,T') has an (ε_0,C_0)-canonical neighbourhood. First consider the case where (x,T') is the centre of an (ε_0,C_0)-cap U. We use Remark 4.2.9 (iv). Since being the centre of an (ε_0,C_0)-cap is an open property in the $\mathcal{C}^{[\varepsilon_0^{-1}]+1}$-topology, U is still an (ε_0,C_0)-cap with respect to $g(t_k)$, with centre x_k, for sufficiently large k. This is a contradiction.

Suppose now that (x,T') is the centre of a strong ε_0-neck N. Let ψ be a parametrisation of N. By definition, there exists a parabolic rescaling $\bar{g}$ of g with factor Q at time T' such that

$$\sup_{t\in[-1,0]} \|\psi^*(\bar{g}(t)) - g_{\mathrm{cyl}}(t)\|_{\mathcal{C}^{[\varepsilon_0^{-1}]+1}} < \varepsilon_0.$$

Note that the inequality is strict. Let $\bar{g}_k(\cdot)$ be the parabolic rescaling of g with factor Q at time t_k. Since $t_k \to T'$, the same inequality holds with $\bar{g}(\cdot)$ replaced by $\bar{g}_k(\cdot)$, for sufficiently large k. Hence (x, t_k) is the centre of a strong ε_0-neck. We conclude using the already remarked fact that being the centre of strong ε_0-neck is an open property. □

It follows that $\{g(t)\}_{t\in[0,T'+\alpha')}$ is a Ricci flow with (r, δ)-bubbling-off. By Proposition C, it is a Ricci flow with (r, δ, κ)-bubbling-off. This contradicts maximality of T'.

Case 2. $R_{\max}(T') = \Theta$. Proposition A yields a metric g_+, which has curvature pinched toward positive at time T'. Applying Theorem C.1.1 with initial metric g_+, we obtain a positive number α and an evolving metric $\{g(t)\}_{t\in(T',T'+\alpha)}$ whose limit from the right as t tends to T' is equal to g_+. We also may assume that $R_{\max}$ remains bounded above by Θ. By Proposition 4.4.5 it has curvature pinched toward positive. Hence the resulting evolving metric $\{g(t)\}_{t\in[0,T'+\alpha)}$ satisfies the first two properties of the definition of Ricci flow with (r, δ)-bubbling-off.

Lemma 5.3.3. *There exists an* $\alpha' \in (0, \alpha]$ *such that Condition* $(\mathrm{CN})_r$ *holds for* $\{g(t)\}_{t\in[0,T'+\alpha')}$.

Proof. Arguing as in the proof of Lemma 5.3.2, consider sequences $x_k \to x \in M$ and $t_k \to T'$ such that for each k, $R(x_k, t_k) \geqslant r^{-2}$ but (x_k, t_k) does not have a canonical neighbourhood. Set $\tilde{g} := h^{-2} g_+(T')$. We will distinguish several cases according to the distance between x and the set

$$\Sigma_{T'} := \overline{\{x \in M \mid g_+(x, T') \neq g(x, T')\}}.$$

By definition of (r, δ)-surgery, this set is a union of 3-balls which are cores of δ-almost standard caps of $g_+(T')$.

Case (a). $d_{\tilde{g}}(x, \Sigma_{T'}) > (2C_0)^{3/2}$.

Claim. *The point* (x, T') *has an* (ε_0, C_0)*-canonical neighbourhood* U' *such that* $\overline{U'} \cap \Sigma_{T'} = \emptyset$.

Once the claim is proved, Case (a) follows by the arguments used to prove Lemma 5.3.2.

Let us prove the claim: since $t \mapsto g(t)$ is smooth on $[T', T'+\alpha)$ in a neighbourhood of x we have $R(x, T') = \lim R(x_k, t_k) \geqslant r^{-2}$. Hence (x, T') has an (ε_0, C_0)-canonical neighbourhood U'. Suppose that $\overline{U'} \cap \Sigma_{T'} \neq \emptyset$. Pick a point $z' \in \overline{U'} \cap \Sigma_{T'} \neq \emptyset$. Since $g_+(T')$ is obtained from $g(\cdot)$ by (r, δ)-surgery, there is a δ-almost standard cap U whose core V contains z'.

Let z be a point of $\overline{U'} \cap \partial V$. We have (see Section 3.2)

$$1 - f_3(\delta') \leqslant \frac{h^{-2}}{R(z, T')} \leqslant 1 + f_3(\delta').$$

Besides, we have $R(z, T') < C_0 R(x, T')$, thus

$$R(x, T')^{-\frac{1}{2}} < R(z, T')^{-\frac{1}{2}} \cdot (2C_0)^{\frac{1}{2}}.$$

It follows that

$$d_{g(T')}(x, z) \leqslant 2C_0 R(x, T')^{-\frac{1}{2}} \leqslant (2C_0)^{\frac{3}{2}} h.$$

We conclude that $d_{\tilde{g}}(x, \Sigma_{T'}) \leqslant (2C_0)^{\frac{3}{2}}$, contradicting our hypothesis.

Case (b). $d_{\tilde{g}}(x, \Sigma_{T'}) \leqslant (2C_0)^{3/2}$.

In this case, there exists a δ-almost standard cap U with core V and tip p such that $d_{\tilde{g}}(x, V) \leqslant (2C_0)^{3/2}$. In particular, $B_{\tilde{g}}(x, 2\varepsilon_0^{-1}) \subset U$ by choice of δ. If $B_{\tilde{g}}(x, 2\varepsilon_0^{-1}) \cap V = \emptyset$, then (x, T') is the centre of a strong ε_0-neck disjoint from $\Sigma_{T'}$, and we can apply the argument used in the proof of Lemma 5.3.2. Hence we assume that $B_{\tilde{g}}(x, 2\varepsilon_0^{-1}) \cap V \neq \emptyset$. Choose an ε_0-neck N in U at distance, say $5{\varepsilon_0}^{-1}$ from p. Then $U' := B_{\tilde{g}}(p, 6\varepsilon_0^{-1}) \cup N$ is an (ε_0, C_0)-cap in $(M, g_+(T'))$, with core $U' \backslash N$ and centre x. As $g(\cdot)$ is smooth of $(T', T' + \alpha)$ and $g_+(T') = \lim_{t \to T', t > T'} g(t)$, we can conclude as in Lemma 5.3.2. □

Hence $\{g(t)\}_{t \in [0, T' + \alpha')}$ is a Ricci flow with (r, δ)-bubbling-off. By Proposition C, it is a Ricci flow with (r, δ, κ)-bubbling-off. Again this contradicts the assumption that T' should be maximal.

□

5.4 Long-time existence of Ricci flow with bubbling-off

This section is not needed for the Poincaré Conjecture and can be skipped on first reading.

By iteration, Theorem 1.2.3 immediately extends to an existence theorem for Ricci flow with bubbling-off defined on $[0, +\infty)$ on our nonspherical manifold M. However, we need to extend the particular construction of Ricci flow with (r, δ, κ)-bubbling-off. More precisely, we need the following statement:

Theorem 5.4.1 (Long-time existence). *There exist decreasing sequences of positive numbers r_i, κ_i and, for every continuous positive function $t \mapsto \bar{\delta}(t)$, a decreasing sequence of positive numbers δ_i with $\delta_i < \min_{i < t \leqslant i+1} \bar{\delta}(t)$ such that the following holds: for every normalised metric g_0 on M, there exists a Ricci flow with bubbling-off $g(\cdot)$ defined on $[0, +\infty)$ such that $g(0) = g_0$ and for all i, $g(\cdot)$ is a Ricci flow with $(r_i, \delta_i, \kappa_i)$-bubbling-off on $(i, (i+1)]$.*

Remark 5.4.2. The presence of the function $\bar{\delta}(\cdot)$ allows us to choose δ small compared to r. This will be used in the proof of the thin-thick decomposition theorem (Theorem 1.3.4), in Part III of this book. In order to get estimates for the long-time behaviour of Ricci flow with bubbling-off, we will need to refine the choice of the parameters.

Theorem 5.4.1 will follow from a version of Theorem 5.3.1 suited for iteration. Before giving the statement, we need a definition.

Definition 5.4.3. Let $Q > 0$. One says that a Riemannian manifold (M, g) is Q-normalised if

(i) $(\operatorname{tr} \operatorname{Rm}^2)^{1/2} \leqslant Q$ on M,

(ii) $\operatorname{vol} B(x, 1) \geqslant \frac{Q^{-1}}{2} v_0(1)$ for all $x \in M$, where $v_0(1)$ denotes the volume of the unit Euclidean ball.

By compactness, any metric g is Q-normalised for some constant $Q > 0$. Moreover we have:

Assertion 5.4.4. *Let $g(\cdot)$ be a Ricci flow with (r, δ, κ)-bubbling-off on some interval $[a, b]$. Then for any $t \in [a, b]$, $g(t)$ is Q-normalised for some constant $Q = Q(r, \delta, \kappa)$.*

Proof. Recall that $R \leqslant \Theta = 2Dh^{-2}$ where D and h are functions of (r, δ), hence $|\operatorname{Rm}| \leqslant \Theta$ by Pinching Lemma 4.4.7. It follows that $(\operatorname{tr} \operatorname{Rm}^2)^{1/2} \leqslant \sqrt{3}\Theta$. On the other hand, setting $s = \Theta^{-1/2}$, one has $|\operatorname{Rm}| \leqslant s^{-2}$ on $P(x, t, s, -s^2)$ for all $(x, t) \in M \times [a, b]$. As $g(\cdot)$ is κ-noncollapsed at scales $\leqslant 1$, we deduce that

$$\operatorname{vol} B(x, t, 1) \geqslant \operatorname{vol} B(x, t, s) \geqslant \kappa s^3 = \frac{2\kappa\Theta^{-3/2}}{v_0(1)} \frac{v_0(1)}{2}.$$

Thus

$$Q = \max\left\{\sqrt{3}\Theta, \frac{v_0(1)}{2\kappa\Theta^{-3/2}}\right\}$$

fits. □

We then have a Q-normalised version of the finite-time existence theorem:

Theorem 5.4.5 (Finite-time existence for Q-normalised manifolds). *For all $Q > 0$, $0 \leqslant T_A < T_\Omega$, there exists $\kappa > 0$, such that for all $\bar{r} \in (0, 10^{-3})$ there exists $r \in (0, \bar{r})$, for all $\bar{\delta}$ there exists $\delta \in (0, \bar{\delta})$ and $Q' > 0$ with the following property. Let g be a Riemannian metric on M such that (M, g) is Q-normalised and has curvature pinched toward positive at time T_A. Then there exists a Ricci flow with (r, δ, κ)-bubbling-off $g(\cdot)$ defined on $[T_A, T_\Omega]$, such that $g(T_A) = g$, and such that $g(\cdot)$ is Q'-normalised.*

We leave it to the reader check that a proof of this theorem can be deduced, as the proof of Theorem 5.3.1 in Section 5.3.2, of Proposition A and the two following propositions.

Proposition BB. *For all $Q, \kappa > 0$ there exist $r(Q, \kappa) \in (0, 10^{-3})$, $\bar{\delta}_B = \bar{\delta}_B(Q, \kappa) \in (0, \delta_0)$ with the following property: let $\delta \leqslant \bar{\delta}_B$, $0 \leqslant a < b$ and $\{g(t)\}$ be a Ricci flow with bubbling-off defined on $[a, b]$, such that $g(a)$ is Q-normalised. Assume that the restriction of $g(\cdot)$ to the half-open interval $[a, b)$ is a Ricci flow with (r, δ, κ)-bubbling-off. Then it satisfies $(\mathrm{CN})_r$ on all of $[a, b]$.*

Proposition CC. *For all $Q, T > 0$ there exists $\kappa = \kappa(Q, T) \in (0, \kappa_0)$ such that for all $r \in (0, 10^{-3})$ there exists $\bar{\delta}_C = \bar{\delta}_C(Q, T, r) \in (0, \delta_0)$ with the following property. For all $0 < \delta \leqslant \bar{\delta}_C$ and $0 \leqslant a < b \leqslant T$, every Ricci flow with (r, δ)-bubbling-off defined on $[a, b)$, such that $g(a)$ is Q-normalised, satisfies* $(\mathrm{NC})_\kappa$.

Applying inductively Theorem 5.4.5 proves the long-time existence theorem.

Definition 5.4.6. In order to simplify the notation, given decreasing sequences r_i and δ_i we introduce step functions $t \mapsto r(t), t \mapsto \delta(t)$, whose values are respectively r_i and δ_i on $(i, i+1]$. We will say that a Ricci flow with bubbling-off defined on some interval $I \subset \mathbb{R}$ is a *Ricci flow with $r(\cdot), \delta(\cdot)$-bubbling-off* if, in restriction to $I \cap (i, i+1]$ it is a Ricci flow with (r_i, δ_i)-bubbling-off. .

Ricci flows with $r(\cdot), \delta(\cdot)$-bubbling-off will appear in Theorems 8.1.2 and 10.4.1 of Part II, which are needed in Part III where the function $r(\cdot)$ given by Theorem 5.4.1 will be fixed. The choice of $\delta(\cdot)$ should be refined to obtain better estimates for the Ricci flow with $r(\cdot), \delta(\cdot)$-bubbling-off.

Definition 5.4.7. For simplicity we shall let $h(t) := h(r(t), \delta(t))$, $D(t) = D(r(t), \delta(t))$ and $\Theta(t) = D(t)h^{-2}(t)$ denote the associated cutoff parameters.

Remark 5.4.8. The functions $r(\cdot)$, $\delta(\cdot)$ given by Theorem 5.4.1, and hence $h(\cdot)$, are nonincreasing and go to zero as $t \to \infty$. The functions $D(\cdot)$ and $\Theta(\cdot)$ are nondecreasing and go to $+\infty$ as $t \to \infty$.

Notes. The concept of metric surgery for Ricci flows is due to Hamilton [Ham86]. The cutoff parameter theorem (Theorem 5.2.4) is a variant of a result of Perelman ([Per03b], Lemma 4.3).

Propositions A, B, C correspond roughly to Section 4, Section 5.4 and Lemma 5.2 of [Per03b], respectively.

Lemma 5.2.9 is new. It is not expected to hold in Perelman's construction. A similar estimate holds for the Huisken–Sinestrari surgically modified mean curvature flow (see [HS09])

Theorem 5.4.1 is similar to [Per03b], Proposition 5.1.

Part II

Ricci flow with bubbling-off: existence

In this part we prove the three main propositions A, B and C. Let us be more precise. Chapter 6 is devoted to presenting some technical material. We prove the Curvature-Distance Theorem (Theorem 6.1.1) which asserts that the curvature on a given ball is comparable to the curvature at the centre provided that we have the canonical neighbourhood property and that the curvature at the centre is large enough. From this we deduce the Cutoff Parameters Theorem 5.2.4.

In Chapter 7 we state a precise definition for the standard solution and describe our version of the metric surgery, proving Theorem 5.2.2. From that we prove Proposition A which shows that we can perform the (r, δ)-surgery when the threshold for the scalar curvature is attained. It is shown to preserve the curvature pinched toward positive assumption.

Chapter 8 is a technical chapter. The main result is the so-called persistence theorem. It shows that near the added almost-standard cap the evolving metric looks like the standard solution for an amount of time close to 1, after rescaling. This is used to show that the canonical neighbourhood property persists after surgery.

In Chapter 9 we prove Proposition B. It shows that the canonical neighbourhood property is a closed condition in time; that is, if it is satisfied on an interval open in the future, it is also true on the closure, again if the parameters are well chosen.

Finally Chapter 10 is devoted to the proof of Proposition C. We prove several versions of the κ-non-collapsing result. We give an n-dimensional version for smooth Ricci flow. We also prove a version on a given interval for the Ricci flow with bubbling-off in dimension 3. Finally we give a useful version adapted to the long-time study of the flow which is used in Part III. It applies in particular in the description of the thick part.

Chapter 6
Choosing cutoff parameters

The main goal of this chapter is to prove the Cutoff Parameters Theorem 5.2.4 which allows us to get h, D, and therefore Θ, from r, δ. Let us recall that Θ is the scalar curvature threshold used to trigger the surgery process (cf. Definitions 5.2.5 and 5.2.7). The surgeries are made in strong δ-necks centred at points of scalar curvature $\approx h^{-2}$. Theorem 5.2.4 proves existence of such δ-necks on any geodesic joining a point of scalar curvature $\leqslant 2r^{-2}$ to a point of scalar curvature $\geqslant Dh^{-2} = \Theta/2$.

Along the way we prove another crucial result, the Curvature-Distance Theorem 6.1.1. Essentially, this theorem says that points at bounded distance from a base point have scalar curvature comparable to that of the base point, provided this latter is large enough. This will be used repeatedly in order to have curvature bounds on balls of rescaled Ricci flows with bubbling-off, which is the first condition to apply the Compactness Theorem C.3.1. This technique will appear as the first step of numerous arguments by contradiction used in this book.

The results on the structure of sets obtained by gluing together ε-necks, described in Chapter 3, will be used in the proofs of Theorems 5.2.4 and 6.1.1.

6.1 Bounded curvature at bounded distance

In this section we prove the following theorem. Let us recall that the constant ε_0 has been fixed in Chapter 3 so that the topology of the intersection of two ε_0-necks is understood.

Theorem 6.1.1 (Curvature-distance). *For all $A, C_1 > 0$ and all $\varepsilon_1 \in (0, 2\varepsilon_0]$, there exists $Q = Q(A, \varepsilon_1, C_1) > 0$ and $\Lambda = \Lambda(A, \varepsilon_1, C_1) > 0$ with the following property. Let $I \subset [0, +\infty)$ be an interval and $\{g(t)\}_{t \in I}$ be a Ricci flow with bubbling-off with curvature pinched toward positive. Let $(x_0, t_0) \in M \times I$ be such that the following holds:*

(i) $(1 + t_0)R(x_0, t_0) \geqslant Q$.

(ii) *For each point $y \in B(x_0, t_0, AR(x_0, t_0)^{-1/2})$, if $R(y, t_0) \geqslant 2R(x_0, t_0)$, then (y, t_0) has an (ε_1, C_1)-canonical neighbourhood.*

Then for all $y \in B(x_0, t_0, AR(x_0, t_0)^{-1/2})$, we have

$$\frac{R(y, t_0)}{R(x_0, t_0)} \leqslant \Lambda.$$

We emphasize that we do not require the Ricci flow with bubbling-off to satisfy the $(\mathrm{CN})_r$ assumption. In particular, the centre (x_0, t_0) may have no canonical neighbourhood, even if its scalar curvature is large. Instead, we assume only that points at bounded distance from the centre with curvature at least twice its curvature have canonical neighbourhoods. This particular situation appears, after a point-picking argument, in the proof of the canonical neighbourhood theorem ([Per02], Theorem 12.1) and in the proof of Proposition B in this book (cf. Lemma 9.2.3). In these proofs, the goal is to *show* that the centre *has* a canonical neighbourhood.

6.1.1 Preliminaries

Let us explain how to obtain local curvature estimates in spacetime. If a point has a neighbourhood where the local derivative estimates (4.2), (4.5) (see p. 40) hold, it suffices to integrate them. This happens in particular if the point is the centre of an (ε, C)-canonical neighbourhood. In fact, it is sufficient that these derivative estimates are satisfied above some curvature threshold, as shown in the lemmas below.

Lemma 6.1.2 (Local curvature-distance lemma). *Let (U, g) be a Riemannian manifold. Let $Q \geqslant 1$, $C > 0$, $x, y \in U$ and set $Q_x = |R(x)| + Q$. Assume that $[xy]$ is a geodesic segment of length $\leqslant \frac{Q_x^{-1/2}}{2C}$, on which*

$$|\nabla R| < C R^{3/2} \tag{6.1}$$

holds at any point of scalar curvature $\geqslant Q$. Then $R(y) < 2Q_x$.

Proof. We can assume that $R(y) \geqslant Q_x$ and hence that there exists a maximal segment $[x'y] \subset [xy]$ such that $R \geqslant Q_x$ on $[x'y]$. In particular $R(x') \geqslant Q_x$. Integrating Inequality (6.1) on $[x'y']$, we find

$$|R^{-1/2}(x') - R^{-1/2}(y)| < \frac{1}{2} C d(x', y) \leqslant \frac{1}{4} Q_x^{-1/2}.$$

Hence

$$R^{-1/2}(y) > R^{-1/2}(x') - \frac{1}{4} Q_x^{-1/2} \geqslant {Q_x}^{-1/2} - \frac{1}{4} Q_x^{-1/2} = \frac{3}{4} Q_x^{-1/2}.$$

Therefore $R(y) < \frac{16}{9} Q_x < 2Q_x$. □

Lemma 6.1.3 (Local curvature-time lemma). *Let $(U, g(\cdot))$ be a Ricci flow defined on $[t_1, t_2]$. Let $Q \geqslant 1$, $C > 0$, $x \in U$ and set $Q_x = |R(x, t_2)| + Q$. Assume that on $[t_2 - (2CQ_x)^{-1}, t_2]$, the inequality*

$$\left| \frac{\partial R}{\partial t} \right| < C R^2 \tag{6.2}$$

holds at (x, t) if $R(x, t) \geqslant Q$. Then $R(x, \cdot) < 2Q_x$ on $[t_2 - (2CQ_x)^{-1}, t_2]$.

Proof. Let $t \in [t_1, t_2]$. We can assume that $R(x,t) \geqslant Q$ and define $t_+ \in [t, t_2]$ maximal such that $R(x, \cdot) \geqslant Q$ on $[t, t_+]$. Note that $R(x, t_+) \leqslant Q_x$. Integrating (6.2) on $\{x\} \times [t, t_+]$ gives

$$|R(x,t_+)^{-1} - R(x,t)^{-1}| < C(t_+ - t) \leqslant (2Q_x)^{-1}.$$

hence $R(x,t)^{-1} > R(x,t_+)^{-1} - (2Q_x)^{-1} = Q_x{}^{-1} - (2Q_x)^{-1} = (2Q_x)^{-1}$. We conclude that $R(x,t) < 2Q_x$. □

Lemma 6.1.4 (Local curvature-control lemma). *Let $Q > 0$, $C > 0$, $\varepsilon \in (0, 2\varepsilon_0]$, and $\{g(t)\}_{t \in I}$ be a Ricci flow with bubbling-off on M. Let $(x_0, t_0) \in M \times I$ and set $Q_0 = |R(x_0, t_0)| + Q$. Suppose that $P = P(x_0, t_0, \frac{1}{2C\sqrt{Q_0}}, -\frac{1}{8CQ_0})$ is unscathed and that each $(x,t) \in P$ with $R(x,t) \geqslant Q$ has an (ε, C)-canonical neighbourhood. Then for all $(x,t) \in P$,*

$$R(x,t) \leqslant 4Q_0.$$

Proof. First we note that $R(x, t_0) \leqslant 2Q_0$. This follows from the local curvature-distance lemma (Lemma 6.1.2). Indeed, any point of $B(x_0, t_0, \frac{1}{2C\sqrt{Q_0}})$ of scalar curvature larger than Q is the centre of an (ε, C)-canonical neighbourhood, hence the hypothesis (6.1) of Lemma 6.1.2 is satisfied. Let us show that $R(x,t) \leqslant 4Q_0$. If $R(x,t) \leqslant Q$ we are done. Otherwise, let $t_+ \in [t, t_0]$ be maximal such that $R(x,s) \geqslant Q$ for all $s \in [t, t_+]$. Observe that $R(x, t_+) = Q \leqslant 2Q_0$ or $t_+ = t_0$ and $R(x, t_0) \leqslant 2Q_0$. As above, integrating $|\partial R / \partial t| \leqslant C_1 R^2$ on $[t, t_+]$, we obtain

$$\begin{aligned} R(x,t) &\leqslant (R(x,t_+)^{-1} - C_1(t_+ - t))^{-1} \\ &\leqslant ((2Q_0)^{-1} - (8Q_0)^{-1})^{-1} \leqslant \frac{8}{3} Q_0 \leqslant 4Q_0. \end{aligned}$$ □

Remark 6.1.5. In particular, this lemma applies to any Ricci flow with bubbling-off satisfying $(\mathrm{CN})_r$, setting $Q = r^{-2}$.

We shall use repeatedly the following property:

Proposition 6.1.6. *Let $(U_k, g_k(\cdot), *_k)$ be a sequence of pointed evolving metrics defined on intervals $I_k \subset \mathbb{R}_+$, and having curvature pinched toward positive. Set $(x_k, t_k) \in U_k \times I_k$ be such that $(1 + t_k) R(x_k, t_k)$ goes to $+\infty$. Let $\bar{g}_k(t) := R(x_k, t_k) g_k(t_k + \frac{t}{R(x_k, t_k)})$. Then we have the following:*

(i) *$R_{\min}(\bar{g}_k(\cdot))$ tends to 0 as $k \to +\infty$.*

(ii) *If $(U_k, \bar{g}_k(\cdot), *_k)$ converges in the pointed $\mathcal{C}^2$ sense, then the limit has nonnegative curvature operator.*

Proof. (i) Set $Q_k := R(x_k, t_k)$ and pick $t \in \mathbb{R}$ such that $t'_k := t_k + t Q_k{}^{-1} \in I_k$.

Then we have

$$\begin{aligned} R_{\min}(\bar{g}_k(t)) &= Q_k{}^{-1} R_{\min}(g_k(t'_k)) \\ &\geqslant \frac{-6}{(4t'_k+1)Q_k} \\ &= \frac{-6}{4t_k Q_k + 4t + Q_k} \\ &\geqslant \frac{-6}{(1+t_k)Q_k} \to 0 \end{aligned}$$

uniformly in t.

(ii) Let us call $(U_\infty, g_\infty(\cdot), *_\infty)$ the limit. Let $y_\infty \in U_\infty$ be the limit of $y_k \in U_k$. Since the convergence is $\mathcal{C}^2$, we now have

$$\begin{aligned} \operatorname{Rm}(y_\infty, t)) &= \lim_{k\to\infty} \operatorname{Rm}_{\bar{g}_k}(y_k, t) \\ &= \lim_{k\to\infty} Q_k{}^{-1} \operatorname{Rm}_{g_k}(y_k, t'_k) \\ &\geqslant -\frac{\phi_{t'_k}(R(y_k, t'_k))}{Q_k} \\ &= -\frac{\phi_0((1+t'_k)R(y_k, t'_k))}{(1+t'_k)Q_k}. \end{aligned}$$

Note that $(1+t'_k)Q_k = (1+t_k)Q_k + t \to +\infty$. If, up to extracting a subsequence, $(1+t'_k)R(y_k, t'_k) < C$ for some constant C, then the above expression is bounded from below by $-\frac{\phi_0(C)}{(1+t'_k)Q_k}$, which goes to 0 as k tends to $+\infty$. Now if $(1+t'_k)R(y_k, t'_k) \to \infty$, the expression above is bounded below by

$$-\frac{\phi_0((1+t'_k)R(y_k, t'_k))}{(1+t'_k)R(y_k, t'_k)} \frac{R(y_k, t'_k)}{Q_k},$$

which tends to 0 by Lemma 4.4.6 (ii), since by assumption, $\frac{R(y_k, t'_k)}{Q_k}$ converges to $R(y_\infty, t) \in \mathbb{R}$. □

To conclude the proof of Theorem 6.1.1, we shall also use Lemma C.5.1 stated in Appendix C.

6.1.2 Proof of Curvature-Distance Theorem 6.1.1

The proof is done by contradiction. Assume that there exist constants $A, C_1 > 0$, $\varepsilon_1 \in (0, 2\varepsilon_0]$, sequences $Q_k \to +\infty$, $\Lambda_k \to +\infty$, and a sequence $(g_k(\cdot), (x_k, t_k))$ of pointed Ricci flows with bubbling-off with curvature pinched toward positive, such that the following is satisfied:

(i) $(1+t_k)R(x_k,t_k) \geqslant Q_k$;
(ii) for all $y \in B(x_k,t_k,AR(x_k,t_k)^{-1/2})$, if $R(y,t_k) \geqslant 2R(x_k,t_k)$, then y has an (ε_1, C_1)-canonical neighbourhood;
(iii) for each k there exists a point z_k such that

$$\frac{R(z_k,t_k)}{R(x_k,t_k)} \geqslant \Lambda_k \quad \text{and} \quad d_{t_k}(x_k,z_k) \leqslant AR(x_k,t_k)^{-1/2}.$$

For every k we let

$$\bar{g}_k(\cdot) := R(x_k,t_k)g_k\Big(t_k + \frac{\cdot}{R(x_k,t_k)}\Big)$$

be the Ricci flow with bubbling-off resulting from $g_k(\cdot)$ by parabolic rescaling at time t_k with factor $R(x_k,t_k)$.

We shall use the following notation: when computing geometric quantities at $x \in M$ with the rescaled metrics, the point will be denoted by $\bar{x}$. For example, the points above satisfy $R(\bar{x}_k,0)=1$, $R(\bar{z}_k,0) \geqslant \Lambda_k$ and $d_0(\bar{x}_k,\bar{z}_k) \leqslant A$. Observe that points of $B(\bar{x}_k,0,A)$ of scalar curvature greater than or equal to 2 have (ε_1,C_1)-canonical neighbourhoods.

Step 1. There exists $\rho > 0$ such that the sequence $(B(\bar{x}_k,0,\rho),\bar{g}_k(0),\bar{x}_k)$ subconverges in the $\mathcal{C}^{1,\alpha}$ topology to $(M_\infty,\bar{g}_\infty,\bar{x}_\infty)$, where M_∞ is a smooth 3-manifold and $\bar{g}_\infty$ is an incomplete $\mathcal{C}^{1,\alpha}$-smooth Riemannian metric.

Proof. This will follow from Gromov's compactness theorem (Theorem B.1.4, see also Remark B.1.5). To apply this result, we have to show that, maybe for a subsequence,

1) there exists $\rho > 0$ such that for all $s \in (0,\rho)$, sectional curvatures of $\bar{g}_k(0)$ are bounded in absolute value on $B(\bar{x}_k,0,s)$, independently of k,

2) the injectivity radius of $\bar{g}_k(0)$ at x_k is positively bounded below independently of k.

Define

$$\rho := \sup\big\{s>0 \mid \exists C(s)>0,\ \forall k \in \mathbb{N},\ \forall \bar{y} \in B(\bar{x}_k,0,s),\ R(\bar{y},0) \leqslant C(s)\big\}.$$

First note that ρ is well defined. Indeed, Curvature-Distance Lemma 6.1.2, applied with $Q=2$ and $C=C_1$, gives $R \leqslant 6$ on $B(\bar{x}_k,0,\frac{3^{-1/2}}{2C_1})$. This shows that $\rho \geqslant \frac{3^{-1/2}}{2C_1}$. On the other hand, $\rho \leqslant A$ by assumption. Since $g_k(\cdot)$ has curvature pinched toward positive, we can apply Pinching Lemma 4.4.7. Let $0 < s < \rho$ and $\bar{y} \in B(\bar{x}_k,0,s)$. One has $R(y,t_k) \leqslant R(x_k,t_k)C(s)$, and we can assume that $(1+t_k)R(x_k,t_k)C(s) \geqslant \bar{s}$ for all $k \in \mathbb{N}$ (let $C(s) \geqslant 1$ and $Q_k \geqslant \bar{s}$). Pinching Lemma 4.4.7 then implies that $|\mathrm{Rm}(y,t_k)| \leqslant R(x_k,t_k)C(s)$. It follows that $|\mathrm{Rm}(\bar{y},0)| \leqslant C(s)$ on $B(\bar{x}_k,0,s)$ for all $k \in \mathbb{N}$.

Now we bound from below the injectivity radius at $(\bar{x}_k,0)$.

First note that, up to extracting a subsequence, there exists $\bar{y}_k \in B(\bar{x}_k, 0, \rho)$ such that

$$R(\bar{y}_k, 0) \xrightarrow[k\to+\infty]{} +\infty \quad \text{and} \quad d_0(\bar{x}_k, \bar{y}_k) \xrightarrow[k\to+\infty]{} \rho.$$

Indeed, if the scalar curvature stays below C on $B(\bar{x}_k, 0, \rho)$, then Lemma 6.1.2 implies that it is below $2(C+2)$ on $B(\bar{x}_k, 0, \rho + \frac{(C+2)^{-1/2}}{2C_1})$. This contradicts the definition of ρ.

Fix $k \in \mathbb{N}$, and consider a ball $B(\bar{x}_k, 0, s)$ of maximal radius such that $R(\bar{y}, 0) \leqslant 2$ for all $\bar{y} \in B(\bar{x}_k, 0, s)$. It follows from the curvature-distance lemma that $R \leqslant 8$ on $B(\bar{x}_k, 0, s + (4C_1)^{-1})$. On the other hand, let $x \in \partial \overline{B(\bar{x}_k, 0, s)}$ such that $R(\bar{x}, 0) = 2$. By assumption (ii) (x, t_k) is the centre of an (ε_1, C_1)-canonical neighbourhood U, not necessarily contained in $B(\bar{x}_k, 0, s + (4C_1)^{-1})$. On this neighbourhood the scalar curvature, hence $|\mathrm{Rm}|$, is bounded by $2C_1$ for the metric $\bar{g}_k(0)$. Then the ball $B(\bar{x}, 0, (4C_1)^{-1}) \subset U$ satisfies inequality (4.1). We deduce a positive lower bound for $\operatorname{vol} B(\bar{x}_k, 0, s + (4C_1)^{-1})$ independent of k. Using that $|\mathrm{Rm}| \leqslant 8$ on this ball, the local injectivity radius estimate (see Theorem B.1.2) gives a positive bound from below for the injectivity radius at $(\bar{x}_k, 0)$.

This allows us to apply Gromov's compactness theorem (Theorem B.1.4, see Remark B.1.5).

Then there is a convergent subsequence in the pointed $\mathcal{C}^{1,\alpha}$ topology to a Riemannian manifold $(M_\infty, \bar{g}_\infty(0), \bar{x}_\infty)$, where $M_\infty = B(x_\infty, \rho)$ is incomplete. □

Up to extracting a subsequence, we may assume that the sequence $[\bar{x}_k \bar{y}_k]$ converges to some geodesic $\gamma_\infty \colon [0, \rho) \to M_\infty$ such that $\gamma_\infty(0) = \bar{x}_\infty$.

We now show that the end of γ is contained in a horn (see Definition 3.2.7 and Figure 6.2 below), that the metric is smooth there and is the final time slice of a partial Ricci flow (see Definition 2.2.3).

Step 2. There exists $a \in [0, \rho)$ and an open set $U_\infty \subset M$ such that $U_\infty \supset \gamma([a, \rho))$, and such that $(U_\infty, g_\infty(0))$ is a union of $2\varepsilon_1$-necks and is the final time slice $U_\infty \times \{0\}$ of a partial Ricci flow with nonnegative curvature operator.

Proof. This will follow from the local compactness theorem for flows, Theorem C.3.3, applied to a suitable covering of a subset of $[\bar{x}_k \bar{y}_k]$ by strong necks.

For k large enough we pick points $\bar{x}'_k, \bar{y}'_k$ such that $R(\bar{x}'_k, 0) = 2C_1$, $R(\bar{y}'_k, 0) = R(\bar{y}_k, 0)/2C_1$, and $[\bar{x}'_k \bar{y}'_k] \subset [\bar{x}_k \bar{y}_k]$ is a maximal subsegment on which

$$2C_1 \leqslant R(\cdot, 0) \leqslant \frac{R(\bar{y}_k, 0)}{2C_1},$$

with $\bar{x}'_k$ closest to $\bar{x}_k$ (see Figure 6.1 below). This is possible as soon as $R(\bar{y}_k, 0) \geqslant 4C_1^2$, which holds if k is large enough. Note that $R(\bar{y}'_k, 0) = \frac{R(\bar{y}_k, 0)}{2C_1}$ goes to $+\infty$, so $d(\bar{x}_k, \bar{y}'_k) \to \rho$ as $k \to \infty$. Since $R \geqslant 2$ on $[\bar{x}'_k \bar{y}'_k]$, each point of this segment has an (ε_1, C_1)-canonical neighbourhood.

Let us show that each of them is a strong ε_1-neck. Let $\bar{z} \in [\bar{x}'_k \bar{y}'_k]$ and $U(\bar{z})$ be an (ε_1, C_1)-canonical neighbourhood centred at $\bar{z}$. Since $R(\bar{z}, 0) \in [2C_1, \frac{R(\bar{y}_k,0)}{2C_1}]$, the scalar curvature on $U(\bar{z})$ lies in $[2, \frac{R(\bar{y}_k,0)}{2}]$, by Property (ii) of an (ε_1, C_1)-canonical neighbourhood (Definition 4.2.10). It follows that neither $\bar{x}_k$ nor $\bar{y}_k$ belong to $U(\bar{z})$ (Figure 6.1). By Corollary 3.3.3, $U(\bar{z})$ cannot be a cap. Hence it is a strong ε_1-neck.

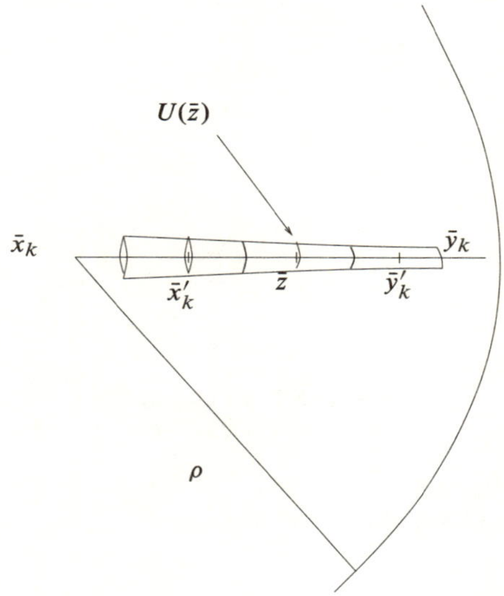

Figure 6.1. $[x'_k y'_k]$ in necks.

For each $\bar{z} \in [\bar{x}'_k \bar{y}'_k]$ we pick a strong ε_1-neck $U(\bar{z})$ centred at $(\bar{z}, 0)$. Let U_k be the union of the $U(\bar{z})$'s. For each $0 < s < \rho$, we have a uniform upper bound on the scalar curvature on $U_k \cap B(\bar{x}_k, 0, s)$. This implies a uniform bound for the curvature operator on $(U_k \cap B(\bar{x}_k, 0, s) \times [-\tau_s, 0])$, for some $\tau_s > 0$.

Besides, $d_0(\bar{x}'_k, \bar{y}'_k) \geqslant (\varepsilon_1 \sqrt{2C_1})^{-1}$ for k large enough, hence $d_0(\bar{x}_k, \bar{x}'_k) \leqslant \rho - (\varepsilon_1 \sqrt{2C_1})^{-1}$. This implies that $\bar{x}'_k$ tends to some point $\bar{x}'_\infty \in M_\infty$. Up to an extraction, the sequence $[\bar{x}_k \bar{y}_k]$ converges to a geodesic segment $\gamma_\infty \colon [0, \rho) \to M_\infty$ such that $\gamma_\infty(0) = \bar{x}_\infty$, and $\bar{x}'_\infty = \gamma_\infty(a)$ for $a \in (0, \rho)$. On the other hand, Theorem C.3.3 implies that $(U_k, \bar{g}_{k(\cdot)})$, with the appropriate base points, subconverges smoothly to a partial Ricci flow $(U_\infty, g_\infty(\cdot))$ (Figure 6.2 below). In particular, $g_\infty(0)$ is smooth on U_∞. By Proposition 6.1.6 (ii), $g_\infty(\cdot)$ has nonnegative curvature operator on U_∞. The final time slice $U_\infty \times \{0\} \subset M_\infty$ is a union of strong $2\varepsilon_1$-necks which covers $\gamma([a, \rho))$. Moreover $R(\gamma_\infty(s), 0) \to +\infty$ as $s \to \rho$. By Lemma 3.2.8, U_∞ is a $20\varepsilon_1$-horn.

Note that for each $z \in U_\infty$, the evolving metric $g_\infty(z, \cdot)$ is defined at least on $[-\frac{R(z,0)^{-1}}{2}, 0]$. □

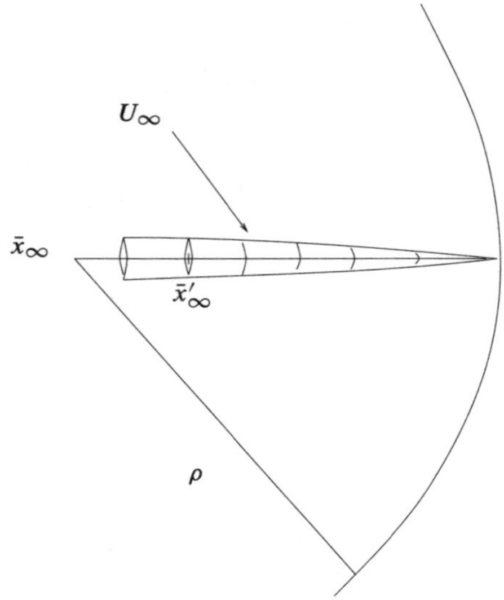

Figure 6.2. γ in a horn.

The horn U_∞ is diffeomorphic to, say, $S^2 \times (a, \rho)$. We do a local completion of U_∞ at ρ by identifying $S^2 \times \{\rho\}$ to a single point, called y_∞. Let us call $(\widehat{U}_\infty, \hat{d})$ the resulting metric space.

Step 3. The pointed metric space $(\widehat{U}_\infty, \hat{d}, y_\infty)$ is a locally complete Alexandrov space with nonnegative curvature. Its tangent cone at y_∞ admits outside the origin a partial Ricci flow whose curvature operator is nonnegative and equal to zero in the radial direction. Furthermore it is not identically zero.

For the definition of an Alexandrov space and of its tangent cone, see Appendix D.

Proof. It is easy to check that $(\widehat{U}_\infty, \hat{d}, y_\infty)$ is locally geodesic (see [BBI01]). Pick three points $a, b, c \in \widehat{U}_\infty$.

Suppose that a, b, c are all different from y_∞, and that c is closest to y_∞ among all three. Let S be the middle sphere of a neck centred at c. Choose a transverse sphere S' such that $d(c, S') = 4\,\mathrm{diam}(S)$ and which separates the convex hull of the triangle Δabc from y_∞. Then the proof of the Toponogov theorem in the Riemannian setting shows that Δabc is thicker than its comparison triangle in Euclidean space. The case where $c = y_\infty$ follows by considering a sequence $c_i \to c$, $c_i \neq c$. This shows that $(\widehat{U}_\infty, \hat{d}, y_\infty)$ is an Alexandrov space of nonnegative curvature.

Let C_∞ be the tangent cone of $\widehat{U}_\infty$ at y_∞. By definition we have

$$(C_\infty, d, 0) = \lim_{\lambda \to \infty} (\widehat{U}_\infty, \lambda \hat{d}, y_\infty),$$

where the limit is taken in the Gromov–Hausdorff topology (see Appendix A).

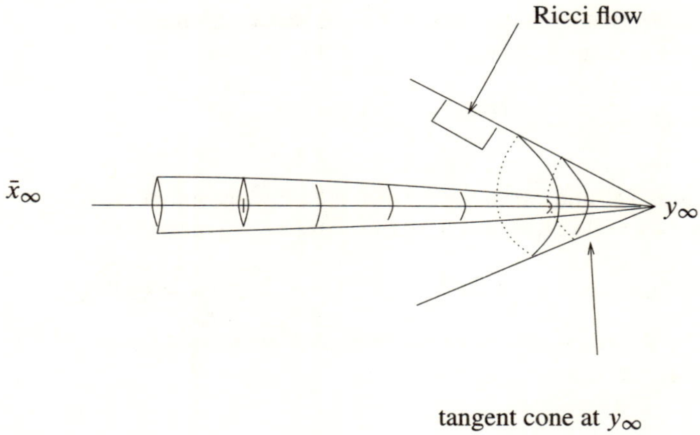

Figure 6.3. A Ricci flow on the tangent cone at y_∞.

Lemma 6.1.7. (C_∞, d) *is a nonnegatively curved metric cone. It is smooth outside* 0 *and not flat. Moreover* $C_\infty \setminus \{0\}$ *is the final time slice of some partial Ricci flow with nonnegative curvature, and every point is the centre of a* $3\varepsilon_1$*-neck.*

Proof. Recall that a cone has a family of homotheties of positive ratio. Let z be a point of $C_\infty \setminus \{0\}$. Without loss of generality, we can assume that $d(0, z) = 1$. With the obvious notation, we have

$$B\left(z, \frac{1}{2}\right) = \lim_{G-H} B_{i\hat{d}}\left(z_i, \frac{1}{2i}\right),$$

where $i \in \mathbb{N}$, $z_i \in \widehat{U}_\infty$, and $\hat{d}(y_\infty, z_i) = \frac{1}{i}$, by definition of the tangent cone.

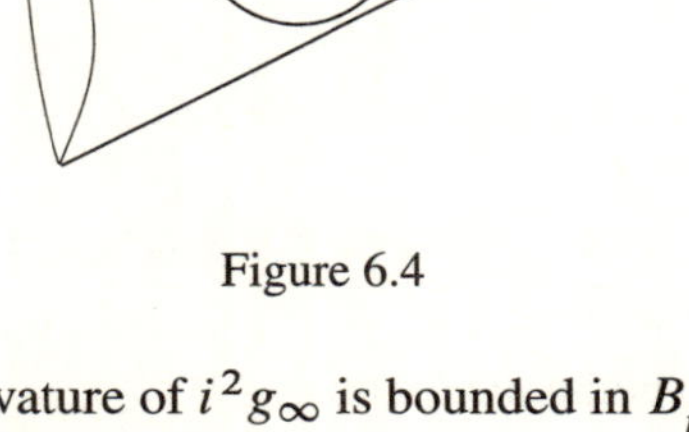

Figure 6.4

Let us show that the curvature of $i^2 g_\infty$ is bounded in $B_{i\hat{d}}(z_i, \frac{1}{2i})$. More precisely, there exist $A, B > 0$, such that

$$0 < A \leqslant R_\infty(z_i)\hat{d}^2(z_i, y_\infty) \leqslant B.$$

1) We know that z_i is the centre of a $2\varepsilon_1$-neck of length $(2\varepsilon_1\sqrt{R_\infty(z_i)})^{-1}$. This neck does not contain y_∞, hence

$$\hat{d}(z_i, y_\infty) \geqslant (2\varepsilon_1\sqrt{R_\infty(z_i)})^{-1},$$

which gives the lower bound, with $A := (4{\varepsilon_1}^2)^{-1}$.

2) Fix $i_0 \in \mathbb{N}^\star$ and two points p, q on the middle sphere at z_{i_0} maximally distant from one another. By closeness with the standard $2\varepsilon_1$-neck we have

$$\hat{d}(p_{i_0}, q_{i_0}) = \operatorname{diam}(S(z_{i_0})) \geqslant (1-2\varepsilon_1)\pi\sqrt{2}(R_\infty(z_{i_0}))^{-1/2}.$$

Let p_i, q_i be the respective intersection points of the middle sphere $S(z_i)$ with minimising geodesics from p_{i_0} and q_{i_0} to y_∞.

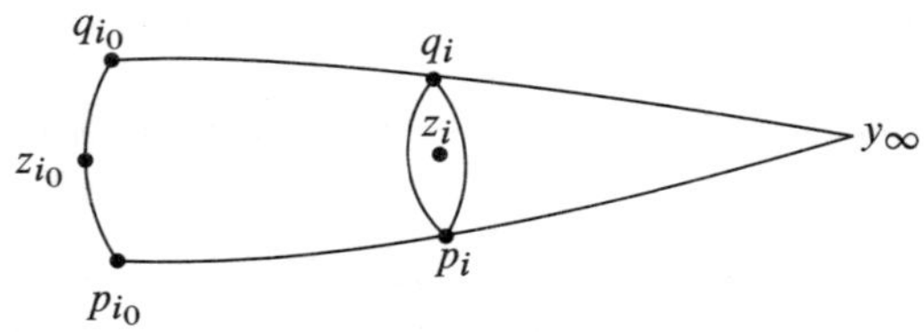

Figure 6.5. Geodesic triangle $\Delta y_\infty p_{i_0} q_{i_0}$.

The triangle inequality gives:

$$\frac{1}{i_0} - \frac{(1+2\varepsilon_1)\pi\sqrt{2}}{\sqrt{R(z_{i_0})}} \leqslant \hat{d}(p_{i_0}, y_\infty) \leqslant \frac{1}{i_0} + \frac{(1+2\varepsilon_1)\pi\sqrt{2}}{\sqrt{R(z_{i_0})}}.$$

Let $\Delta\bar{y}_\infty\bar{p}_{i_0}\bar{q}_{i_0} \subset \mathbb{R}^2$ be a comparison triangle for $\Delta y_\infty p_{i_0} q_{i_0}$.

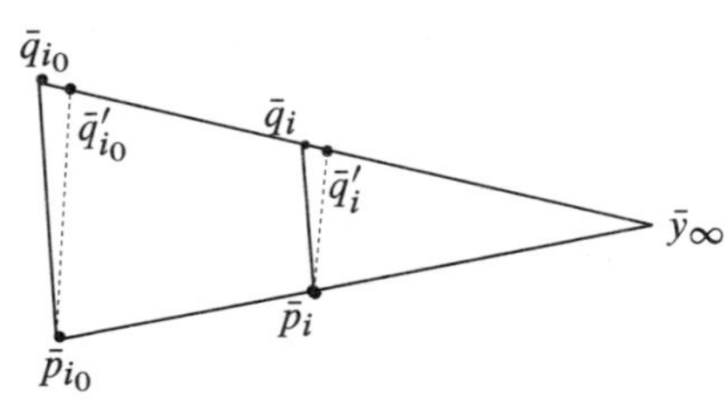

Figure 6.6. Comparison triangle $\Delta\bar{y}_\infty\bar{p}_{i_0}\bar{q}_{i_0}$.

Pick points $\bar{p}_i, \bar{q}_i$ in this triangle such that $|\bar{p}_i\bar{y}_\infty| = \hat{d}(p_i, y_\infty)$ and $|\bar{q}_i\bar{y}_\infty| = \hat{d}(q_i, y_\infty)$. For simplicity assume that $\hat{d}(p_i, y_\infty) \leqslant \hat{d}(q_i, y_\infty)$ and $\hat{d}(p_{i_0}, y_\infty) \leqslant \hat{d}(q_{i_0}, y_\infty)$, the other cases being analogous.

Let q'_{i_0} (resp. $\bar{q}'_{i_0}$) be the point of $[y_\infty q_{i_0}]$ (resp. $[\bar{y}_\infty\bar{q}_{i_0}]$) such that

$$\hat{d}(q'_{i_0}, y_\infty) = \hat{d}(p_{i_0}, y_\infty) \quad (\text{resp. } |\bar{q}'_{i_0}\bar{y}_\infty| = |\bar{p}_{i_0}\bar{y}_\infty|),$$

and define similarly points $q_i', \bar{q}_i'$. By construction,

$$|\bar{p}_i\bar{q}_i| \geqslant |\bar{p}_i\bar{q}_i'|.$$

Then one has

$$|\bar{p}_i\bar{q}_i'| \geqslant |\bar{p}_{i_0}\bar{q}_{i_0}'|\frac{|\bar{y}_\infty\bar{p}_i|}{|\bar{y}_\infty\bar{p}_{i_0}|}.$$

Now

$$\frac{|\bar{p}_{i_0}\bar{q}_{i_0}'|}{|\bar{p}_{i_0}\bar{q}_{i_0}|} = C > 0.$$

Hence

$$\begin{aligned}\hat{d}(p_i, q_i) = |\bar{p}_i\bar{q}_i| \geqslant |\bar{p}_i\bar{q}_i'| &= |\bar{p}_{i_0}\bar{q}_{i_0}'|\frac{|\bar{y}_\infty\bar{p}_i|}{|\bar{y}_\infty\bar{p}_{i_0}|}\\ &= C|\bar{p}_{i_0}\bar{q}_{i_0}|\frac{|\bar{y}_\infty\bar{p}_i|}{|\bar{y}_\infty\bar{p}_{i_0}|}\\ &= C|\bar{p}_{i_0}\bar{q}_{i_0}|\frac{\hat{d}(y_\infty p_i)}{\hat{d}(y_\infty, p_{i_0})}.\end{aligned}$$

That is,

$$\hat{d}(p_i, q_i) \geqslant C\frac{(1-2\varepsilon_1)\pi\sqrt{2}}{\sqrt{R_\infty(z_{i_0})}}\frac{1/i(1-\pi\sqrt{2}(1+2\varepsilon_1)i(R_\infty(z_i))^{-1/2}}{1/i_0(1+\pi\sqrt{2}(1+2\varepsilon_1)i_0(R_\infty(z_{i_0}))^{-1/2}}.$$

Step 1 shows that $i \leqslant 2\varepsilon_1\sqrt{R_\infty(z_i)}$. Recall that $\varepsilon_1 \leqslant 1/100$. Hence

$$\hat{d}(p_i, q_i) \geqslant \frac{C}{i} > 0.$$

for some constant $C > 0$ depending only of the curvature at z_{i_0}. As a consequence, we have

$$i\operatorname{diam}(S(z_i)) \geqslant C > 0,$$

hence

$$(1+2\varepsilon_1)(R_\infty(z_i))^{-1/2}\pi i\sqrt{2} \geqslant C,$$

therefore

$$\hat{d}(z_i, y_\infty)\sqrt{R_\infty(z_i)} \leqslant B.$$

Remark 6.1.8. The above argument works for all z such that $\hat{d}(z, y_\infty) > 1/i_0$.

This shows that all strong $2\varepsilon_1$-necks centred at z_i, henceforth denoted by $U(z_i)$, which have curvature roughly $R_\infty(z_i)$ also have bounded curvature after rescaling by i^2. By the Local Compactness Theorem C.3.3 we deduce that the rescaled strong necks $\bar{U}(z_i)$ subconverge smoothly to some strong $3\varepsilon_1$-necks, centred at z and denoted by $U_\infty(z)$. Furthermore, these necks have nonnegative curvature operator by Proposition 6.1.6, and they are not flat. We can adapt the above arguments by taking z at arbitrary distance from 0. It then follows that $C_\infty \setminus \{0\}$ has a partial Ricci flow as stated. □

This finishes Step 3. □

The above work contradicts Lemma C.5.1. This concludes the proof of Curvature-Distance Theorem 6.1.1.

6.2 Existence of cutoff parameters

For the convenience of the reader, we restate Theorem 5.2.4.

Theorem 6.2.1 (Cutoff parameters). *For all $r, \delta > 0$, there exist $h \in (0, \delta r)$ and $D > 10$ such that if $g(\cdot)$ is a Ricci flow with bubbling-off on M defined on an interval $[a,b]$, with curvature pinched toward positive and satisfying* $(\mathrm{CN})_r$, *then the following holds:*

Suppose that $x, y, z \in M$ and $t \in [a,b]$ are such that $R(x,t) \leqslant 2r^{-2}$, $R(y,t) = h^{-2}$, $R(z,t) \geqslant Dh^{-2}$, and y lies on a $g(t)$-geodesic segment connecting x to z. Then (y,t) is the centre of a strong δ-neck.

Proof. We again argue by contradiction. Fix constants $r > 0, \delta > 0$, sequences $h_k \to 0$, $D_k \to +\infty$, and Ricci flows with bubbling-off $(M_k, g_k(\cdot))$ satisfying the above hypotheses. Let $x_k, y_k, z_k \in M_k$, $t_k > 0$ and $\gamma_k \subset M_k$ a $g_k(t_k)$-geodesic segment connecting x_k to z_k, containing y_k and such that $R(x_k, t_k) \leqslant 2r^{-2}$, $R(z_k, t_k) \geqslant D_k h_k^{-2}$, and $R(y_k, t_k) = h_k^{-2}$ but (y_k, t_k) is not the centre of any strong δ-neck.

Consider the sequence $(M_k, \bar{g}_k(\cdot))$ defined by the following parabolic rescaling

$$\bar{g}_k(t) = h_k^{-2} g_k(t_k + t h_k^2).$$

Let us recall that we shall put a bar on the points when they are involved in geometric quantities computed with respect to the metric $\bar{g}_k$. Note that $R(\bar{y}_k, 0) = 1$. The contradiction will come from extracting a converging subsequence of the pointed sequence $(M_k, \bar{g}_k(\cdot), (\bar{y}_k, 0))$ and showing that the limit is the standard flow on $S^2 \times \mathbb{R}$, which implies that for k large enough, y_k is the centre of some strong δ-neck, contrary to the hypothesis.

Step 1. $(M_k, \bar{g}_k(0), \bar{y}_k)$ subconverges in the pointed $\mathcal{C}^p$ topology, for large p, to $(S^2 \times \mathbb{R}, \bar{g}_\infty, \bar{y}_\infty)$ where $\bar{g}_\infty$ is a product metric of nonnegative curvature operator and scalar curvature close to 1.

Note that it is not asserted at this stage that the metric on the S^2 factor is round.

Proof. We first control the curvature on balls around $\bar{y}_k$. Since $R(y_k, t_k)$ goes to $+\infty$, Theorem 6.1.1 implies that for all $\rho > 0$, there exists $\Lambda(\rho) > 0$ and $k_0(\rho) > 0$ such that $\bar{g}_k(0)$ has scalar curvature bounded above by $\Lambda(\rho)$ on $B(\bar{y}_k, 0, \rho)$ for $k \geqslant k_0(\rho)$. Since (y_k, t_k) has an (ε_0, C_0)-canonical neighbourhood, $\bar{g}_k(0)$ satisfies the volume estimate (4.1) at $\bar{y}_k$. As before, we deduce by Theorem B.1.2 a uniform positive lower bound for the injectivity radius at $\bar{y}_k$. By Gromov's compactness theorem (Theorem B.1.4), the sequence $(M_k, \bar{g}_k(0), \bar{y}_k)$ subconverges in the $\mathcal{C}^{1,\alpha}$ topology to a complete Riemannian manifold $(M_\infty, \bar{g}_\infty, \bar{y}_\infty)$.

Let us show that for all sufficiently large k, the ball $B(\bar{y}_k, 0, \rho)$ is covered by (ε_0, C_0)-canonical neighbourhoods. Each metric $g_k(t_k)$ satisfies

$$|\nabla R| < C_0 R^{3/2}$$

at points of scalar curvature $\geqslant r^{-2}$. Choose a point y such that $R(y, t_k) \leqslant 3r^{-2}$ and integrate the previous inequality on the portion of $[y_k y]$ where $R \geqslant 3r^{-2}$. An easy computation yields

$$d(\bar{y}, \bar{y}_k) \geqslant \frac{1}{h_k} \frac{2}{C_0} \Big(\frac{r}{\sqrt{3}} - h_k \Big), \tag{6.3}$$

which goes to ∞ with k. As a consequence, for all $\rho > 0$ there exists $k_1(\rho) \geqslant k_0(\rho)$ such that for every integer $k \geqslant k_1(\rho)$, the scalar curvature of $g_k(t_k)$ is above $3r^{-2}$ on $B(y_k, 0, \rho)$. Hence the ball is covered by (ε_0, C_0)-canonical neighbourhoods for large k.

Note that $d(\bar{y}_k, \bar{x}_k)$ goes to $+\infty$. Likewise, $d(\bar{y}_k, \bar{z}_k)$ goes to $+\infty$, because $R(\bar{z}_k, 0) = h_k^2 R(z_k, t_k) \geqslant D_k \geqslant \Lambda(\rho)$ for large k. In particular, for each $\rho > 0$, there exists $k_2(\rho) \geqslant k_1(\rho)$ such that for every integer $k \geqslant k_2(\rho)$ we have $R(\bar{z}_k, 0) > \Lambda(\rho)$.

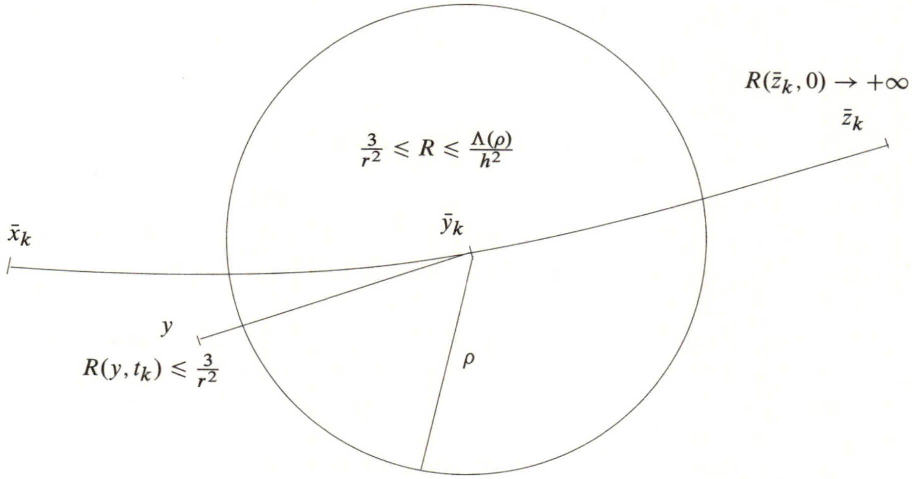

Figure 6.7. Covering $B(\bar{y}_k, 0, \rho)$ by canonical neighbourhoods.

As a result of Corollary 3.3.3, each canonical neighbourhood U centred on $[\bar{x}_k \bar{z}_k] \cap B(\bar{y}_k, 0, \rho)$ is a strong ε_0-neck. It follows from Proposition 3.2.3 that $U_{\rho,k}$ is a tube, which contains $B(\bar{y}_k, 0, \rho)$.

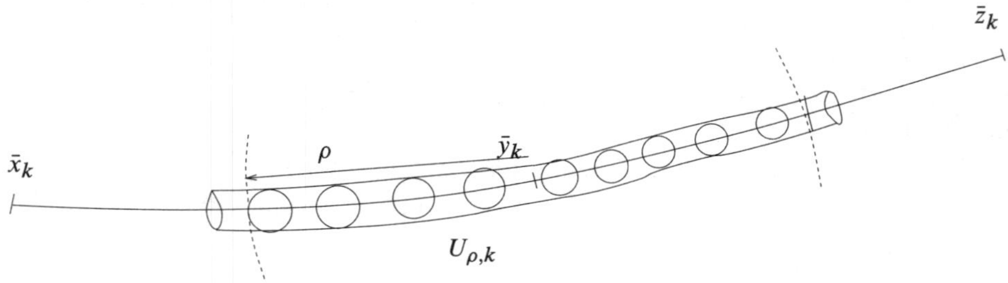

Figure 6.8. Covering $B(\bar{y}_k, 0, \rho)$ by necks.

When $k \geqslant k_0(\rho)$, the scalar curvature on $B(\bar{y}_k, 0, \rho)$ is less than $\Lambda(\rho)$. It follows that on each strong ε_0-neck U of $U_{k,\rho}$, $\bar{g}_k(t)$ is smoothly defined at least on $[-\frac{1}{2\Lambda(\rho)}, 0]$, and has curvature bounded above by $2\Lambda(\rho)$. Hence for each $\rho > 0$, the parabolic balls $P(\bar{y}_k, 0, \rho, -\frac{1}{2\Lambda(\rho)})$ are unscathed, with scalar curvature bounded above by $2\Lambda(\rho)$ for all $k \geqslant k_1(\rho)$. Since $g_k(\cdot)$ has curvature pinched toward positive, this implies a control on the curvature operator there, which is uniform in k, but depends on ρ.

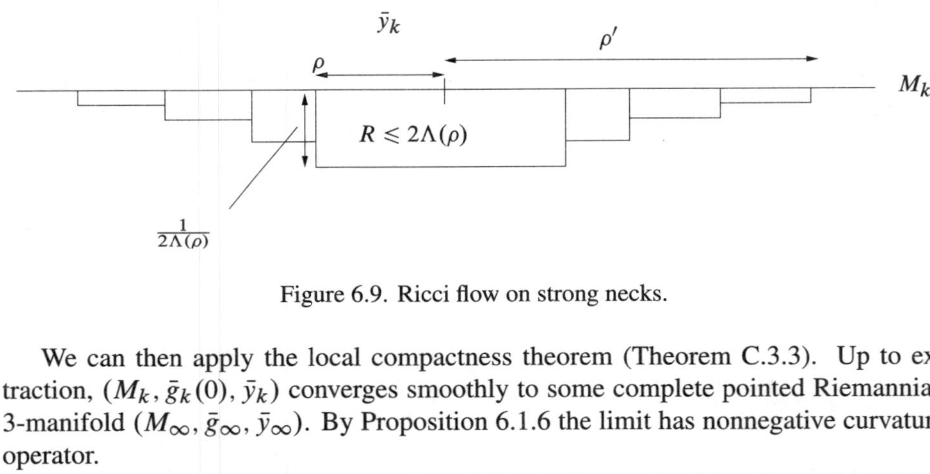

Figure 6.9. Ricci flow on strong necks.

We can then apply the local compactness theorem (Theorem C.3.3). Up to extraction, $(M_k, \bar{g}_k(0), \bar{y}_k)$ converges smoothly to some complete pointed Riemannian 3-manifold $(M_\infty, \bar{g}_\infty, \bar{y}_\infty)$. By Proposition 6.1.6 the limit has nonnegative curvature operator.

Passing to the limit we get a covering of M_∞ by $2\varepsilon_0$-necks. Then Proposition 3.2.9 shows that M_∞ is diffeomorphic to $S^2 \times \mathbb{R}$. In particular, it has two ends, so it contains a line, and Toponogov's theorem implies that it is the metric product of some (possibly nonround) metric on S^2 with $\mathbb{R}$.

As a consequence, the spherical factor of this product must be $2\varepsilon_0$-close to the round metric on S^2 with scalar curvature 1. Hence the scalar curvature is close to 1 everywhere. This finishes the proof of Step 1. □

Henceforth we take a subsequence $(M_k, \bar{g}_k(0), \bar{y}_k)$ satisfying the conclusion of Step 1. At this stage we can conclude, using the closeness to the limit and properties of strong necks, that there exists $k_2(\rho) \in \mathbb{N}$ for each $\rho > 0$, such that for $k \geqslant k_2(\rho)$, the parabolic neighbourhoods $P(\bar{y}_k, 0, \rho, -(2\Lambda(\rho))^{-1})$ are unscathed and satisfy $\frac{1}{2} \leqslant R \leqslant 2$.

Step 2. The sequence of pointed evolving metrics $(M_k, \bar{g}_k(\cdot), (y_k, 0))$ converges smoothly to the cylindrical flow on $S^2 \times \mathbb{R}$.

Proof. The Local Compactness Theorem C.3.3 implies that $(M_k, \bar{g}_k(\cdot), (\bar{y}_k, 0))$ converges to some complete Ricci flow $\bar{g}_\infty(\cdot)$ on M_∞. This flow is defined on $(-\frac{1}{2}, 0]$, it satisfies $R \leqslant 2$, and also $\mathrm{Rm} \geqslant 0$ by Proposition 6.1.6. For each $t \in (\frac{1}{2}, 0]$, the metric $\bar{g}_\infty(t)$ is also a product. We now have to extend the convergence to $(-\infty, 0]$. Set

$$\tau_0 := \sup\{\tau > 0 \mid \forall \rho > 0\,, \exists C(\rho,\tau) > 0, \exists k(\rho)\,, \forall k \geqslant k(\rho)\,, P(\bar{y}_k, 0, \rho, -\tau) \text{ is unscathed and } C(\rho,\tau)^{-1} \leqslant R \leqslant C(\rho,\tau) \text{ there }\}.$$

By the above argument, $\tau_0 \geqslant 1/2$. Let us show that $\tau_0 = +\infty$.

The convergence extends and gives a flow $\bar{g}_\infty(\cdot)$ defined on $(-\tau_0, 0]$, with each time-slice being a split nonnegatively curved metric. Let $(p, s) \in M_\infty \times (-\tau_0, 0]$. We shall prove that $\frac{\partial R_\infty}{\partial t}(p, s) > 0$. Here R_∞ denotes the scalar curvature of $\bar{g}_\infty$.

With this goal in mind, let $\rho > 0$ and $\tau \in (0, \tau_0)$ such that $(p, s) \in P(\bar{y}_\infty, 0, \rho, -\tau]$. Pick a sequence $(\bar{p}_k, s_k) \in P(\bar{y}_k, 0, \rho, \tau)$ whose limit is $(\bar{p}, s)$. Set $u_k := t_k + s_k h_k^2$. Since $h_k^2 R(p_k, u_k)$ goes to $R_\infty(\bar{p}, s) \in [C(\rho,\tau)^{-1}, C(\rho,\tau)]$, it follows that $R(p_k, u_k) \to +\infty$ with k. Let us recall that here p_k is the point $\bar{p}_k$ but considered on the manifold M_k endowed with the unscaled metric g_k; similarly, u_k corresponds, after rescaling, to the time s_k. Hence $(\bar{p}_k, s_k)$ has a canonical neighbourhood, which can only be a strong neck because $(\bar{p}_k, s_k)$ tends to $(\bar{p}, s)$. Denote by $\tilde{g}_k(\cdot)$ the parabolic rescaling of $g_k(\cdot)$ with factor $R(p_k, u_k)$ at time u_k and $\widetilde{R}$ its scalar curvature. Then

$$\frac{\partial \widetilde{R}}{\partial t}(p_k, 0) \sim 1,$$

which is the time zero value on the cylindrical flow. An easy computation shows that

$$\frac{\partial R_\infty}{\partial s}(\bar{p}, s) = R_\infty(\bar{p}, s)^2 \lim_{k\to\infty} \frac{\partial \widetilde{R}}{\partial t}(p_k, 0) > 0.$$

Therefore, the scalar curvature of $\bar{g}_\infty$ is bounded above by 1.5 on $M_\infty \times (-\tau_0, 0]$. It follows that for every $\rho > 0$ and every $\tau \in (0, \tau_0)$, there exists $k(\rho,\tau) \in \mathbb{N}$ such that for all $k \geqslant k(\rho,\tau)$, the scalar curvature stays below 2 on $P(\bar{y}_k, 0, \rho, -\tau)$.

For every point $(\bar{p}, t) \in P(\bar{y}_k, 0, \rho, -\tau)$, the strong ε_0-neck centred in $(\bar{p}, t)$ has then a Ricci flow defined at least on $[t - 1/3, t]$, with scalar curvature in $[\frac{C^{-1}}{3}, 3]$. If $\tau_0 < +\infty$, we can consider $\tau = \tau_0 + 1/6$ and $t = -\tau$. Then for all $\rho > 0$, for $k \geqslant k(\rho,\tau)$ the parabolic neighbourhoods $P(\bar{y}_k, 0, \rho, -\tau_0 - 1/6)$ are then unscathed with scalar curvature in $[\frac{C^{-1}}{3}, 3]$ (see Figure 6.10). This contradicts the definition of τ_0.

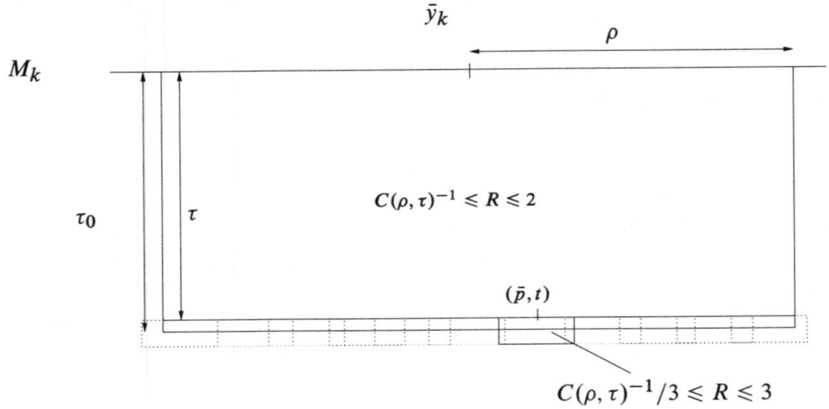

Figure 6.10. Extending the flow on $(-\infty, 0]$.

Hence $\tau_0 = +\infty$. Our limit is a κ-solution which splits at time 0. The classification of complete non-compact nonnegatively curved 3-dimensional Ricci flows, given by W. X. Shi (see [Shi89a]), shows that $\bar{g}_\infty(t)$ is the product of $\mathbb{R}$ with a positively curved Ricci flow on a surface. Since $\bar{g}_\infty(t)$ is a κ-solution the 2-dimensional flow is also a κ-solution. We conclude the proof of Step 2 by quoting Theorem 4.2.4, that is the 2-dimensional κ-solution is the round sphere. □

The conclusion of Step 2 implies that for k large enough, the point (y_k, t_k) is in fact centre of a strong δ-neck. This contradiction brings the proof of Theorem 5.2.4 to an end. □

Remark 6.2.2. In the proof of the result above, we have used the crucial assumption that canonical neighbourhoods which are ε-necks are in fact strong ε-necks. This has been used to extend the flow backward.

Notes. The results of Section 6.1 are close to Claim 2 in [Per02], Theorem 12.1, and to [Per03b], Section 4.2.

For the proof of the Cutoff Parameters Theorem 6.2.1 we follow closely Perelman (see [Per03b], Lemma 4.3). For the third step in the proof of Theorem 6.1.1 a more explicit treatment can be found in R. Bamler's Diploma thesis [Bam07].

A version of the Curvature-Distance Theorem 6.1.1, stated as an independent result, first appeared in [MT07] (Theorem 10.2). Our proof follows [KL08].

Chapter 7

Metric surgery and proof of Proposition A

The goal of this chapter is to prove the metric surgery theorem (Theorem 5.2.2) and Proposition A. This is done in Sections 7.2 and 7.3. The operation we call 'metric surgery' should not be confused with the topological concept of surgery: the manifold M does not change, only its metric g is modified. There is a 3-ball $B \subset M$ where the metric is decreased, and made close to a metric ball in some standard model, the so-called *standard solution*, which we already discussed in Section 4.3. One important feature of this operation is that it preserves the property of having curvature pinched toward positive.

In the actual surgery operation, this theorem is applied on a finite collection of suitably chosen distinct 3-balls. This is the content of Proposition A. The main point here is that the new metric g_+ satisfies $R_{\max}(g_+) \leqslant R_{\max}(g)/2$. This allows us to prevent singularities in Ricci flow from appearing, while preserving the estimates coming from the maximum principle. The fact that the metric decreases is only used to control the variation of the Colding–Minicozzi width, as explained in Section 1.2.

The fact that after surgery, the part that has changed is close to some model is used in several ways: first it is used to show that surgery does not disrupt the canonical neighbourhood property and the κ-noncollapsing property. It is also useful to control the behaviour of Ricci flow for a short time afterwards in this region (cf. Chapter 8).

The first thing we need to do is give a precise definition of the standard solution. This is done in Section 7.1 below.

7.1 The standard solution II

Let $r_0 := \frac{\pi\sqrt{2}}{2}$ be the radius of the hemisphere of sectional curvature $1/2$. Note that $r_0 < 3$.

Fact. There exists a number $\alpha \in (0, 10^{-2})$ and a smooth function $u\colon [0,+\infty) \to [0,\sqrt{2}]$ such that

$$\begin{cases} u(r) = \sqrt{2}\sin(\frac{r}{\sqrt{2}}) & \text{if } r \leqslant r_0 - \alpha, \\ 0 < u' < 10^{-1},\ -\frac{1}{\sqrt{2}} - \alpha < u'' < 0 & \text{if } r_0 - \alpha \leqslant r < r_0, \\ u(r) = \sqrt{2} & \text{if } r_0 \leqslant r. \end{cases}$$

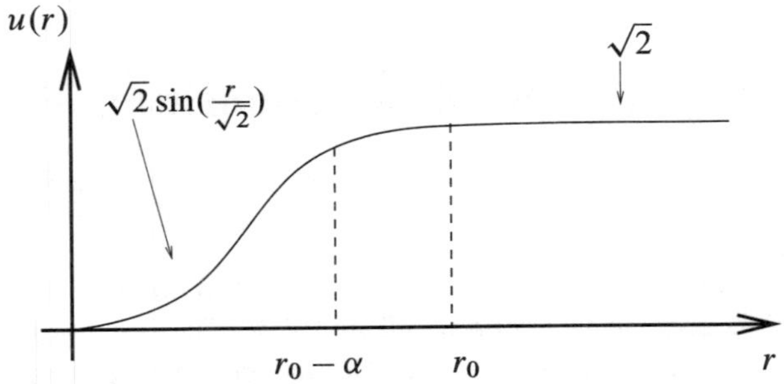

In the sequel, we fix α, u satisfying the above conclusion. We also choose a smooth function $\chi_0\colon [0,+\infty) \to [0,1]$ such that

$$\begin{cases} \chi_0 \equiv 1 & \text{on } [0,\alpha], \\ \chi_0' < 0 & \text{on } (\alpha, 2\alpha), \\ \chi_0 \equiv 0 & \text{on } [2\alpha, +\infty). \end{cases}$$

For all $\Lambda > 0$, let $f_\Lambda\colon [0,+\infty) \to [0,+\infty)$ be the smooth function defined by the formula

$$\begin{cases} f_\Lambda(r) := \chi_0(r)e^{-\frac{\Lambda}{5-\alpha}} + (1-\chi_0(r))e^{-\frac{\Lambda}{5-r}} & \text{if } r \leqslant 5, \\ f_\Lambda(r) := 0 & \text{if } r \geqslant 5. \end{cases}$$

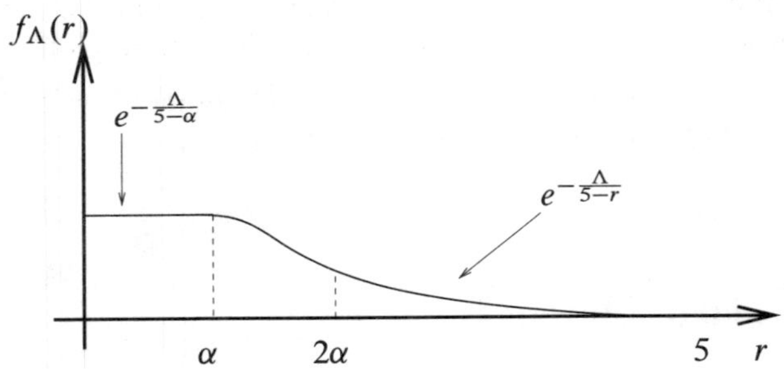

In particular, we have $f_\Lambda' < 0$ on $(\alpha, 2\alpha]$, and $f_\Lambda'' > 0$ on $[2\alpha, 5)$.

Fact. For all $0 < \beta < 1$, there exists $\Lambda(\beta) > 0$ such that for all $\Lambda \geqslant \Lambda(\beta)$, the function f_Λ satisfies the following estimates:

- On $[0,\infty)$, $|f_\Lambda|_{\mathcal{C}^2} < \beta$, and $|\frac{f_\Lambda' u'}{u}| \leqslant \beta$.
- On $[0,\infty)$, $|1 - e^{-2f_\Lambda}| \leqslant \beta |f_\Lambda''|$.
- On $[2\alpha, +\infty)$, $\max(|f_\Lambda'|, |f_\Lambda'|^2) \leqslant \beta |f_\Lambda''|$.

The proofs are left to the reader. We set once and for all $\Lambda_0 := \Lambda(\frac{\alpha}{500})$ and $f := f_{\Lambda_0}$. For notational convenience we set $\beta := \frac{\alpha}{500}$ and $\Lambda := \Lambda_0$.

Let $d\theta^2$ denote the round metric of sectional curvature 1 on S^2.

Definition 7.1.1. The *standard initial metric* is the Riemannian manifold $\mathcal{S}_0 := (\mathbb{R}^3, \bar{g}_0)$, where the metric $\bar{g}_0$ is given in polar coordinates by

$$\bar{g}_0 := e^{-2f(r)} g_u,$$

where

$$g_u = dr^2 + u^2(r)d\theta^2.$$

The origin of $\mathbb{R}^3$, which is also the centre of spherical symmetry, will be denoted by p_0.

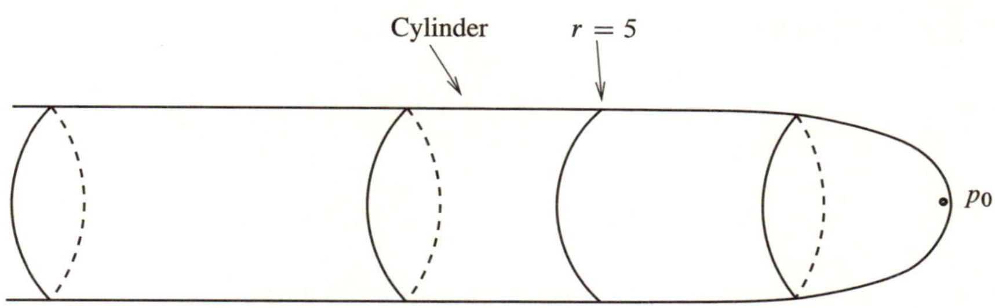

Standard initial metric

Lemma 7.1.2. *For all $r \in [0, 5)$, all sectional curvatures of $\bar{g}_0$ lie in $(0, 1]$.*

Proof. Let $\{e_1, e_2, e_3\}$ be a g_u-orthonormal basis such that $e_1 = \partial/\partial r$. Since g_u is a warped product, the Bishop–O'Neil formulas [Sak96] give sectional curvatures

$$K_{12} = K_{13} = -\frac{u''}{u},$$

which is positive on $[0, r_0)$ and vanishes on $[r_0, +\infty)$, and

$$K_{23} = \frac{1 - u'^2}{u^2},$$

which is greater than or equal to $\frac{1-10^{-2}}{2}$ on $[0, +\infty)$.

We deduce the Ricci curvatures with similar notation:

$$\mathrm{Ric}_{g_u}(e_1, e_1) = K_{12} + K_{13} = -\frac{2u''}{u},$$

$$\mathrm{Ric}_{g_u}(e_2, e_2) = \mathrm{Ric}_{g_u}(e_3, e_3) = -\frac{u''}{u} + \frac{1 - u'^2}{u^2}.$$

We now compute the curvatures of $\bar{g}_0 = e^{-2f} g_u$ using the formulas given on p. 59 of [Bes87] (where f should be replaced by $-f$),

$$\mathrm{Ric}_{\bar{g}_0} = \mathrm{Ric}_{g_u} - \Delta f g_u + (n-2)(Ddf + df \circ df - |df|^2 g_u),$$

the right-hand side quantities being computed using g_u.

To fix convention, here $\Delta = -\mathrm{trace}\, Dd$. It is easily checked that

$$Ddf\left(\frac{\partial}{\partial r}, \frac{\partial}{\partial r}\right) = f''(r),$$

and

$$Ddf(X, X) = \frac{f'u'}{u} g_u(X, X)$$

if $X \perp \partial/\partial r$. It follows that

$$\mathrm{Ric}_{\bar{g}_0}(e_1, e_1) = \mathrm{Ric}_{g_u}(e_1, e_1) + f'' + 2\frac{f'u'}{u} + f'',$$

and

$$\mathrm{Ric}_{\bar{g}_0}(\bar{e}_1, \bar{e}_1) = e^{2f}\Big(\mathrm{Ric}_{g_u}(e_1, e_1) + 2f'' + 2\frac{f'u'}{u}\Big),$$

where $\bar{e}_1 = e^f e_1$. For $i = 2, 3$ we get, with similar notation,

$$\mathrm{Ric}_{\bar{g}_0}(\bar{e}_i, \bar{e}_i) = e^{2f}\Big(\mathrm{Ric}_{g_u}(e_i, e_i) + f'' + 3\frac{f'u'}{u} - f'^2\Big).$$

We deduce the values of the sectional curvatures of $\bar{g}_0$, with the obvious notation:

$$\bar{K}_{12} = \bar{K}_{13} = e^{2f}\Big(K_{12} + f'' + \frac{f'u'}{u}\Big).$$

By choice of f and u, we have for all $r \in [0, r_0 - \alpha]$:

$$\bar{K}_{12} = \bar{K}_{13} \geqslant e^{2f}\Big(\frac{1}{2} - 2\beta\Big) > 0.$$

Let $r \in [r_0 - \alpha, 5)$. Since $2\alpha \leqslant r_0 - \alpha$, $f'' > 0$, we get

$$\bar{K}_{12} = \bar{K}_{13} \geqslant e^{2f}(K_{12} + (1-\beta)f'') \geqslant e^{2f}(1-\beta)f'' > 0.$$

Similarly, it is easily checked that

$$\begin{aligned}\bar{K}_{23} &= e^{2f}\Big(K_{23} + \frac{1}{2}\Big(4\frac{f'u'}{u} - 2f'^2\Big)\Big) = e^{2f}\Big(K_{23} + 2\frac{f'u'}{u} - f'^2\Big)\\ &\geqslant e^{2f}\Big(\frac{0.99}{2} - 3\beta\Big) > 0.\end{aligned}$$

Observe that there is a constant $K_0 > 0$ such that for all $r \in [3, 4]$ and all $i \neq j$,

$$\bar{K}_{ij} \geqslant K_0 > 0. \tag{7.1}$$

To conclude, notice that the maximum of the sectional curvature is close to $1/2$, hence less than 1 by choice of the constants. □

7.2 Proof of the metric surgery theorem

We recall that if ϕ is a nonnegative function, then a Riemannian manifold (N, g) is said to have ϕ-almost nonnegative curvature if it satisfies $\mathrm{Rm} \geqslant -\phi(R)$. This is trivially true at points where $\nu \geqslant 0$, i.e., all sectional curvatures are nonnegative. For convenience we restate the metric surgery theorem:

Theorem 7.2.1. *There exists $\delta_0 > 0$ and a function $\delta'\colon (0, \delta_0] \ni \delta \mapsto \delta'(\delta) \in (0, \varepsilon_0/10]$ tending to zero as $\delta \to 0$, with the following property:*

Let ϕ be a nondecreasing, nonnegative function, let $\delta \leqslant \delta_0$, let (M, g) be a Riemannian 3-manifold with ϕ-almost nonnegative curvature, and $U \subset M$ be a δ-neck whose middle sphere bounds a 3-ball $B \subset M$.

Then there exists a Riemannian metric g_+ on M with the following properties:

(i) *$g_+ = g$ on $M \setminus \mathrm{Int}\, B$ and $g_+ < g$ on $\mathrm{Int}\, B$;*
(ii) *$U \cup B$ is a $(\delta, \delta'(\delta))$-almost standard cap with respect to g_+ and B is a core thereof;*
(iii) *g_+ has ϕ-almost nonnegative curvature.*

First we introduce some notation. For $0 \leqslant r_1 \leqslant r_2$, we let $C[r_1, r_2]$ denote the annular region of $\mathbb{R}^3$ defined by the inequalities $r_1 \leqslant r \leqslant r_2$ in polar coordinates. Define $\mathcal{S}_u = (\mathbb{R}^3, g_u)$. Observe that for all $3 \leqslant r_1 < r_2$, the restriction of g_u to $C[r_1, r_2]$ is isometric to the cylinder $S^2 \times [r_1, r_2]$ with scalar curvature 1. We recall that the origin of $\mathbb{R}^3$, which is the centre of spherical symmetry of g_u and $\bar{g}_0$, is denoted by p_0.

Summary of the proof. Let ϕ be a nondecreasing nonnegative function, δ be a positive number, $U \subset M$ be a δ-neck whose middle sphere S bounds a 3-ball $B \subset M$ (see Figure 7.1, p. 81). Let y be a point of S. Our first task is to construct a $(1+\delta)$-Lipschitz diffeomorphism

$$\psi\colon (B, R(y)g) \to \overline{B(p_0, 5)} \subset \mathcal{S}_u$$

which is almost an isometry near $C[3, 5]$. With this we push forward $R(y)g$ in a metric $\bar{g}$ on $B(p_0, 5)$. Then we interpolate the metrics $e^{-2f}\bar{g}$ and $e^{-2f}g_u = \bar{g}_0$ to get a metric $\bar{g}_+$ on $B(p_0, 5)$. The conformal factor will guarantee that the resulting metric has ϕ-almost nonnegative curvature; we will then explain how to choose δ small enough so that $\bar{g}_+ < \bar{g}$ on $B(p_0, 5)$, and finally pull back $R(y)^{-1}\bar{g}_+$ to Int B using ψ. To do the interpolation we shall use the following.

Choose a smooth function $\chi\colon [0,5] \to [0,1]$ such that

$$\begin{cases} \chi \equiv 1 & \text{on } [0,3], \\ \chi' < 0 & \text{on } (3,4), \\ \chi \equiv 0 & \text{on } [4,5]. \end{cases}$$

Proposition 7.2.2. *There exists $\delta_1 > 0$ and a function $\delta'\colon (0,\delta_1] \to (0, \frac{\varepsilon_0}{10}]$ with limit zero at zero, having the following property: let ϕ be a nondecreasing positive function, $0 < \delta \leqslant \delta_1$ and $\bar{g}$ be a metric on $B(p_0,5) \subset \mathbb{R}^3$, with ϕ-almost nonnegative curvature, such that $\|\bar{g} - g_u\|_{\mathcal{C}^{[\delta^{-1}]}} < \delta$ on $C(3,5)$. Then the metric*

$$\bar{g}_+ = e^{-2f}\,(\chi g_u + (1-\chi)\bar{g})$$

has ϕ-almost nonnegative curvature, and is $\delta'(\delta)$-close to $\bar{g}_0$ on $B(p_0,5)$.

Proof of Theorem 7.2.1 *assuming Proposition* 7.2.2. Set $\delta_2 := e^{2e^{-\Lambda}} - 1$ and $\delta_0 := \min(\delta_1, \delta_2)$. We begin with the construction of the $(1+\delta)$-Lipschitz diffeomorphism from $B \subset M$ to $\overline{B(p_0,5)} \subset \mathcal{S}_u$.

Lemma 7.2.3. *There exists a diffeomorphism $\psi\colon B \to \overline{B(p_0,5)}$ such that the metric $\bar{g} := R(y)\psi_* g$ satisfies $g_u \leqslant (1+\delta)\bar{g}$ on $\overline{B(p_0,5)}$ and $\|\bar{g} - g_u\|_{[\delta^{-1}]} \leqslant \delta$ on $C[3,5]$.*

Proof. By definition of the δ-neck U there exists a diffeomorphism $\psi_U\colon U \to S^2 \times (-\delta^{-1}, \delta^{-1})$ such that $\psi_U(S) = S^2 \times \{0\}$ and $\|R(y)\psi_{U*}g - g_{\text{cyl}}\|_{[\delta^{-1}]} \leqslant \delta$. We may assume that $U \cap B$ is sent to $S^2 \times [0, \delta^{-1})$. Set $U_{[0,5]} := {\psi_U}^{-1}(S^2 \times [0,5])$ and let $\psi_{[0,5]}$ be the restriction of ψ_U to $U_{[0,5]}$. Let s be the map

$$s\colon S^2 \times [0,5] \to S^2 \times [3,8], \quad (\theta, z) \mapsto (\theta, 8-z).$$

Let i be the map that identifies $S^2 \times [3,8]$ with $C[3,8] \subset \mathcal{S}_u$ via polar coordinates. Define a map $\bar{\psi}_{[0,5]}\colon U_{[0,5]} \to C[3,8] \subset \mathcal{S}_u$ by setting $\bar{\psi}_{[0,5]} := i \circ s \circ \psi_{[0,5]}$. Observe that $\bar{\psi}_{[0,5]}$ maps S onto $S^2 \times \{8\}$. Using a theorem of Smale [Sma59] we extend $\bar{\psi}_{[0,5]}$ to a diffeomorphism $\bar{\psi}_B\colon B \to \overline{B(p_0,8)} \subset \mathcal{S}_u$.

Clearly, on $U_{[0,5]}$ we still have

$$\|R(y)\bar{\psi}_{B*}g - g_u\|_{\mathcal{C}^{[\delta^{-1}]}} \leqslant \delta,$$

since this involves only the cylindrical part of g_u. Besides, by compactness, there is a constant $C > 0$ such that $\bar{\psi}_B\colon (B \setminus U_{[0,5]}, R(y)g) \to (B(p_0,3), g_u)$ is C-Lipschitz.

Let $\eta > 0$ be small enough so that $\sup_{0<r\leqslant 3} \frac{u(\frac{\eta r}{6})}{u(r)} \leqslant \frac{1}{C}$ and $\frac{\eta}{6} \leqslant \frac{1}{C}$. Define a map $T\colon \overline{B(p_0,8)} \to \overline{B(p_0,5)}$ in polar coordinates by $T(\theta, r) := (\theta, t(r))$, where

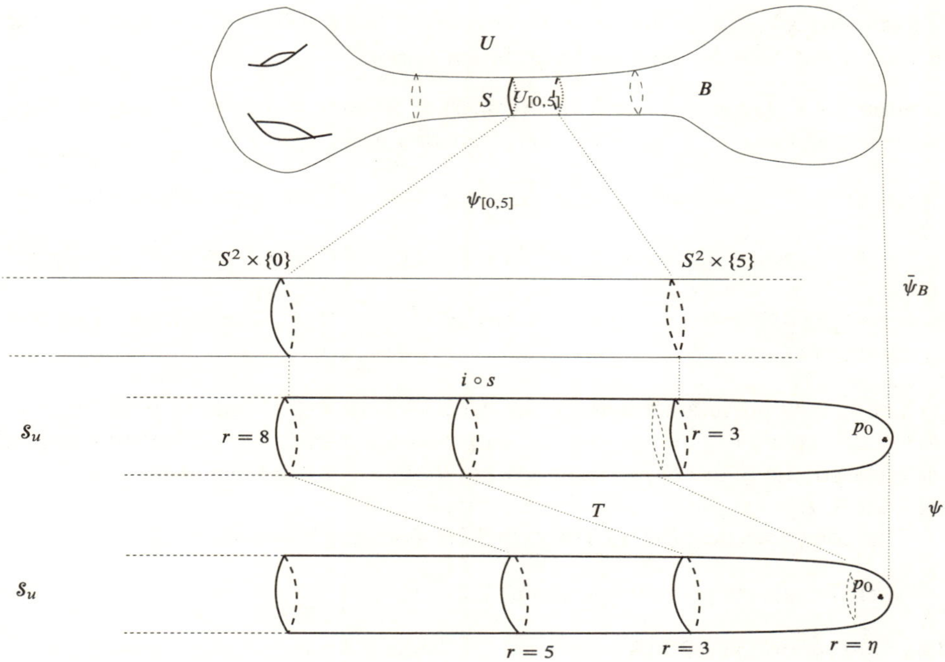

Figure 7.1. $(1 + \delta)$-Lipschitz diffeomorphism from B to $\overline{B(p_0, 5)}$.

$t\colon [0, 8] \to [0, 5]$ is a smooth function satisfying

$$\begin{cases} t(r) = \frac{\eta}{6} r & \text{on } [0, 3], \\ 0 < t' < 1 & \text{on } [3, 3 + \eta], \\ t(r) = r - 3 & \text{on } [3 + \eta, 8]. \end{cases}$$

In particular, T is $\frac{1}{C}$-Lipschitz on $(B(p_0, 3), g_u)$. Indeed, the dilatation in the radial direction is $\frac{\eta}{6}$ while in the orthoradial directions it is $u(\frac{\eta r}{6})/u(r)$. On $(C[3, 6], g_u)$, T is 1-Lipschitz since u is nondecreasing. Lastly, T restricts to an isometry from $C[6, 8]$ onto $C[3, 5]$ endowed with g_u.

Define a map $\psi := T \circ \bar{\psi}_B$ from B onto $\overline{B(p_0, 5)}$. For the metric $\bar{g} = R(y)\psi_* g$, we have by composition $g_u \leqslant (1+\delta)\bar{g}$ on $B(p_0, 3)$. On $C[3, 5]$ we have $\|\bar{g} - g_u\|_{[\delta^{-1}]} \leqslant \delta$, and the proof of Lemma 7.2.3 is complete. □

We turn to the proof of assertion (i) of Theorem 7.2.1: on $M \setminus \operatorname{Int} B$, set $g_+ := g$. Then apply Proposition 7.2.2 to the metric $\bar{g} = R(y)\psi_* g$ on $B(p_0, 5)$ and the rescaled pinching function $\bar{\phi}(\cdot) := R(y)^{-1}\phi(R(y)\cdot)$. Observe that $\bar{g}$ is $\bar{\phi}$-pinched. We get a $\bar{\phi}$-pinched metric $\bar{g}_+$, and the metric $g_+ := R(y)^{-1}\psi^* \bar{g}_+$ is ϕ-pinched on $\operatorname{Int} B$.

The property $g_+ < g$ on this set is a consequence of the following lemma and of the inequality $\delta \leqslant \delta_0 \leqslant \delta_2$, since we have chosen $\delta_2 = e^{2e^{-\Lambda}} - 1$.

Lemma 7.2.4. *Let $\bar{g}$ be a metric on $B(p_0, 5) \subset \mathbb{R}^3$ such that $g_u \leqslant (1+\delta_2)\bar{g}$. Then the metric $\bar{g}_+ := e^{-2f}(\chi g_u + (1-\chi)\bar{g})$ satisfies $\bar{g}_+ < \bar{g}$ on $B(p_0, 5)$.*

Proof. On $C[4,5)$ we have $\bar{g}_+ = e^{-2f}\bar{g} < \bar{g}$. On $[0,4)$, we argue as follows:

$$\bar{g}_+ = e^{-2f}(\chi g_u + (1-\chi)\bar{g}) \leqslant e^{-2f}(\chi(1+\delta_2)\bar{g} + (1-\chi)\bar{g}) \leqslant e^{-2f}(1+\delta_2)\bar{g}.$$

Now on $C[0,4)$, we have $f > e^{-\Lambda}$; hence $e^{-2f} < e^{-2e^{-\Lambda}}$. The conclusion follows by choice of δ_2. □

Let us now prove assertions (ii) and (iii). Assertion (iii) follows from Proposition 7.2.2 applied to $\bar{g} = R(y)\psi^* g$. To prove assertion (ii), it suffices to combine the $\delta'(\delta)$-proximity of $(B(p_0,5), \bar{g}_+)$ with $(B(p_0,5), \bar{g}_0)$ and that of the half δ-neck $(U \setminus \operatorname{Int} B, R(y)g_+ = R(y)g)$ with $S^2 \times [5, \frac{1}{\delta} + 5)$.

This completes the proof of Theorem 7.2.1 assuming Proposition 7.2.2. □

There remains to prove Proposition 7.2.2.

Proof of Proposition 7.2.2. The delicate part is the ϕ-almost nonnegativity. Set $\tilde{g} := e^{-2f}\bar{g}$. Let $a \leqslant b \leqslant c$ be the eigenvalues of $\operatorname{Ric}_{\bar{g}}$ and $\{e_1, e_2, e_3\}$ be a $\bar{g}$-orthonormal basis consisting of eigenvectors of $\operatorname{Ric}_{\bar{g}}$. Remark that a is close to 0, whereas b, c are close to $1/2$ on $C(3,5)$. Let D_i denote the covariant derivative along e_i with respect to $\bar{g}$.

Assertion 7.2.5. *There exists $\delta_1' > 0$ such that for all $0 < \delta < \delta_1'$ and on $C(3,5)$,*

$$\begin{aligned} |D_1 r - 1| &\leqslant \beta, \\ |D_i r| &\leqslant \beta \quad \textit{for } i = 2,3, \\ |D_i D_j r| &\leqslant \beta \quad \textit{for } i,j = 1,2,3. \end{aligned}$$

This follows immediately from the fact that $\bar{g}$ is δ-close to the cylindrical metric, for which the corresponding quantities are $D_1 r = 1$, $D_2 r = D_3 r = 0$.

Assertion 7.2.6. *For all $0 < \delta < \delta_1'$ we have*

$$\begin{aligned} |D_i f| &\leqslant 2\beta f'', \\ |D_i D_j f| &\leqslant 3\beta f'' \quad \textit{if } i,j \neq 1, \\ |D_1 D_1 f - f''| &\leqslant 5\beta f''. \end{aligned}$$

Proof. We have $D_i f = \frac{\partial f}{\partial r} D_i r = f' D_i r$, so

$$|D_i f| \leqslant \beta f'' |D_i r| \leqslant \beta f''(1+\beta) \leqslant 2\beta f''.$$

The first inequality comes from the properties of f and the second one from Assertion 7.2.5. Furthermore,

$$D_i D_j f = f'' D_i r D_j r + f' D_i D_j r.$$

For $i, j \neq 1$, we derive

$$|D_i D_j f| \leqslant f'' \beta(1+\beta) + \beta f'' \beta \leqslant 3\beta f'' \leqslant 3\beta f''.$$

On the other hand,

$$D_1 D_1 f = D_1(f' D_1 r) = f''(D_1 r)^2 + f' D_1 D_1 r.$$

Therefore,

$$\begin{aligned} |D_1 D_1 f - f''| &\leqslant f''|(D_1 r)^2 - 1| + f'|D_1 D_1 r| \\ &\leqslant f''|D_1 r - 1||D_1 r + 1| + f'|D_1 D_1 r| \\ &\leqslant f'' \beta(2+\beta) + \beta f''(1+\beta) \\ &\leqslant 5\beta f'', \end{aligned}$$

which proves the claim. □

For simplicity we denote by $\widetilde{\operatorname{Ric}}$ the Ricci tensor of $\tilde{g} = e^{-2f} \bar{g}$, and $\tilde{a} \leqslant \tilde{b} \leqslant \tilde{c}$ its eigenvalues.

Assertion 7.2.7. *For all $0 < \delta < \delta_1'$, we have*

$$\begin{aligned} |\tilde{a} - e^{2f}(a + 2D_1 D_1 f)| &\leqslant \alpha D_1 D_1 f, \\ |\tilde{b} - e^{2f}(b + D_1 D_1 f)| &\leqslant \alpha D_1 D_1 f, \\ |\tilde{c} - e^{2f}(c + D_1 D_1 f)| &\leqslant \alpha D_1 D_1 f. \end{aligned}$$

Proof. Set $\beta' := 5\beta$ and $\tilde{e}_i := e^f e_i$. The basis $\{\tilde{e}_1, \tilde{e}_2, \tilde{e}_3\}$ is $\tilde{g}$-orthonormal. Let $\tilde{A}$ be the matrix of $\widetilde{\operatorname{Ric}}$ in this basis. First we show that the eigenvalues of $\widetilde{\operatorname{Ric}}$ are close to the diagonal terms $\tilde{A}_{ii}$ of the matrix $\tilde{A}$. By [Bes87], p. 59 (applied with $-f$ instead of f), we have

$$\widetilde{\operatorname{Ric}} = \operatorname{Ric} - \Delta f \bar{g} + (n-2)(Ddf + df \circ df - |df|^2 \bar{g}),$$

where the right-hand side operators are those of $\bar{g}$. Evaluating this formula on a pair $(\tilde{e}_i, \tilde{e}_j)$ for $i \neq j$, we get

$$\tilde{A}_{ij} = e^{2f}(D_i D_j f + D_i f D_j f),$$

so

$$|\tilde{A}_{ij}| \leqslant 2(\beta' f'' + \beta'^2 f''^2) \leqslant 4\beta' f'' \leqslant 4\beta' \frac{1}{1 - 5\beta'} D_1 D_1 f \leqslant 8\beta' D_1 D_1 f.$$

Using the following elementary lemma, whose proof is left to the reader, we get an upper bound on the distance between the $\tilde{A}_{ii}$'s and the eigenvalues of $\tilde{A}$ by $24\beta' D_1 D_1 f \leqslant \frac{\alpha}{2} D_1 D_1 f$.

Lemma 7.2.8. *Let A, B be symmetric $n \times n$ matrices, of eigenvalues $a_1 \leqslant a_2 \leqslant \cdots \leqslant a_n$ and $b_1 \leqslant b_2 \leqslant \cdots \leqslant b_n$ respectively. Define $\|M\| = \sup_{i,j} |m_{ij}|$. Then for all $i = 1, \ldots, n$, we have*

$$|a_i - b_i| \leqslant n\|B - A\|.$$

There remains to show that for $i = 1, 2, 3$, we have

$$\begin{aligned} |\tilde{A}_{11} - e^{2f}(a + 2D_1 D_1 f)| &\leqslant \frac{\alpha}{2} D_1 D_1 f, \\ |\tilde{A}_{22} - e^{2f}(b + D_1 D_1 f)| &\leqslant \frac{\alpha}{2} D_1 D_1 f, \\ |\tilde{A}_{33} - e^{2f}(c + D_1 D_1 f)| &\leqslant \frac{\alpha}{2} D_1 D_1 f. \end{aligned}$$

We deduce

$$\begin{aligned} \tilde{A}_{11} &= \widetilde{\mathrm{Ric}}(\tilde{e}_1, \tilde{e}_1) \\ &= e^{2f}(a - \Delta f + D_1 D_1 f + D_1 f D_1 f - |df|^2) \\ &= e^{2f}(a + 2D_1 D_1 f + D_2 D_2 f + D_3 D_3 f + (D_1 f)^2 - |df|^2), \end{aligned}$$

hence

$$\begin{aligned} |\tilde{A}_{11} - e^{2f}(a + 2D_1 D_1 f)| &\leqslant e^{2f}(\beta' f' + \beta' f'' + 2(\beta f'')^2) \\ &\leqslant 2(4\beta' f'') \leqslant 16\beta' D_1 D_1 f \leqslant \frac{\alpha}{2} D_1 D_1 f. \end{aligned}$$

Now

$$\begin{aligned} \tilde{A}_{22} &= e^{2f}(b - \Delta f + D_2 D_2 f + D_2 f D_2 f - |Df|^2) \\ &= e^{2f}(b + D_1 D_1 f + 2D_2 D_2 f + D_3 D_3 f - (D_1 f)^2 - (D_3 f)^2), \end{aligned}$$

thus

$$\begin{aligned} |\tilde{A}_{22} - e^{2f}(b + D_1 D_1 f)| &\leqslant e^{2f}(3\beta' f'' + 2(\beta' f'')^2) \\ &\leqslant 10\beta' f'' \leqslant 20\beta' D_1 D_1 f \leqslant \frac{\alpha}{2} D_1 D_1 f. \end{aligned}$$

The proof of the third term is similar. □

Assertion 7.2.9. *For all $0 < \delta \leqslant \delta_1'$, on $C(4, 5)$ we have*

(i) *$\tilde{a} \geqslant a$, $\tilde{b} \geqslant b$, $\tilde{c} \geqslant c$ and $\widetilde{R} \geqslant R$, where $\widetilde{R}$ (resp. R) is the scalar curvature of $\tilde{g}$ (resp. $\bar{g}$),*

(ii) $\bar{g}_+ = \tilde{g}$ *has* ϕ*-almost nonnegative curvature.*

Proof. By Assertion 7.2.7,

$$\tilde{a} \geqslant e^{2f}(a + 2D_1D_1f) - \alpha D_1D_1f \geqslant a + (2-\alpha)D_1D_1f \geqslant a + (2-\alpha)(1-\alpha)f''.$$

Now on $[4,5]$, f is convex, so $\tilde{a} \geqslant a$. We can argue similarly for b and c. The result on the scalar curvature then follows from the formula $R = a + b + c$.

Let $\lambda \geqslant \mu \geqslant \nu$ be the eigenvalues of the curvature operator of $\bar{g}$ and $\tilde{\lambda} \geqslant \tilde{\mu} \geqslant \tilde{\nu}$ those of the curvature operator of $\tilde{g}$. Recall that

$$\nu = \frac{a+b-c}{2}, \quad \mu = \frac{a+c-b}{2}, \quad \lambda = \frac{b+c-a}{2}$$

and similarly for $\tilde{g}$. By 7.2.7,

$$|\tilde{\nu} - e^{2f}(\nu + D_1D_1f)| \leqslant \frac{3\alpha}{2}D_1D_1f \leqslant \frac{1}{2}D_1D_1f.$$

In particular,

$$\tilde{\nu} \geqslant e^{2f}(\nu + D_1D_1f) - \frac{1}{2}D_1D_1f \geqslant \nu + \frac{1}{2}D_1D_1f \geqslant \nu + \frac{1}{2}(1-\alpha)f'' \geqslant \nu.$$

Since the metric $\bar{g}$ has ϕ-almost nonnegative curvature, and ϕ is nondecreasing, we deduce:

$$\tilde{\nu} \geqslant \nu \geqslant -\phi(R) \geqslant -\phi(\tilde{R}).$$

Hence in this interval, $\bar{g}_+$, which coincides with $\tilde{g}$, has ϕ-almost nonnegative curvature. □

Recall that on $C(3,4)$, the metric $\bar{g}_+$ is obtained by blending $\bar{g}_0 = e^{-2f}g_u$ and $\tilde{g} = e^{-2f}\bar{g}$. We show that if δ is sufficiently small, this metric has positive (hence *a fortiori* ϕ-almost nonnegative) curvature.

Assertion 7.2.10. *There exists* $\delta_1'' > 0$ *such that for* $\delta \leqslant \delta_1''$*, the metric* $\bar{g}_+$ *has positive curvature on* $C(3,4)$.

Proof. Recall that

$$\bar{g}_+ = \chi\bar{g}_0 + (1-\chi)\tilde{g}.$$

By hypothesis $\bar{g}$ is δ-close to g_u on $C(3,5)$. The metric $\bar{g}_+ = e^{-2f}(\chi g_u + (1-\chi)\bar{g})$ is therefore $\delta'(\delta)$-close to $\bar{g}_0$ on $C(3,5)$ for some $\delta'(\delta)$ depending on f and on χ, and tending to 0 when δ does.

Since on $C(3,4)$, the metric $\bar{g}_0$ has sectional curvatures bounded from below by $K_0 > 0$ (see inequality (7.1) on p. 79). We can choose δ_1'' sufficiently small so that $\bar{g}_+$ has positive curvature on $C(3,4)$ provided $\delta \leqslant \delta_1''$. □

On $B(p_0,3)$ we have $\bar{g}_+ = \bar{g}_0$, which also has positive curvature. Moreover, $\bar{g}_+$ is $\delta'(\delta)$-close to $\bar{g}_0$ on $B(p_0,5)$. Proposition 7.2.2 is proven by setting $\delta_1 := \min(\delta_1', \delta_1'')$. □

7.3 Proof of Proposition A

Let us first recall the statement of this proposition.

Proposition A. *There exists a universal constant $\bar{\delta}_A > 0$ having the following property: let r, δ be surgery parameters, a, b be positive numbers with $a < b$, and $\{g(t)\}_{t \in (a,b]}$ be a Ricci flow with (r,δ)-bubbling-off on M. Suppose that $\delta \leqslant \bar{\delta}_A$, and $R_{\max}(b) = \Theta$.*

Then M admits a metric g_+ which is obtained from $g(\cdot)$ by (r,δ)-surgery at time b, and in addition satisfies that

(i) *g_+ has ϕ_b-almost nonnegative curvature;*

(ii) *$R_{\min}(g_+) \geqslant R_{\min}(g(b))$.*

Throughout this section we shall work in the Riemannian manifold $(M, g(b))$. In particular all curvatures and distances are taken with respect to this metric. Let $\mathcal{G}$ (resp. $\mathcal{O}$, resp. $\mathcal{R}$) be the set of points of M of scalar curvature less than $2r^{-2}$, (resp. $\in [2r^{-2}, \Theta/2)$, resp. $\geqslant \Theta/2$).

Step 1. There exists a compact submanifold $Y \subset \mathcal{G}$ such that every component of ∂Y is a 2-sphere that bounds a 3-ball disjoint from the interior of Y.

Proof. Let $X \subset M$ be a compact submanifold containing $\mathcal{O} \cup \mathcal{R}$ and such that every point of X has scalar curvature at least, say $\frac{3}{2}r^{-2}$. Then every point of X is the centre of an ε_0-neck or an ε_0-cap, so we can apply Corollary 3.2.6. Since M is irreducible and not spherical, we get a finite collection $N_1, \dots, N_p$ of $10\varepsilon_0$-caps and $10\varepsilon_0$-tubes which cover X and have disjoint closures.

Then $N := \overline{\bigcup_i N_i}$ is a compact submanifold of M whose complement is contained in $\mathcal{G}$. Since each component of ∂N is a sphere and M is irreducible and not diffeomorphic to S^3, it follows that there is exactly one component Y of $M \setminus \operatorname{Int} N$ having the required property. □

Note that M is the union of Y and the 3-balls bounded by its boundary components. In particular, the 'bad' set $\mathcal{R}$ is contained in a union of 3-balls. Our next goal is to find δ-necks separating $\mathcal{R}$ from Y. For technical convenience, we first choose a compact submanifold $\tilde{\mathcal{R}} \subset M$ containing $\mathcal{R}$, and such that every point of $\tilde{\mathcal{R}}$ has curvature at least $\frac{1}{2}Dh^{-2} = \Theta/4$. In particular, $\tilde{\mathcal{R}}$ has finitely many connected components, which we denote by $R_1, \dots, R_p$. We also assume that each component of $\tilde{\mathcal{R}}$ contains a point of $\mathcal{R}$.

Step 2. For every i, there exists a point $y_i \in \mathcal{O}$ of scalar curvature h^{-2}, and a strong δ-neck U_i centred at y_i whose middle sphere S_i separates R_i from Y.

Proof. Let x_i be a point of the component of ∂Y closest to R_i. Let $z_i \in R_i \cap \mathcal{R}$ be an arbitrary point. Choose a geodesic segment $[x_i z_i]$ and a point $y_i \in [x_i z_i]$ of curvature h^{-2}. By hypothesis $R(x_i) \leqslant 2r^{-2}$ and $R(z_i) \geqslant Dh^{-2}$. Applying the Cutoff Parameters Theorem 5.2.4, we obtain a strong δ-neck U_i centred at y_i. Call S_i its middle sphere. Since M is irreducible, S_i is separating in M. Together with the fact that U_i is traversed by a geodesic from x_i to z_i, Lemma 3.3.2 implies that S_i separates x_i from z_i, and therefore Y from R_i. □

For each i, the sphere S_i is contained in a unique component B_i' of $M \setminus \operatorname{Int} Y$. Since B_i' is a 3-ball, by Alexander's theorem, S_i bounds a 3-ball $B_i \subset B_i'$. The fact that S_i separates R_i from Y implies that $R_i \subset B_i$.

Step 3. There exists a set of indices $1 \leqslant i_1, \dots, i_q \leqslant p$ such that the 3-balls $B_{i_1}, \dots, B_{i_q}$ are disjoint and their union contains $\mathcal{R}$.

Proof. First remove among the B_i's those which are not maximal for inclusion. The only problem left is that some of the S_i's might intersect each other. Now if there exist two distinct indices i, j such that $S_i \cap S_j \neq \emptyset$, then S_j is contained in U_i and isotopic to S_i there. Hence R_j is also separated from Y by S_i, and we may safely remove B_j. After finitely many of these operations, we are left with a finite subcollection $B_{i_1}, \dots, B_{i_q}$ which has the desired properties. □

To conclude, it suffices to apply Theorem 5.2.2 finitely many times, replacing at each step the cap B_{i_k} by a δ-almost standard cap. Call g_+ the resulting metric. Since all points in a δ-almost standard cap have scalar curvature comparable to h^{-2}, we have $R_{\max}(g_+) \leqslant Dh^{-2} = \Theta/2$. Hence g_+ is obtained from $g(\cdot)$ by (r, δ)-surgery, and satisfies the pinching assumption. Moreover, note that $R_{\min}(g_+)$ is attained in $M \setminus \bigcup B_i = Y'$, where $g_+ = g(b)$. Indeed, $Y' \supset \mathcal{G}$ where $R \leqslant 2r^{-2}$, and on $\bigcup B_i$, R is comparable to h^{-2}. Hence

$$R_{\min}(g_+) = R_{\min|Y'}(g_+) = R_{\min|Y'}(g(b)) \geqslant R_{\min}(g(b)).$$

Thus the proof of Proposition A is complete.

Chapter 8
Persistence

In this chapter, we show that the behaviour of almost standard caps which have been added by surgery is modelled on the standard solution for some time after surgery (Theorem 8.1.2 below). We call this phenomenon 'persistence' of the closeness to some model solution. Recall that the standard solution is defined on $[0, 1)$. Duration of the persistence, denoted by θ, can be fixed arbitrarily close to 1. In order to have persistence for this time length, it is then needed to choose the parameter δ small enough. Similarly, the size A of the part of the almost standard cap which persists can also be chosen arbitrarily large. Again persistence occurs on this part if δ is small enough.

This persistence theorem is crucial in two parts of the proof of the existence of Ricci flows with (r, δ)-bubbling-off, namely in the proofs of Proposition B and Proposition C. In the proof of Proposition B, the goal is to extend in time the canonical neighbourhood property $(\mathrm{CN})_r$. The parts of the flow that "persist" are modelled on the standard solution, and hence inherit canonical neighbourhoods by closeness. One can then consider parts of space-time far away from surgeries.

In the proof of Proposition C, the goal is to extend in time the noncollapsing property $(\mathrm{NC})_\kappa$. Here the persistence theorem is used in the core of a technical argument to compute a kind of energy for curves that approach the singularities (Section 10.3.2).

In order to clarify the proof of the persistence theorem, we have split it into several parts. One of them is an n-dimensional persistence result for Ricci flows with bounded curvature (Theorem 8.1.3 below).

Finally, let us emphasize that persistence of almost standard caps should not be confused with persistence of hyperbolic limits for large time as in Hamilton's article [Ham99].

8.1 Introduction

First we need a few technical definitions:

Definition 8.1.1. If $g(\cdot)$ is a piecewise $\mathcal{C}^1$ evolving metric on some interval $I \subset \mathbb{R}$ and $[a, b] \subset I$, one calls *restriction of* g *to* $[a, b]$ the map

$$t \mapsto \begin{cases} g_+(a) & \text{if } t = a, \\ g(t) & \text{if } t \in (a, b]. \end{cases}$$

We shall still denote by $g(\cdot)$ the restriction.

If $(U, \{g(t)\}_{t\in I})$ is a Ricci flow, then for all $(x,t) \in U \times I$, $r > 0$ and $\Delta t > 0$ we define the forward parabolic neighbourhood $P(x,t,r,\Delta t)$ as the set $B(x,t,r) \times [t, t+\Delta t] \subset U \times I$. When we consider $g(\cdot)$ is restricted to some $[a,b] \subset I$, note that $P(x,a,r,\Delta t)$ is defined using $B_{g_+}(x,a,r)$.

Given two Ricci flows $(U, g(\cdot))$ and $(U_0, g_0(\cdot))$, we say that a parabolic neighbourhood $P(x,t,r,\Delta t)$ of $(U, g(\cdot))$ is *ε-close* to a parabolic neighbourhood $P(x_0,t,r_0,\Delta t)$ of $(U_0, g_0(\cdot))$ if $(B(x,t,r), g(\cdot))$ is ε-close to $(B(x_0,t,r_0), g_0(\cdot))$ on $[t, t+\Delta t]$. We say that $P(x,t,r,\Delta t)$ is *ε-homothetic* to $P(x_0,t_0,r_0,\lambda\Delta t)$ if it is ε-close after a parabolic rescaling by λ.

Here is the main theorem of this chapter:

Theorem 8.1.2 (Persistence of almost standard caps). *For all $A > 0$, $\theta \in [0,1)$ and $\hat{r} > 0$, there exists $\bar{\delta} = \bar{\delta}_{\text{per}}(A, \theta, \hat{r})$ with the following property. Let $g(\cdot)$ be a Ricci flow with bubbling-off on M defined on some interval $[a,b]$, which is a Ricci flow with $r(\cdot), \delta(\cdot)$-bubbling-off on $[a,b)$, where $r(\cdot) \geqslant \hat{r}$ and $\delta(\cdot) \leqslant \bar{\delta}$. Let $t_0 \in [a,b)$ be a singular time and consider the restriction $g(\cdot)$ to $[t_0,b]$. Let $p \in (M, g(t_0))$ be the tip of some δ-almost standard cap of scale h. Let $t_1 \leqslant \min\{b, t_0 + \theta h^2\}$ be maximal subject to this inequality such that $P(p,t_0,Ah,t_1-t_0)$ is unscathed. Then the following holds:*

(i) *The parabolic neighbourhood $P(p,t_0,Ah,t_1-t_0)$ is A^{-1}-homothetic to $P(p_0, 0, A, (t_1-t_0)h^{-2})$.*
(ii) *If $t_1 < \min\{b, t_0+\theta h^2\}$, then $B(p,t_0,Ah)$ is contained in the singular set at time t_1.*

Let T_0 be a positive real number and $\mathcal{M}_0 = (X_0, g_0(\cdot), p_0)$ (the 'model') be a complete n-dimensional pointed Ricci flow defined on $[0, T_0]$. Assume that the quantity

$$\Lambda_N := \|\operatorname{Rm}(g_0(\cdot))\|_{N, X_0\times[0,T_0], g_0(\cdot)}$$

is finite for all $N \in \mathbb{N}$. The norm here is the one defined on p. 28.

Theorem 8.1.3 (Persistence under bounded curvature). *For all $A, \Lambda > 0$, there exists $\rho = \rho(\mathcal{M}_0, A, \Lambda) > A$ with the following property. Let U be an open subset of X_0 and $T \in (0, T_0]$. Let $g(\cdot)$ be a Ricci flow defined on $U \times [0,T]$, such that the ball $B(p_0, 0, \rho) \subset U$ is relatively compact. We assume that*

(i) $\|\operatorname{Rm}(g(\cdot))\|_{0, U\times[0,T], g(\cdot)} \leqslant \Lambda$,
(ii) *$g(0)$ is ρ^{-1}-close to $g_0(0)$ on $B(p_0, 0, \rho)$.*

Then $g(\cdot)$ is A^{-1}-close to $g_0(\cdot)$ on $B(p_0, 0, A) \times [0, T]$.

Remark 8.1.4. This result fails for $T_0 = \infty$. Indeed, one can approximate an arbitrary large metric ball in the standard cylinder by an almost cylindrical ball in the cigar soliton.

Remark 8.1.5. Notice also that ρ may have to be much larger than A. One can think for example of a flat disk in a sphere. It cannot remain almost flat for a very long time, unless it is contained at time zero in a much larger almost flat disk.

8.2 Persistence of a model

In this section we prove Theorem 8.1.3 and a few corollaries. We begin with a 'local' persistence result.

Lemma 8.2.1 (Local persistence). *For all $\alpha, K > 0$, there exist $T_{\mathrm{loc}} = T_{\mathrm{loc}}(\alpha, K) > 0$ and $\delta_{\mathrm{loc}} = \delta_{\mathrm{loc}}(\alpha, K) > 0$ with the following property. Let U be an n-manifold, and $T \in (0, T_{\mathrm{loc}})$. Let $g_0(\cdot)$ and $g_1(\cdot)$ be two Ricci flows defined on $U \times [0, T]$. Suppose that*

(a) $\| \mathrm{Rm}(g_i(\cdot))\|_{[\alpha^{-1}], U\times[0,T], g_i(\cdot)} \leqslant K$ *for* $i = 0, 1$;
(b) $g_1(0)$ *is* δ*-close to* $g_0(0)$ *on* U.

Then $g_1(\cdot)$ is α-close to $g_0(\cdot)$ on $U \times [0, T_{\mathrm{loc}}]$.

Proof. First we replace time-dependent norms by one involving $g_0(0)$, covariant derivatives being still taken with respect to $g_0(t)$. Let p be an integer in $[0, [\alpha^{-1}]]$ representing the order of derivative of Rm. For $i = 0, 1$ the bound $a)$ for $p = 0$ implies:

$$e^{-2(n-1)Kt} g_i(0) \leqslant g_i(t) \leqslant e^{2(n-1)Kt} g_i(0), \tag{8.1}$$

for all $x \in U$. In the sequel we omit reference to x. Hence we have, at order 0:

$$|g_1(t) - g_0(t)|_{g_0(t)} \leqslant e^{2(n-1)Kt} |g_1(t) - g_0(t)|_{g_0(0)},$$

and likewise

$$|\nabla^p_{g_0(t)} g_1(t)|_{g_0(t)} \leqslant e^{(p+2)(n-1)Kt} |\nabla^p_{g_0(t)} g_1(t)|_{g_0(0)}.$$

To estimate $|\nabla^p_{g_0(t)} g_1(t)|_{g_0(0)}$ we shall show the existence of $\Lambda_p > 0$, for all $p \leqslant [\alpha^{-1}]$, such that

$$\frac{d}{dt} |\nabla^p_{g_0(t)} g_1(t)|_{g_0(0)} \leqslant \Lambda_p.$$

Indeed, it suffices to integrate and use closeness of the metrics at time 0 to obtain closeness at time $t > 0$. We have for $p > 0$:

$$\frac{d}{dt} |g_1(t) - g_0(t)|_{g_0(0)} \leqslant \left| \frac{d}{dt} (g_1(t) - g_0(t)) \right|_{g_0(0)},$$

and

$$\frac{d}{dt} |\nabla^p_{g_0(t)} g_1(t)|_{g_0(0)} \leqslant \left| \frac{d}{dt} \nabla^p_{g_0(t)} g_1(t) \right|_{g_0(0)}.$$

To simplify the exposition, we only do the cases $p = 0$ and 1 and leave the general case to the reader. Note that

$$\begin{aligned} |\frac{d}{dt}(g_1(t) - g_0(t))|_{g_0(0)} &= 2|\mathrm{Ric}(g_1(t)) - \mathrm{Ric}(g_0(t))|_{g_0(0)} \\ &\leqslant 2|\mathrm{Ric}(g_1(t))|_{g_0(0)} + 2|\mathrm{Ric}(g_0(t))|_{g_0(0)}. \end{aligned}$$

Closeness of $g_0(0)$ and $g_1(0)$, the estimate (8.1) and Hypothesis (a) together imply that

$$\begin{aligned}|\mathrm{Ric}(g_i(t))|_{g_0(0)} &\leqslant (1+\delta)|\mathrm{Ric}(g_i(t))|_{g_i(0)}\\ &\leqslant (1+\delta)e^{2(n-1)Kt}|\mathrm{Ric}(g_i(t))|_{g_i(t)}\\ &\leqslant (n-1)K(1+\delta)e^{2(n-1)Kt}.\end{aligned}$$

We deduce that

$$\left|\frac{d}{dt}(g_1(t)-g_0(t))\right|_{g_0(0)} \leqslant 4(n-1)K(1+\delta)e^{2(n-1)Kt},$$

and by integration,

$$\begin{aligned}|g_1(t)-g_0(t)|_{g_0(0)} &\leqslant |g_1(0)-g_0(0)|_{g_0(0)} + 4(n-1)K(1+\delta)e^{2(n-1)Kt}t\\ &\leqslant \delta + 4(n-1)K(1+\delta)e^{2(n-1)Kt}t.\end{aligned}$$

For $p=1$ we have

$$\begin{aligned}\frac{d}{dt}\nabla_{g_0(t)}g_1(t) &= \frac{d}{dt}[(\nabla_{g_0(t)}-\nabla_{g_1(t)})g_1(t)]\\ &= (\dot\nabla_{g_0(t)}-\dot\nabla_{g_1(t)})g_1(t) - 2(\nabla_{g_0(t)}-\nabla_{g_1(t)})\,\mathrm{Ric}(g_1(t))\end{aligned} \tag{8.2}$$

where $\dot\nabla_{g_0(t)}$ (resp. $\dot\nabla_{g_1(t)}$) denotes the time-derivative of the Levi-Civita connection of $g_0(t)$ (resp. $g_1(t)$). Recall that if T is a 2-tensor and X_1, X_2, X_3 are vector fields, then

$$\nabla_{X_1}T(X_2,X_3) = X_1\cdot T(X_2,X_3) - T(\nabla_{X_1}X_2,X_3) - T(X_2,\nabla_{X_1}X_3).$$

The difference of two connections is a 1-form with values in the Lie algebra of $\mathrm{GL}(n,\mathbb{R})$; in particular $\nabla_{g_0}-\nabla_{g_1}$, when applied to a tensor, does not differentiate it. The same holds for $\dot\nabla_{g_0(t)}-\dot\nabla_{g_1(t)}$. The formula (8.2) shows that it is enough to control the norms of the 1-forms $\nabla_{g_0(t)}-\nabla_{g_1(t)}$ and $\dot\nabla_{g_0(t)}-\dot\nabla_{g_1(t)}$. Now

$$|\dot\nabla_{g_0(t)}-\dot\nabla_{g_1(t)}|_{g_0(0)} \leqslant C_0|\dot\nabla_{g_0(t)}|_{g_0(t)} + C_1|\dot\nabla_{g_1(t)}|_{g_1(t)},$$

where as before, C_0, C_1 depend on K,δ,t; Formula (18.3) from [KL08] gives

$$2\langle\dot\nabla_{X_1}X_2,X_3\rangle = -2(\nabla_{X_1}\mathrm{Ric})(X_2,X_3) - 2(\nabla_{X_2}\mathrm{Ric})(X_1,X_3) + 2(\nabla_{X_3}\mathrm{Ric})(X_1,X_2).$$

Hence we can estimate $|\dot\nabla_{g_i(t)}|_{g_i(t)}$ from above by $|\nabla_{g_i(t)}\mathrm{Ric}_{g_i(t)}|_{g_i(t)}$ (up to a constant). In the sequel, we denote by Λ a generic constant which may change from line to line. We conclude that

$$|\dot\nabla_{g_0(t)}-\dot\nabla_{g_1(t)}|_{g_0(0)} \leqslant \Lambda(\delta,K,t),$$

and, integrating with respect to t,

$$|\nabla_{g_0(t)} - \nabla_{g_1(t)}|_{g_0(0)} \leqslant |\nabla_{g_0(0)} - \nabla_{g_1(0)}|_{g_0(0)} + \Lambda(\delta, K, t).$$

Proximity of the metrics at time 0 yields

$$|\nabla_{g_0(t)} - \nabla_{g_1(t)}|_{g_0(0)} \leqslant \Lambda(\delta, K, t),$$

and, using (8.2),

$$\frac{d}{dt}|\nabla_{g_0(t)} g_1(t)|_{g_0(0)} \leqslant \Lambda(\delta, K, t).$$

Generalising to arbitrary $p \geqslant 1$, one can show that there exist $0 < \Lambda_p = \Lambda_p(\delta, K, T)$ such that, for all $t \leqslant T$,

$$\frac{d}{dt}|\nabla^p_{g_0(t)} g_1(t)|_{g_0(0)} \leqslant \Lambda_p.$$

By integration and proximity at time 0, we get

$$|\nabla^p_{g_0(t)} g_1(t)|_{g_0(0)} \leqslant \delta + \Lambda_p t,$$

and

$$|\nabla^p_{g_0(t)} g_1(t)|_{g_0(t)} \leqslant e^{(p+2)(n-1)Kt}(\delta + \Lambda_p t).$$

Summing up all terms, we conclude that

$$\sup_{x \in U} \Big\{ \textstyle\sum_{p=1}^{[\alpha^{-1}]} |\nabla^p_{g_0} g_1|^2_{g_0}(x) + |g_1 - g_0|^2_{g_0}(x) < \alpha^2 \Big\},$$

if δ, t are small enough. □

Proof of Theorem 8.1.3. We argue by contradiction. Let us fix $A > 0$ and $\Lambda > 0$ and assume that there exist a sequence of Ricci flows $g_k(\cdot)$ defined on $U_k \times [0, T_k]$ ($T_k \leqslant T_0$), a sequence $\rho_k \to +\infty$ when $k \to \infty$, such that $B_k(p_0, 0, \rho_k) \subset U_k$ are relatively compact, and

(i) $|\mathrm{Rm}_{g_k(t)}|_{g_k(t)} \leqslant \Lambda$ on $U_k \times [0, T_k]$,

(ii) $g_k(0)$ is ρ_k^{-1}-close to $g_0(0)$ on $B(p_0, 0, \rho_k)$.

Furthermore there exists $t_k \in [0, T_k]$ such that $g_k(t_k)$ is not A^{-1}-close to $g_0(t_k)$. We may assume that t_k is minimal for this property.

Set $N := [A]$. By closeness at time 0, the quantities $|\nabla^p \mathrm{Rm}_{g_k(0)}|$, for $p \in \{0, \dots, N\}$, are bounded independently of k, say by $2\Lambda_N$, on $B(p_k, 0, \rho_k)$. Then, by the local Shi estimates, there exists $T_{\mathrm{Shi}} > 0$ and $\Lambda_{\mathrm{Shi}} > 0$ such that if $\rho_k \geqslant A + 1$ the inequalities $|\nabla^p \mathrm{Rm}_{g_k(t)}|_{g(t)} \leqslant \Lambda_{\mathrm{Shi}}$ hold on $B(p_k, 0, \rho_k - 1) \times [0, \min\{t_k, T_{\mathrm{Shi}}\}]$.

Indeed, let us consider a ball $B(x, 0, 1) \subset B(p_k, 0, \rho_k)$ and set $K := \max\{\Lambda, 2\Lambda_N\}$. By Theorem 3.28 of [MT07] applied with $T_{\mathrm{Shi}} = K^{-1}$ and $C'_p := C'_p(K, 1, 1, n)$ for $p \in \{1, \dots, N\}$, one has for all $y \in B(x, 0, 1/2)$ and $t \in [0, \min\{T_{\mathrm{Shi}}, T_k\}]$,

$$|\nabla^p \mathrm{Rm}_{g_k(t)}(y, t)|_{g_k(t)} \leqslant C'_p.$$

This proves the above fact with $\Lambda_{\mathrm{Shi}} = \sup\{K, C'_p \mid 0 \leqslant p \leqslant N\}$. Now the local persistence lemma, applied with $\alpha = A^{-1}$ and Λ_{Shi}, gives numbers $T_{\mathrm{loc}}(A^{-1}, \Lambda_{\mathrm{Shi}})$ and $\delta_{\mathrm{loc}}(A^{-1}, \Lambda_{\mathrm{Shi}})$ such that $B(p_k, 0, \rho_k^{-1}) \times [0, \min\{T_k, T_{\mathrm{Shi}}, T_{\mathrm{loc}}\}]$ is A^{-1}-close to $B(p_0, 0, \rho_k^{-1}) \times [0, \min\{T_k, T_{\mathrm{Shi}}, T_{\mathrm{loc}}\}]$, for all k large enough such that $\rho_k^{-1} < \delta_{\mathrm{loc}}$. In particular $t_k \geqslant \min\{T_{\mathrm{Shi}}, T_{\mathrm{loc}}\}$ for k large enough.

Up to taking a subsequence of the t_k's we may assume that it converges; we then define $t_\infty := \lim_{k\to\infty} t_k \geqslant \min\{T_{\mathrm{Shi}}, T_{\mathrm{loc}}\} > 0$. By assumption, when k goes to infinity, $g_k(0)$ is arbitrarily close to $g_0(0)$ on balls of arbitrarily large radius. This implies the closeness of curvatures, Jacobi fields, lengths and exponential maps and hence the injectivity radius. This shows that the metrics $g_k(0)$ have a uniform positive lower bound for their injectivity radius at p_k. Applying Hamilton's compactness theorem with extra curvature bounds ([Ham95a], Theorem 2.3, see [HH97], Theorem 13, for the local version) and arguing as in [Ham95a], Section 2, one obtains that

$$(B(p_0, 0, \rho_k), g_k(\cdot), p_0) \times [0, t_\infty) \to (X_\infty, g_\infty(\cdot), p_\infty) \times [0, t_\infty)$$

when k goes to infinity. Here the right-hand side is a complete Ricci flow with bounded curvature. As $g_k(0) - g_0(0)$ is going to zero on arbitrarily large balls, it follows from the proof of the compactness theorem that the diffeomorphism involved in the convergence above can be taken as the identity map. This implies that $X_\infty = X_0$ and $g_\infty(0) = g_0(0)$. By Chen–Zhu uniqueness theorem [CZ06c], we get that

$$(X_\infty, g_\infty(\cdot), p_\infty) = (X_0, g_0(\cdot), p_0).$$

Let $\sigma > 0$ to be chosen later. For k big enough $t_k > t_\infty - \sigma$ and $P(p_k, 0, A+1, t_\infty - \sigma)$ converges towards $P(p_0, 0, A+1, t_\infty - \sigma)$; in particular, $|\nabla^p \mathrm{Rm}| \leqslant 2\Lambda_N$ for $p \in \{1, \dots, N\}$ on $B(p_k, 0, A+1)$ at time $t_\infty - \sigma$. Now, on $[t_\infty - \sigma, t_k]$, $|\mathrm{Rm}| \leqslant \Lambda$ by hypothesis. Then, by Shi's Theorem there exists $T'_{\mathrm{Shi}} > 0$ and $\Lambda'_{\mathrm{Shi}} > 0$ such that

$$|\nabla^p \mathrm{Rm}| \leqslant \Lambda'_{\mathrm{Shi}} \quad \text{on } P(p_k, 0, A, \min\{t_k, t_\infty - \sigma + T'_{\mathrm{Shi}}\}).$$

We choose $2\sigma < T'_{\mathrm{Shi}}$, then $t_k \leqslant t_\infty - \sigma + T'_{\mathrm{Shi}}$. Indeed, for k big enough, we have $t_k - (t_\infty - \sigma) \leqslant 2\sigma$.

Now, we can apply the local persistence lemma on $B(p_k, 0, A)$ between $t_\infty - \sigma$ and $\min(t_k, t_\infty - \sigma + T'_{\mathrm{loc}})$ where $T'_{\mathrm{loc}} = T_{\mathrm{loc}}(A^{-1}, \Lambda'_{\mathrm{Shi}})$. If we furthermore choose $\sigma \leqslant T'_{\mathrm{loc}}/2$, this gives a contradiction since it shows that the persistence holds up to t_k for k big enough. □

From Theorem 8.1.3 we deduce the two following corollaries, one in dimension n and the second one specific to dimension 3.

Corollary 8.2.2 (Persistence under curvature estimate). *Let $A > 0$ and $C > 0$, there exists $\rho = \rho(\mathcal{M}_0, A, C) > A$ with the following property. Let X be an n-manifold, U be an open subset of X, $p \in U$ and $T \in (0, T_0]$. Let $g(\cdot)$ be a Ricci flow defined on $U \times [0, T]$, such that the ball $B(p, 0, \rho) \subset U$ is relatively compact. We assume that* (i′) $|\frac{\partial}{\partial t}\mathrm{Rm}_{g(t)}(x)|_{g(t)} \leqslant C|\mathrm{Rm}_{g(t)}|^2_{g(t)}(x)$ *when* $|\mathrm{Rm}(x)|_{g(t)} \geqslant C$,

(ii) $B(p,0,\rho)$ *is* ρ^{-1}*-close to* $B(p_0,0,\rho)$.

Then $P(p,0,A,t)$ *is* A^{-1}*-close to* $P(p_0,0,A,t)$.

Remark 8.2.3. Notice that inequality (i′) is the n-dimensional analogous of (4.5).

Proof of Corollary 8.2.2. First we define, for some $A_1 > 0$ to be determined later, $\rho = \rho(\mathcal{M}_0, A_1, 4\Lambda_0)$ given by Theorem 8.1.3. We set $g_1(\cdot) = \psi^* g(\cdot)$, where $\psi\colon B(p_0,0,\rho) \to B(p,0,\rho)$ is a diffeomorphism given by the assumption that $g(0)$ is ρ^{-1}-close to $g_0(0)$. By closeness again, $|\mathrm{Rm}_{g_1(0)}| \leqslant 2\Lambda_0$ on $B(p_0,0,\rho)$. By integrating the inequality (i′) we get $|\mathrm{Rm}| \leqslant 4\Lambda_0$ on $B(p_0,0,\rho) \times [0,T_{\times 2}]$, where $T_{\times 2} = (4\Lambda_0 C)^{-1}$ is the doubling time (we may assume $\Lambda \geqslant C$). Applying Theorem 8.1.3 gives that $g_1(\cdot)$ is ρ_1^{-1}-close to $g_0(\cdot)$ on $B(p_0,0,A_1) \times [0,T_{\times 2}]$ and, in particular, $|\mathrm{Rm}| \leqslant 2\Lambda_0$ on this set (if ρ_1 is big enough). We now can make the same argument with $A_1 = \rho(\mathcal{M}_0, A_2, 4\Lambda_0)$ and A_2 to be chosen later. Then $g_1(\cdot)$ is A_2^{-1}-close to $g_0(\cdot)$ on $B(p,0,A_2) \times [0,2T_{\times 2}]$. One iterates this procedure, defining inductively $A_i = \rho(\mathcal{M}_0, A_{i+1}, 4\Lambda_0)$ up to $A_m = \rho(\mathcal{M}_0, A, 4\Lambda_0)$, where $m \approx T/T_{\times 2} \leqslant T_0/T_{\times 2}$.

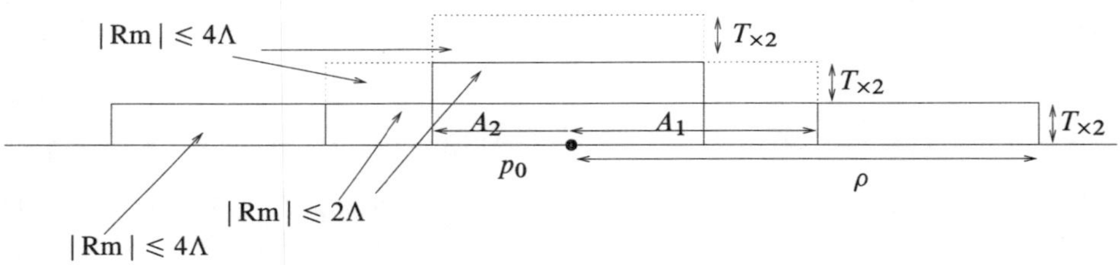

□

For the next corollary we suppose that the model X_0 is 3-dimensional. We assume as usual that M is a closed, irreducible, nonspherical 3-manifold. For applications to sub-intervals of a given Ricci flow with bubbling-off, we use a property which could be given by the $(\mathrm{CN})_r$ assumption (see equation 4.5).

Corollary 8.2.4 (Persistence in a Ricci flow with bubbling-off). *Let* $A > 0$, *there exists* $\rho = \rho(\mathcal{M}_0, A) > A$ *with the following property. Let* $\{g(t)\}_{t\in[0,T]}$ *be a Ricci flow with bubbling-off on* M *with* $T \leqslant T_0 < +\infty$. *Suppose that*

(a) $g(\cdot)$ *has curvature pinched toward positive,*

(b) $|\frac{\partial}{\partial t}R(x,t)| \leqslant C_0 R^2$ *if* $R(x,t) \geqslant 1$.

Let $(p,t) \in M \times (0,T]$ *such that*

(c) $B(p,0,\rho)$ *is* ρ^{-1}*-close to* $B(p_0,0,\rho) \subset X_0$,

(d) $P(p,0,\rho,t)$ *is unscathed.*

Then $P(p,0,A,t)$ *is* A^{-1}*-close to* $P(p_0,0,A,t)$.

Proof of Corollary 8.2.4. The idea is the same as for the proof of Corollary 8.2.2. To estimate the doubling time for Rm, one now uses the derivative estimate on the scalar curvature (4.5) and the curvature pinched toward positive assumption. □

8.3 Application: persistence of almost standard caps

In this section, we prove Theorem 8.1.2.

Proof of Theorem 8.1.2. We now consider as our model, $\mathcal{M}_0 = (\mathcal{S}_0, g_0(\cdot), p_0)$, which is the standard solution restricted to $[0, \theta]$. Let us assume for simplicity that $b \geqslant t_0 + \theta h^2$, so that $t_1 \geqslant t_0 + \theta h^2$. For any nonnegative integer N, we now set

$$\Lambda_N := \max_{\mathcal{S}_0 \times [0,\theta]} \{|\nabla^p \operatorname{Rm}|, |R|; 0 \leqslant p \leqslant N\}.$$

In the sequel we consider the restriction of $g(\cdot)$ to $[t_0, b]$ and we define

$$\bar{g}(t) := h^{-2} g(t_0 + th^2) \quad \text{for } t \in [0, \theta].$$

Note that $\bar{g}(\cdot)$ satisfies assumptions (a) and (b) of Corollary 8.2.4. Indeed, if $g(\cdot)$ satisfies $(\mathrm{CN})_r$ on $[a, b)$, it follows easily by a continuity argument that any (x, b) with $R(x, b) \geqslant 2r(b)^{-2}$ satisfies the estimate $|\frac{\partial R}{\partial t}| \leqslant C_0 R^2$ at (x, b). After rescaling by $h(b)^{-2} \gg 8r^{-2}$, this property holds at points with scalar curvature above $1/4$.

Let us choose $A > 0$ and, for some $\rho_1 > A$ to be defined later, let $\rho := 4\rho(\mathcal{M}_0, \rho_1) \gg C$ be the parameter given by Corollary 8.2.4. Assume that $P(p, 0, A, t)$ is unscathed for some $t \in [0, \theta]$. By assumption we can choose $\delta \ll 1$ such that $B(p, 0, \delta^{-1})$ is $\delta'(\delta)$-close to $B(p_0, 0, \delta^{-1}) \subset \mathcal{S}_0$. In order to apply Corollary 8.2.4 we need to show that $P(p, 0, \rho, t)$ is unscathed.

Let $T_{\times 2} = \min(\theta, 4\Lambda_0^{-1} C_0^{-1})$.

Claim. $P(p, 0, \rho, T_{\times 2})$ is unscathed or $B(p, 0, \rho) \subset \Sigma_{t'}$ for some $t' \in [0, T_{\times 2})$.

Proof of the claim. As usual we argue by contradiction. If $P(p, 0, \rho, t)$ is scathed then $P(p, 0, \rho, t) \cap \Sigma_{t'} \neq \emptyset$ for some $t' \in [0, T_{\times 2})$. If $B(p, 0, \rho) \not\subset \Sigma_{t'}$ then $B(p, 0, \rho)$ intersects the middle sphere S of some δ-neck N for $g(t')$. Let $s \colon N \to \mathbb{R}$ denote the radial coordinate of the neck. We shall prove that $p \in s^{-1}((-\delta^{-1}/10, \delta^{-1}/10))$.

By closeness at time 0 with $g_0(0)$, we have that the scalar curvature of $\bar{g}(0)$ satisfies

$$R_{\bar{g}} \in [1/2, 2\Lambda_0] \tag{8.3}$$

on $B_{\bar{g}(0)}(p, 0, \delta^{-1})$. Now using the derivative estimate $(\partial R/\partial t < C_0 R^2)$ on the unscathed neighbourhood $P(p, 0, \rho, t')$ and the definition of $T_{\times 2}$ we get that $R_{\bar{g}} \in [1/4, 4\Lambda_0]$ on $P(p, 0, \rho, t')$. From the curvature pinched toward positive property we then have (cf. Remark 4.4.2) that $|\mathrm{Rm}| \leqslant 2\Lambda_0 + 2\phi_0(4\Lambda_0)$ on this set. Hence on $B_{\bar{g}}(p, 0, \rho)$ and for $t \in [0, t']$,

$$\frac{1}{2C} \leqslant \frac{\bar{g}(t)}{\bar{g}(0)} \leqslant 2C,$$

for some constant $C := C(\Lambda_0) = C(\theta)$. From that we deduce $\operatorname{diam}_{\bar{g}(t)} B_{\bar{g}}(p, 0, \rho) \leqslant 4\rho C$.

This implies that for $x \in B_{\bar{g}}(p, 0, \rho) \cap S$, $B_{\bar{g}}(p, 0, \rho) \subset B_{\bar{g}}(x, t', 8C\rho)$.

Furthermore, if δ is small enough,

$$\frac{\delta^{-1}}{10} R_{\bar{g}}(x, t')^{-1/2} \geqslant \frac{\delta^{-1}}{10} \left(\frac{1}{4\Lambda_0}\right)^{1/2} > 8C\rho,$$

then $B_{\bar{g}}(x, t'8C\rho) \subset s^{-1}((-\delta^{-1}/10, \delta^{-1}/10))$, and so $s(p) \in (-\delta^{-1}/10, \delta^{-1}/10)$. We define S' to be the transversal sphere of the neck $s^{-1}(s(p))$ that contains p. Note that the diameter of S' for the metric $\bar{g}(t')$ is approximately $R_{\bar{g}}(x, t')^{-1/2} \leqslant 4$. On the other hand, by the distance distortion

$$d_{t'}(p, \partial B(p, 0, \rho/2)) \geqslant \frac{\rho}{4C}.$$

It then follows that $S' \subset B_{\bar{g}}(p, 0, \rho/2)$. Now a geodesic segment for $\bar{g}(t')$, γ, which traverses the neck N must intersect S', hence it must also traverse twice the cylinder $B_{\bar{g}}(p, 0, \rho) \setminus B_{\bar{g}}(p, 0, \rho/2)$. Now the estimates

$$\operatorname{diam}_{t'} \partial B(p, 0, \rho) \leqslant 4C$$

and

$$d_{t'}(\partial B(p, 0, \rho), \partial B(p, 0, \rho/2)) \geqslant \frac{\rho}{4C}.$$

prove that γ cannot be minimising. □

Note that if $B_{\bar{g}}(p, 0, \rho) \subset \Sigma_{t'}$ then $B(p, 0, A) \subset \Sigma_{t'}$. From the assumptions this cannot happen if $t' < t_0$. Then for $\bar{t} = \min(t, T_{\times 2})$, we have that $P(p, 0, \rho, \bar{t})$ is unscathed. Applying Corollary 8.2.4 it follows that $B(p, 0, \rho_1, \bar{t})$ is A'-close to $P(p_0, 0, \rho_1, \bar{t}) \subset \mathcal{S}_0 \times [0, \theta]$. In particular (8.3) holds on $B(p_0, 0, \rho_1)$ at time $\bar{t}$. We can make the same argument on $[\bar{t}, \bar{t} + T_{\times 2}]$ with $\rho_1 := \rho(\mathcal{M}_0, 4\rho_2)$, where $\rho_2 > 0$ is yet to be defined. We iterate this procedure which stops at a finite number of steps. □

Notes. Theorem 8.1.2 is similar to [Per03b], Lemma 4.5.

The local persistence lemma (Lemma 8.2.1) seems to be new, although there are similar results in the literature. Theorem 8.1.3 is new, and relies in an essential way on the Chen–Zhu uniqueness theorem, which was not available to Perelman.

Chapter 9

Canonical neighbourhoods and the proof of Proposition B

This chapter is about extending the canonical neighbourhoods property $(\mathrm{CN})_r$ forward in time. The definition of this property was formulated so as to make it easy to prove that it is open in time (cf. Lemmas 5.3.2 and 5.3.3). There remains to show that it is closed, provided the surgery parameters are small enough. This is the content of Proposition B, whose statement we now recall:

Proposition B. *For all $\kappa > 0$ there exist $r = r(\kappa) \in (0, 10^{-3})$ and $\bar{\delta}_B = \bar{\delta}_B(\kappa) \in (0, \delta_0)$ with the following property: let $0 < \delta \leqslant \bar{\delta}_B$, $b > 0$ and $g(\cdot)$ be a Ricci flow with bubbling-off defined on $[0, b]$ with normalised initial condition. Assume that the restriction of $g(\cdot)$ to the half-open interval $[0, b)$ is a Ricci flow with (r, δ, κ)-bubbling-off. Then it satisfies $(\mathrm{CN})_r$ on all of $[0, b]$.*

The proof is by contradiction, considering a sequence of pointed Ricci flows with bubbling-off such that the basepoint does not have an (ε_0, C_0)-canonical neighbourhood. By a continuity argument, it does have a canonical neighbourhood with slightly different parameters. This argument is performed in Section 9.1 as a preliminary lemma. The argument by contradiction itself follows the same outline as the proof of the cutoff parameters theorem, i.e., backward extension using the compactness theorem for Ricci flows. It is harder because of possible surgeries occurring in the near past of the basepoint. This is where the persistence theorem is used. Another key ingredient is Hamilton's Harnack type inequality, which is used for backward extension of curvature bounds.

We remark that the parameter κ is fixed throughout the proof. This is essential in order to apply the compactness theorem. The (even harder) work needed to propagate the κ-noncollapsing property forward in time will be done in the next chapter.

We begin gently, in Section 9.1 below, showing by a continuity argument that a limit of ε_0-strong necks is a $2\varepsilon_0$-strong neck (see also Lemma 9.2.3).

9.1 Warming up

Lemma 9.1.1. *Let $\varepsilon \in (0, \varepsilon_0)$. Let $g(\cdot)$ be a Ricci flow with bubbling-off and let x be a point of M such that $R(x, 0) > 0$ and there exists a sequence of negative numbers t_i tending to 0 such that for every i, (x, t_i) is the centre of a strong ε-neck. Then $(x, 0)$ is the centre of a strong 2ε-neck.*

Proof. For each i, we fix a strong ε-neck N_i centred at (x, t_i), a parametrisation $\psi_i : S^2 \times (-\varepsilon^{-1}, \varepsilon^{-1}) \to N_i$ and a scaling factor Q_i. We set $\bar{g}_i(t) := Q_i g(t_i + tQ_i^{-1})$. By definition, we have

$$\sup_{t\in[-1,0]} \|\psi_i^* \bar{g}_i(t) - g_{\text{cyl}}(t)\|_{[\varepsilon^{-1}], S^2\times(-\varepsilon^{-1},\varepsilon^{-1}), g_{\text{cyl}}(t)} < \varepsilon. \tag{9.1}$$

By choice of ε_0, we have $|Q_i R(x, t_i) - 1| \leqslant 10^{-1}$. Now the sequence $R(x, t_i)$ converges to the positive number $R(x, 0)$. Hence up to extracting a subsequence, we can assume that Q_i converges to some positive number Q. Set $\bar{g}(t) := Qg(t/Q)$. This is defined for at least $t/Q \geqslant t_i - Q_i^{-1}$, hence for $t > -1$ since $t_i Q - Q/Q_i \to -1$. Hence $\bar{g}$ is defined on $(-1, 0]$.

For all $t \in (-1, 0]$ we have $\bar{g}(t) = \lim \bar{g}_i(t)$ in the $\mathcal{C}^{[\varepsilon^{-1}]}$ topology. Hence we have a uniform upper bound for the $\mathcal{C}^{[\varepsilon^{-1}]+1}$-norm of ψ_i computed with respect to the fixed metrics g_{cyl}, $\bar{g}(0)$.

After passing to a subsequence, we get a map $\psi : S^2 \times (-\varepsilon^{-1}, \varepsilon^{-1}) \to M$ such that $\psi_i \to \psi$ in the $\mathcal{C}^{[\varepsilon^{-1}]}$ sense. For all $y, y' \in S^2 \times (-2\varepsilon^{-1}/3, 2\varepsilon^{-1}/3)$ we have

$$(1 - \varepsilon) d_{g_{\text{cyl}}}(y, y') \leqslant d_{\bar{g}_i(0)}(\psi_i(y), \psi_i(y')) \leqslant (1 + \varepsilon) d_{g_{\text{cyl}}}(y, y'). \tag{9.2}$$

Passing to the limit, we obtain:

$$(1 - \varepsilon) d_{g_{\text{cyl}}}(y, y') \leqslant d_{\bar{g}(0)}(\psi(y), \psi(y')) \leqslant (1 + \varepsilon) d_{g_{\text{cyl}}}(y, y'). \tag{9.3}$$

Hence the restriction of ψ to $S^2 \times (-2\varepsilon^{-1}/3, 2\varepsilon^{-1}/3)$ is injective. From inequality (9.1) we derive

$$\|\psi^* \bar{g}(t) - g_{\text{cyl}}(t)\|_{\mathcal{C}^{[\varepsilon^{-1}]-1}, g_{\text{cyl}}} \leqslant \varepsilon$$

for all $t \in (-1, 0]$. Hence the restriction of ψ to $S^2 \times (-2\varepsilon^{-1}/3, 2\varepsilon^{-1}/3)$ is a $\mathcal{C}^{[\varepsilon^{-1}]}$-diffeomorphism onto its image.

Pick $\lambda > 1$ close enough to 1 such that the metric $\tilde{g}(t) := \lambda \bar{g}(t/\lambda)$ satisfies

$$\|\psi^* \tilde{g}(t) - g_{\text{cyl}}(t)\|_{[\varepsilon^{-1}]-1, S^2\times(-2\varepsilon^{-1}/3, 2\varepsilon^{-1}/3), g_{\text{cyl}}(t)} \leqslant \tfrac{3}{2}\varepsilon < 2\varepsilon$$

on $(-\lambda, 0] \supset [-1, 0]$. Then $\psi(S^2 \times (-(2\varepsilon)^{-1}, (2\varepsilon)^{-1}))$ is a strong 2ε-neck. □

9.2 The proof

We argue by contradiction. Suppose that some fixed number $\kappa > 0$ has the property that for all $r \in (0, r_0)$ and $\bar{\delta}_B > 0$ there exist counterexamples. Then we can consider sequences $r_k \to 0$, $\delta_k \to 0$, $b_k > 0$, and a sequence of Ricci flows with bubbling-off $g_k(\cdot)$ on $[0, b_k]$, which are Ricci flows with (r_k, δ_k, κ)-bubbling-off on $[0, b_k)$, but such that $(\mathrm{CN})_{r_k}$ fails at b_k. This last assertion means that there exists $x_k \in M_k$ such that

$$R(x_k, b_k) \geqslant r_k^{-2},$$

and yet x_k does not have any (ε_0, C_0)-canonical neighbourhood.

Without loss of generality, we assume that

$$\delta_k \leqslant \bar{\delta}_{\text{per}}\Big(k, 1 - \frac{1}{k}, r_k\Big),$$

the right-hand side being the parameter given by Persistence Theorem 8.1.2.

By hypothesis, $\{g_k(t)\}_{t \in [0,b_k)}$ is κ-noncollapsed on scales less than 1. The same is true on the closed interval $[0, b_k]$ by Lemma 4.1.4. The hypothesis that $\{g_k(t)\}_{t \in [0,b_k)}$ has curvature pinched toward positive implies that any blow-up limit (provided it exists) will have nonnegative curvature operator by Proposition 6.1.6.

We shall work with the rescaled evolving metric

$$\bar{g}_k(t) = R(x_k, b_k) g_k\Big(b_k + \frac{t}{R(x_k, b_k)}\Big),$$

which is defined on $[-R(x_k, b_k)b_k, 0]$.

As in Chapter 6, we put a bar on the points when they are involved in geometric quantities computed with respect to the metric $\bar{g}_k$. By the Remark 5.1.9 and the hypothesis that $g_k(0)$ is normalised, we have $b_k \geqslant 1/16$ for all k. This implies that $R(x_k, b_k)b_k$ tends to $+\infty$ as k tends to ∞.

Let us begin with two preliminary lemmas. Set $\rho_0 := 4C_{\text{st}}(\beta_0\varepsilon_0/2)$, where $C_{\text{st}}(\beta_0\varepsilon_0/2)$ is the constant fixed in Definition 5.1.1.

Lemma 9.2.1 (Parabolic balls of bounded curvature are unscathed). *For all $K, \tau > 0$ and $\rho \geqslant \rho_0$, there exists $k_0 := k_0(K, \rho, \tau) \in \mathbb{N}$ such that, if $|\mathrm{Rm}| \leqslant K$ on $B(\bar{x}_k, 0, \rho) \times (-\tau, 0]$ then $P(\bar{x}_k, 0, \rho, -\tau)$ is unscathed for all $k \geqslant k_0$.*

Remark 9.2.2. Note that $B(\bar{x}_k, 0, \rho) \times [-\tau, 0]$ is unscathed when the conclusion holds.

Proof. We prove the lemma by contradiction. Let us fix k and assume that there exist $z_k \in B(\bar{x}_k, 0, \rho)$ and $s_k \in [-\tau, 0)$ such that $\bar{g}_{k,+}(z_k, s_k) \neq \bar{g}_k(z_k, s_k)$. We can take s_k maximal satisfying this property, i.e., the set $B(\bar{x}_k, 0, \rho) \times (s_k, 0]$ is unscathed. In the sequel we consider the restriction of $\bar{g}_k(\cdot)$ to $[s, 0]$, in the sense of Definition 8.1.1. We drop indices for simplicity.

By definition of the (r, δ)-surgery, there exists a marked δ-almost standard cap (U, V, p, y) (see Definition 5.2.3) such that $(U, R(\bar{y}, s)\bar{g}(s))$ is δ'-close to $B(p_0, 5 + \delta^{-1}) \subset \mathcal{S}_0$. Set $\tilde{g}(s) := R(\bar{y}, s)\bar{g}(s)$. We shall put a tilde on the points when they are involved in geometric quantities computed with respect to the metric $\tilde{g}$. We now show that $d(\tilde{x}, \tilde{p})$ is bounded independently of k (if k is large enough).

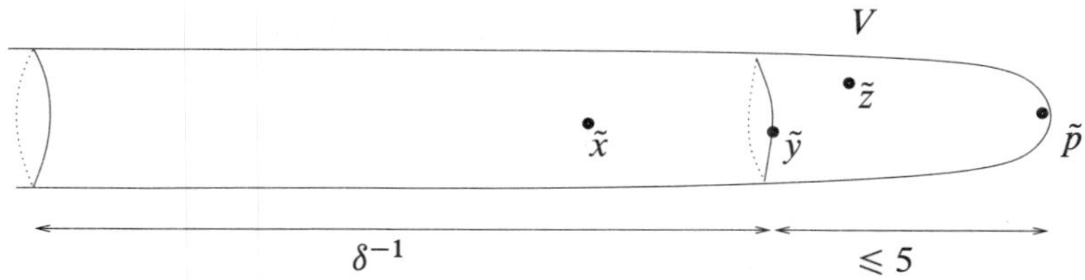

Figure 9.1. Almost standard cap.

As $B(\bar{x}, 0, \rho) \times (s, 0]$ is unscathed with $|\mathrm{Rm}| \leqslant K$, by the Distance-Distortion Lemma 2.2.7 we have on $B(\bar{x}, 0, \rho)$:

$$e^{-4K\tau} \leqslant \frac{\bar{g}(0)}{\bar{g}(s)} \leqslant e^{4K\tau}.$$

The point $\bar{y}$ may not be in $B(\bar{x}, 0, \rho)$ but its scalar curvature is comparable to $\bar{z} \in B(\bar{x}, 0, \rho)$, hence $R(\bar{y}, s) \leqslant 7K$. This implies that $d(\tilde{x}, \tilde{z}) \leqslant \sqrt{7K} e^{2K\tau} \rho$, hence (see Figure 9.2)

$$d(\tilde{x}, \tilde{p}) \leqslant \sqrt{7K} e^{2K\tau} \rho + 5 = \rho'(K, \rho, \tau) =: \rho'.$$

The Persistence Theorem 8.1.2 implies that the set

$$\tilde{P} := B(\tilde{p}, A) \times [0, \min\{\theta, |s| R(\bar{y}, s)\}),$$

endowed with the rescaled flow

$$\tilde{g}(t) := R(\bar{y}, s) \bar{g}(s + t(R(\bar{y}, s))^{-1}),$$

is A^{-1}-close to $P(p_0, 0, A, \min\{\theta, |s| R(\bar{y}, s)\})$. Indeed, the second alternative of the persistence theorem does not occur since in this case, we would have $B(\tilde{p}, 0, A) \subset \Sigma_{t_+}$ for some $t_+ \in (-s, 0]$. As $x \in B(\tilde{p}, 0, A)$, this would imply (x, t_+) is in the singular set for $t_+ > s$, contradicting the choice of s.

We now choose $A = k$ and $\theta = 1 - 1/k$.

Claim. We have $s + \theta R(\bar{y}, s)^{-1} > 0$ for large k.

Proof of the claim. We have $s \geqslant -\tau$ and $\theta \geqslant 1/2$. If $R(\bar{y}, s) > 2\tau$, then

$$s + \theta R_{\bar{g}}(y, s)^{-1} \geqslant -\tau + \frac{1}{2} R(\bar{y}, s)^{-1} > 0.$$

Suppose now that $R(\bar{y}, s) \leqslant 2\tau$. We argue by contradiction. Assume that $s_1 := s + \theta R(\bar{y}, s)^{-1} \leqslant 0$ and apply the Persistence Theorem 8.1.2 up to this time. By Proposition 4.3.4, $R_{\min}(t) \geqslant \mathrm{const}_{\mathrm{st}}(1-t)^{-1}$ on the standard solution, hence we have, by closeness

$$R(\bar{x}, s_1) \geqslant \frac{1}{2} R(\bar{y}, s) \mathrm{const}_{\mathrm{st}}(1-\theta)^{-1} \geqslant \mathrm{const}_{\mathrm{st}}(4\tau(1-\theta))^{-1} = \frac{k \cdot \mathrm{const}_{\mathrm{st}}}{4\tau}.$$

On the other hand, $R(\bar{x}, s_1) \leqslant 6K$, this is a contradiction for sufficiently large k. This proves the claim. □

By the conclusion of the Persistence Theorem 8.1.2, there exists a diffeomorphism $\psi\colon B(p_0, 0, A) \to B(\tilde{p}, 0, A)$ such that $\psi^*\tilde{g}(\cdot)$ is A^{-1}-close to $g_0(\cdot)$ on $B(p_0, 0, A)\times [0, \min\{\theta, |s|R(\bar{y}, s)\}]$. By the above claim, $\min\{\theta, |s|R(\bar{y}, s)\} = |s|R(\bar{y}, s) =: s'$. Let $x' = \psi^{-1}(x)$, the point $(x, 0)$ corresponds to (x', s') by the diffeomorphism. By Proposition 4.3.4, for all $\varepsilon > 0$, there exists $C_{\mathrm{st}}(\varepsilon)$ such that any point (x', t) of the standard solution has an $(\varepsilon, C_{\mathrm{st}}(\varepsilon))$-canonical neighbourhood or $t < 3/4$ and $x' \notin B(p_0, 0, \varepsilon^{-1})$. Let us choose $\varepsilon = \varepsilon_0\beta_0/2 \ll \varepsilon_0$, where β_0 is the parameter fixed in Definition 5.1.1. Let us denote $C_{\mathrm{st}} = C_{\mathrm{st}}(\varepsilon)$. There are again two possibilities.

Case 1. The point (x', s') has an $(\varepsilon, C_{\mathrm{st}})$-canonical neighbourhood,

$$U' \subset B(x', s', 2C_{\mathrm{st}}R(x', s')^{-1/2}).$$

Here $R(x', s')$ is the scalar curvature of the standard solution at (x', s'). The A^{-1}-proximity between $g_0(s')$ and $\psi^*\tilde{g}(s')$ gives

$$R(x', s') \simeq R(\tilde{x}, s') = R(\bar{y}, s)^{-1}R(\bar{x}, 0) = R(\bar{y}, s)^{-1}.$$

On the other hand, we have

$$\frac{\rho}{2}R(\bar{y}, s)^{1/2} \simeq \frac{\rho}{2}R(x', s')^{-1/2} > 2C_{\mathrm{st}}R(x', s')^{-1/2},$$

which implies

$$U' \subset B\Big(x', s', \frac{\rho}{2}R(\bar{y}, s)^{1/2}\Big) \subset \psi^{-1}(B(\bar{x}, 0, \rho)).$$

Hence $\psi(U')$ is a canonical neighbourhood at $(x, 0)$ of parameters $(2\varepsilon, 2C_{\mathrm{st}})$, hence an (ε_0, C_0)-canonical neighbourhood by the choices made in Section 5.1.

Case 2. The point (x', s') has no $(\varepsilon, C_{\mathrm{st}})$-canonical neighbourhood. Then $s' \leqslant 3/4$ and $x' \notin B(p_0, 0, \varepsilon^{-1})$. Hence we have

$$d'_s(\tilde{x}, \tilde{p}) \geqslant \frac{9}{10}\varepsilon^{-1} \geqslant \frac{3}{2}(\varepsilon_0\beta_0)^{-1} > (\varepsilon_0\beta_0)^{-1} + 5.$$

We infer that $(x, -s')$ is the centre of an $(\varepsilon_0\beta_0)$-neck, coming from the strong δ-neck that was there at the singular time. We now apply the neck strengthening lemma (Lemma 4.3.5) which asserts that $(x, 0)$ has a strong ε_0-neck. Indeed the proximity with the standard solution ensures that $P(\bar{x}, -s', (\varepsilon_0\beta_0)^{-1}, s') \subset P(\bar{x}, -s', A, s')$ is unscathed and has $|\mathrm{Rm}| \leqslant 2K_{\mathrm{st}}$. This proves Lemma 9.2.1. □

Lemma 9.2.3. *Each point (x, b_k) of scalar curvature at least $2r_k^{-2}$ for $g_k(b_k)$ has a $(2\varepsilon_0, 2C_0)$-canonical neighbourhood.*

Proof. For each $t < b_k$ close to b_k, $R(x,t) > r_k^{-2}$ hence (x,t) has an (ε_0, C_0)-canonical neighbourhood. If there is a sequence of strong ε_0-necks at times $t_i \to b_k$, the conclusion follows from Lemma 9.1.1. Otherwise it is easy to see that (x, b_k) is the centre of a $(2\varepsilon_0, 2C_0)$-cap. □

Now we continue the proof of Proposition B.

Step 1. The sequence $(M_k, \bar{g}_k(0), \bar{x}_k)$ subconverges to some complete pointed Riemannian manifold $(M_\infty, g_\infty, x_\infty)$ of nonnegative curvature operator.

Proof. We have to show that the sequence satisfies the hypothesis of the Local Compactness Theorem C.3.3. First we wish to apply the Curvature-Distance Theorem 6.1.1 to obtain uniform bounds on balls $B(\bar{x}_k, 0, \rho)$, for all $\rho > 0$. Fix some $\rho > \rho_0$, where ρ_0 is the radius which appears in Lemma 9.2.1. Let us check that the hypothesis of the Curvature-Distance Theorem 6.1.1 are satisfied on $B(\bar{x}_k, 0, \rho)$ for k large enough. Lemma 9.2.3 implies that each point $x \in M_k$ with $R(x, b_k) \geqslant 2R(x_k, b_k)$ has a $(2\varepsilon_0, 2C_0)$-canonical neighbourhood. Let $Q(\rho) := Q(\rho, 2\varepsilon_0, 2C_0)$ and $\Lambda(\rho) := \Lambda(\rho, 2\varepsilon_0, 2C_0)$ be the parameters given by Theorem 6.1.1. Let $k(\rho) \in \mathbb{N}$ be such that $R(x_k, b_k) \geqslant Q(\rho)$ for $k \geqslant (\rho)$. Then Theorem 6.1.1 applies for $k \geqslant k(\rho)$ and implies that the scalar curvature of $\bar{g}_k(0)$ is bounded above on $B(\bar{x}_k, 0, \rho)$ by $\Lambda(\rho)$. Next we obtain similar bounds on parabolic balls $P(\bar{x}_k, 0, \rho, -\tau(\rho))$ for some $\tau(\rho) > 0$, and deduce that they are unscathed.

In order to apply the Curvature-Time Lemma 6.1.3, we set $C(\rho) := \Lambda(\rho) + 2$ and $K(\rho) := 2C(\rho)$. Define $k_1(\rho) := \max\{k_0(K(\rho), \rho, (2C_0C(\rho))^{-1}), k(\rho)\}$, where k_0 is the parameter given by Lemma 9.2.1.

Claim. If $k \geqslant k_1(\rho)$, then the set $P(\bar{x}_k, 0, \rho, -(2C_0C(\rho))^{-1})$ is unscathed and satisfies $|\mathrm{Rm}| \leqslant K(\rho)$.

Proof. Let $s_k \in [-(2C_0C(\rho))^{-1}, 0]$ be minimal such that $B(\bar{x}_k, 0, \rho) \times (s_k, 0]$ is unscathed. By the Curvature-Time Lemma 6.1.3, we have $R \leqslant K(\rho)$ on this set. This implies $|\mathrm{Rm}| \leqslant K(\rho)$ by Pinching Lemma 4.4.7. By Lemma 9.2.1, $P(\bar{x}_k, 0, \rho, s_k)$ is unscathed. By the minimality of s_k we then have $s_k = -(2C_0C(\rho))^{-1}$. □

By hypothesis, the solutions $g_k(\cdot)$ are κ-noncollapsed on scales less than 1. Hence $\bar{g}_k(0)$ is κ-noncollapsed on scales less than $\sqrt{R(x_k, b_k)}$. This, together with the curvature bound implies a positive lower bound for the injectivity radius at $(\bar{x}_k, 0)$ (Theorem B.1.2). Therefore, Compactness Theorem C.3.3 applies to the sequence $(M_k, \bar{g}_k(\cdot), (x_k, 0))$. In particular, it implies that the sequence subconverges to $(M_\infty, g_\infty(\cdot), (x_\infty, 0))$, where M_∞ is a smooth manifold, $g_\infty(0)$ is complete and $g_\infty(\cdot)$ is defined on $B(x_\infty, 0, \rho) \times (-(2C_0C(\rho))^{-1}, 0]$ for each $\rho > 0$.

Lastly, since the metrics $g_k(\cdot)$ are pinched toward positive and the scaling factor $R(x_k, b_k)$ goes to $+\infty$, the metric $g_\infty(0)$ has nonnegative curvature operator by Proposition 6.1.6. This argument completes the proof of Step 1. □

We remark that M_∞ must be noncompact: indeed, if it were compact, it would be diffeomorphic to M_k for large k; this would imply that M_k carries a nonflat metric of nonnegative sectional curvature. By the classification of closed manifolds on nonnegative curvature B.2.5, this is incompatible with our assumption that M_k is irreducible and not spherical.

By the Soul Theorem B.2.1, M_∞ is diffeomorphic to $\mathbb{R}^3$, $S^1 \times \mathbb{R}^2$, or a line bundle over $\mathbb{R}P^2$ or S^2. In particular, every smoothly embedded 2-sphere in M_∞ is separating.

Step 2. The Riemannian manifold $(M_\infty, g_\infty(0))$ has bounded curvature.

Proof. By passing to the limit, we see that every point $p \in M_\infty$ of scalar curvature at least 3 is the centre of a $(3\varepsilon_0, 3C_0)$-cap or a (not necessarily strong) $3\varepsilon_0$-neck. In the sequel, we refer to this fact by saying that the limiting partial flow $g_\infty(\cdot)$ satisfies the *weak canonical neighbourhood property.*

We argue by contradiction. Let $p_k \in M_\infty$ be such that $R_\infty(p_k, 0) \to +\infty$ as $k \to +\infty$; in particular, $d_0(p_k, x_\infty) \to +\infty$. Consider segments $[x_\infty p_k]$ which, after passing to a subsequence, converge to a geodesic ray c starting at x_∞.

Pick a point on $[x_\infty p_k]$ q_k such that $R(q_k, 0) = \frac{R(p_k,0)}{4C_0}$. For sufficiently large k, the point $(q_k, 0)$ is the centre of a weak $(3\varepsilon_0, 3C_0)$-canonical neighbourhood U_k. Then the scalar curvatures on U_k belong to the closed interval $[(3C_0)^{-1} R(q_k, 0), 3C_0 R(q_k, 0)]$. As a consequence, x_∞ and p_k do not belong to U_k, and U_k is traversed by $[x_\infty p_k]$. By Corollary 3.3.3 this neighbourhood is a $3\varepsilon_0$-neck.

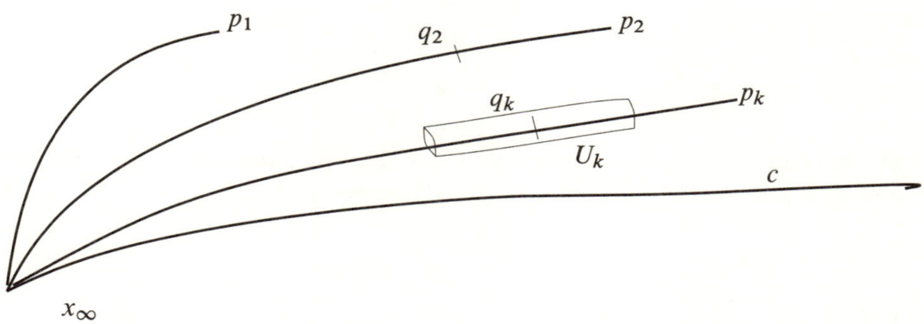

Figure 9.2. Necks in $(M_\infty, g_\infty(0))$.

Lemma 9.2.4. *For k large enough, c traverses U_k.*

Proof. Assume it does not. Recall that $g_\infty(0)$ is nonnegatively curved. Consider a geodesic triangle $\Delta x_\infty q_k c(t)$ for $t \geqslant 10C_0$. Choose k large enough so that the angle at x_∞, denoted by $\measuredangle(p_k x_\infty c(t))$ is less than $\pi/100$. Let S_- (resp. S_+) be the component of ∂U_k which is closest to (resp. farthest from) x_∞. By comparison with a Euclidean triangle $\Delta \bar{x}_\infty \bar{q}_k \bar{c}(t)$, we see that $\liminf_{t\to\infty} \measuredangle(x_\infty q_k c(t)) = \pi$.

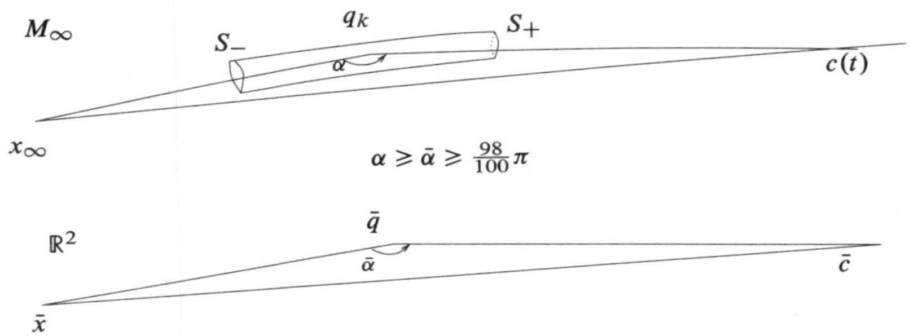

Fix t large enough so that this angle is greater than 0.98π. Then $[q_k c(t)]$ intersects S_+. The loop γ obtained by concatenating $[x_\infty q_k]$, $[q_k c(t)]$ and $[c(t)x_\infty]$ has an odd intersection number with S_+. This implies that S_+ is nonseparating. This contradiction proves Lemma 9.2.4. □

We proceed with the proof of Step 2. Pick k_0 large enough so that U_{k_0} is traversed by c for all $k \geqslant k_0$. Let S_0 be the middle sphere of U_{k_0}. Let $a_0, a'_0 \in S_0$ be two points maximally distant from each other. Call a_k, a'_k the respective intersections of the segments $[a_0 c(t)]$ and $[a'_0 c(t)]$ with the middle sphere S_k of U_k, where t is large enough so that $d_0(a_0, c(t)) \gg d_0(S_0, S_k)$.

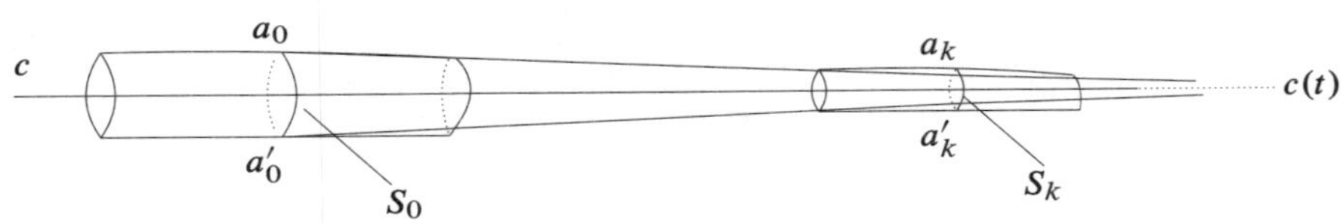

By comparison with Euclidian triangles and Thales theorem, the distance between a_k and a'_k is greater than $\frac{1}{2}d_\infty(a_0, a'_0)$. Thus we have

$$\operatorname{diam}(S_k) \geqslant d_0(a_k, a'_k) \geqslant \frac{1}{2} d_0(a_0, a'_0) = \frac{1}{2}\operatorname{diam}(S_0).$$

Now the diameter of S_k is close to $\pi\sqrt{2}R_\infty(q_k, 0)^{-1}$ and tends to 0 by hypothesis, which gives a contradiction. This completes the proof of Step 2. □

Applying again Lemma 9.2.1 and the Curvature-Time Lemma 6.1.3, there exists $\tau > 0$ such that $(M_k, \bar{g}_k(\cdot), (\bar{x}_k, 0))$ converges to some Ricci flow on $M_\infty \times [-\tau, 0]$. Define

$$\tau_0 := \sup\{\tau \geqslant 0 \mid \exists K(\tau), \forall \rho > 0, \exists k(\rho,\tau) \text{ such that } B(\bar{x}_k, 0, \rho) \times [-\tau, 0] \text{ is unscathed and has curvature bounded by } K(\tau) \text{ for } k \geqslant k(\rho,\tau)\}.$$

We already know that $\tau_0 > 0$ and the Compactness Theorem C.3.1 enables us to construct a Ricci flow $g_\infty(\cdot)$ on $M_\infty \times (-\tau_0, 0]$, which is a limit of $(M_k, \bar{g}_k(\cdot))$ for the pointed convergence of flows. Moreover, the curvature of g_∞ is bounded by $K(\tau)$ on $(-\tau, 0]$ by passing to the limit, for all $\tau \in [0, \tau_0)$.

Step 3. There exists $Q > 0$ such that the curvature of $g_\infty(t)$ is bounded above by Q for all $t \in (-\tau_0, 0]$.

Proof. We know that $g_\infty(t)$ is nonnegatively curved and has the above-mentioned weak canonical neighbourhood property. Let us show that the conclusion of the curvature-distance theorem holds on M_∞, at points of scalar curvature > 2. For this, let $p \in M_\infty$ and $t \in (-\tau_0, 0]$ be such that $R_\infty(p,t) > 2$. Then there exists a sequence $(\bar{p}_k, t)$, where $\bar{p}_k \in M_k$, converging to (p,t) and such that $R(\bar{p}_k, t_k) \geqslant 2$ for k large enough. As a consequence, they satisfy the hypotheses of the curvature-distance theorem as explained in the proof of Step 1. Passing to the limit, we deduce that for every $A > 0$, there exists $\Lambda(A) > 0$ such that for all $q \in M_\infty$,

$$\frac{R(q,t)}{R(p,t)} \leqslant \Lambda(d_t(p,q)R(p,t)^{-1/2}). \tag{9.4}$$

Let us estimate the variation of curvatures and distances on $M_\infty \times (-\tau_0, 0]$. We recall the Harnack inequality for the scalar curvature (see Corollary C.4.2 in the Appendix)

$$\frac{\partial R}{\partial t} + \frac{R}{t + \tau_0} \geqslant 0,$$

Note that when $\tau_0 = +\infty$, this boils down to $\frac{\partial R_\infty}{\partial t} \geqslant 0$. In this case, $R(\cdot, t)$ is pointwise nondecreasing, hence $R(\cdot, t) \leqslant K(0)$ for all $t \leqslant 0$.

Assume now that $\tau_0 < +\infty$. Then

$$R(\cdot, t) \leqslant K(0)\frac{\tau_0}{t + \tau_0}.$$

The Ricci curvature, which is positive, satisfies a similar estimate, which implies

$$\text{const.}\left(\frac{\tau_0}{t+\tau_0}\right) g_\infty \leqslant \frac{\partial g_\infty}{\partial t} \leqslant 0,$$

thus

$$\text{const.}\sqrt{\frac{\tau_0}{t+\tau_0}} \leqslant \frac{\partial d_t(x,y)}{\partial t} \leqslant 0.$$

By integration we obtain

$$|d_t(x,y) - d_0(x,y)| \leqslant C.$$

where $C = C(\tau_0)$. Since $(M_\infty, g_\infty(0))$ is noncompact and nonnegatively curved, it is asymptotically conical (see [BGS85], p. 58–59, and [KL08], Appendix B). Hence

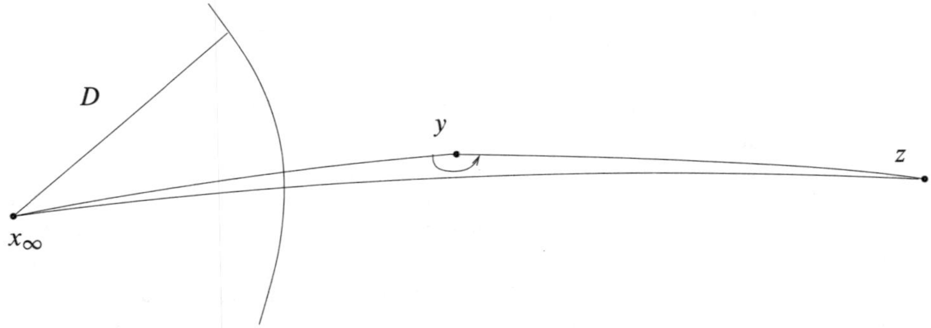

Figure 9.3. Every $y \notin B(x_\infty, 0, D)$ is almost the middle of a segment.

there exists $D > 0$ such that for all $y \in M_\infty$, if $d_0(x_\infty, y) > D$, then there exists $z \in M_\infty$ such that

$$d_0(y,z) = d_0(x_\infty, y) \quad \text{and} \quad d_0(x_\infty, z) \geqslant 1.99 d_0(x,y) \tag{9.5}$$

i.e., the points x_∞, y, z are almost aligned.

By comparison with Euclidean space, if D is large enough we have $\pi - 1/100 \leqslant \measuredangle(x_\infty y z) \leqslant \pi$ for any such y and z. Observe that since $|d_t(\cdot,\cdot) - d_0(\cdot,\cdot)| < C$ uniformly in t, we can choose $D \gg C$ large enough so that, for all $t \in (-\tau_0, 0]$, we have

$$|d_t(y,z) - d_t(x_\infty, y)| < 2C \quad \text{and} \quad d_t(x_\infty, z) \geqslant 1.98 d_t(x_\infty, y),$$

thus $\measuredangle_t(x_\infty y z) \geqslant \pi - 1/50$.

Let us show that the scalar curvature of $g_\infty(t)$ is uniformly bounded above on $M_\infty \setminus B(x_\infty, 0, 2D)$. We argue by contradiction. Suppose that there exists (y_i, t_i) such that $d_0(x_\infty, y_i) > 2D$ and such that $R(y_i, t_i) \to +\infty$ as i tends to ∞. Each (y_i, t_i) has a weak $(3\varepsilon_0, 3C_0)$-canonical neighbourhood. If this neighbourhood is a $(3\varepsilon_0, C_0)$-cap, we let y_i' be a centre of its $3\varepsilon_0$-neck. Since the diameter for $g(t_i)$ of the cap is small (less than $12C_0 R(y_i, t_i)^{-1/2} < C$ for large i), we still have (for large i):

$$d_0(x_\infty, y_i') \geqslant d_{t_i}(x_\infty, y_i') - C \geqslant d_{t_i}(x_\infty, y_i) - 2C \geqslant d_0(x_\infty, y_i) - 3C > D.$$

Furthermore, the curvature $R(y_i', t_i) \geqslant \frac{1}{3C_0} R(y_i, t_i)$ tends to $+\infty$. Up to replacing y_i by y_i', we may assume that there exists a sequence (y_i, t_i) such that $d_0(x_\infty, y_i) > D$, $R(y_i, t_i) \to +\infty$ and (y_i, t_i) is the centre of a $3\varepsilon_0$-neck U_i. For each $i \in \mathbb{N}$, pick $z_i \in M_\infty$ satisfying (9.5). By the above remark, $\measuredangle_{t_i}(x_\infty y_i z_i) \geqslant \pi - \frac{1}{50}$.

The points x_∞ and z_i being outside U_i, we deduce that $[x_\infty y_i]$ and $[y_i z_i]$ each intersect some component ∂U_i. Let S_i be the middle sphere of U_i. It separates M_∞ into two connected components, one containing x_∞, the other z_i.

Now $\mathrm{diam}(S_i, t_i) \to 0$ since $R(y_i, t_i) \to +\infty$. Since $g_\infty(t)$ is nonnegatively curved, distances are nonincreasing in t. As a consequence, $\mathrm{diam}(S_i, 0) \to 0$ with i. At time 0 the curvature bounds and the hypothesis of κ-noncollapsing implies a uniform lower bound on the injectivity radius. Then for large i, the diameter of S_i is less than the injectivity radius of $g_\infty(0)$. This implies that S_i bounds a 3-ball of radius less than the injectivity radius. Since x_∞ and z_i are far from S_i, this is a contradiction.

This implies that the curvature is bounded outside the $g_\infty(0)$-ball of radius $2D$ around x_∞. We deduce a uniform curvature bound on $(-\tau_0, 0]$ in the ball using equation (9.4) (see p. 105). □

Step 4. $\tau_0 = +\infty$.

Proof. Consider a subsequence $(M_k \times (-\tau_0, 0], \bar{g}_k(t), (x_k, 0))$ that converges to $(M_\infty \times (-\tau_0, 0], g_\infty(t), (x_\infty, 0))$. Since the limit has scalar curvature bounded above by Q we deduce that (up to replacing Q by $Q + 1$), for all $0 < \tau < \tau_0$, for all $\rho > 0$, there exists $k'(\tau, \rho) \in \mathbb{N}$ such that for all $k \geqslant k'(\tau, \rho)$, the parabolic neighbourhood $P(\bar{x}_k, 0, \rho, -\tau)$ is unscathed and has scalar curvature $\leqslant Q$.

Suppose that $\tau_0 < +\infty$ and let $0 < \sigma < (4C_0(Q + 2))^{-1}$. Then up to extracting a subsequence, for every $K > 0$, there exists $\rho = \rho(\sigma, K)$ such that $P_k := P(\bar{x}_k, 0, \rho, -(\tau_0 + \sigma))$ is scathed or does not have curvature bounded above by K.

Set $K := 2(Q + 2)$ and $\rho := \rho(\sigma, K)$. If $k \geqslant k'(\tau_0 - \sigma, \rho)$, then P_k is scathed. Indeed, if $k \geqslant k'(\tau_0 - \sigma, \rho)$, we have $R \leqslant Q$ on $P_{\bar{g}_k}(x_k, 0, \rho, -\tau_0 + \sigma))$. If P_k is unscathed, the Curvature-Time Lemma 6.1.3 applied between $-\tau_0 + \sigma$ and $-\tau_0 - \sigma$ (since $2\sigma \leqslant (2C_0(Q + 2))^{-1}$) implies that $R \leqslant 2(Q + 2)$ on P_k, which excludes the second alternative.

Thus there exists $x'_k \in B(x_k, 0, \rho)$, and $t_k \in [-\tau_0 - \sigma, -\tau_0 + \sigma]$, assumed to be maximal, such that $\bar{g}_k(t_k) \neq \bar{g}_{k+}(t_k)$ at x'_k. Since $B(\bar{x}_k, 0, \rho) \times (t_k, 0]$ is unscathed, the above argument shows that $R \leqslant 2(Q + 2)$ on this set, for all sufficiently large k. This implies an upper bound on the Riemann tensor on this set and hence by Lemma 9.2.1 the parabolic neighbourhood $P(\bar{x}_k, 0, \rho, t_k)$ is unscathed. This contradicts the definition of t_k. □

We can now finish the proof of Proposition B: since $\tau_0 = +\infty$, the Ricci flow $(M_\infty, g_\infty(\cdot))$ is defined on $(-\infty, 0]$, and has bounded, nonnegative curvature operator. Moreover, the rescaled evolving metric $\bar{g}_k(\cdot)$ is κ-noncollapsed on scales less than $\sqrt{R(x_k, b_k)}$, so passing to the limit we see that $g_\infty(\cdot)$ is κ-noncollapsed on all scales. The metric $g_\infty(\cdot)$ is not flat since it has scalar curvature 1 at the point $(x_\infty, 0)$.

This shows that $(M_\infty, g_\infty(\cdot))$ is a κ-solution. By Theorem 4.2.11 and the choice of constants in Section 5.1, every point of M_∞ has a $(\frac{\varepsilon_0}{2}, \frac{C_0}{2})$-canonical neighbourhood. Hence for sufficiently large k, $(\bar{x}_k, 0)$ has an (ε_0, C_0)-canonical neighbourhood. This is the final contradiction.

Notes. Our proof of Proposition B is inspired by [Per03b], Section 5.4, and [KL08], Section 80. It also uses arguments from [Per02], Theorem 12.1, and [KL08], Section 52.

Chapter 10

κ-noncollapsing and the proof of Proposition C

The goal of this chapter is to prove Proposition C, whose statement we now recall:

Proposition C. *For all $T > 0$ there exists $\kappa = \kappa(T)$ such that for all $r \in (0, 10^{-3})$ there exists $\bar{\delta}_C = \bar{\delta}_C(T, r) \in (0, \delta_0)$ such that the following holds.*

For all $\delta \leqslant \bar{\delta}_C$, any Ricci flow with (r, δ)-bubbling-off defined on $[0, b]$ with $b \leqslant T$ and having normalised initial condition satisfies $(\mathrm{NC})_\kappa$.

This statement is part of what one can calls the κ-noncollapsing theory of Ricci flow. This notion is defined in all dimensions as follows:

Definition 10.0.1. Let X be a smooth n-dimensional manifold, $\{g(t)\}_{t\in I}$ be a Ricci flow on X and ρ be a positive number. We say that $g(\cdot)$ is *κ-noncollapsed on the scale ρ* if for all $(x, t) \in X \times I$, if $|\mathrm{Rm}| \leqslant \rho^{-2}$ on $P(x, t, \rho, -\rho^2)$, then $\operatorname{vol} B(x, t, \rho) \geqslant \kappa\rho^n$.

We extend the definition of a normalised metric to all dimensions in the obvious way:

Definition 10.0.2. Let X be a smooth n-manifold. A metric g on X is *normalised* if $\operatorname{tr}\mathrm{Rm}^2 \leqslant 1$ and each ball of radius one has volume at least half of the volume of the unit ball in Euclidean n-space.

A general κ-noncollapsing result for Ricci flows is the following:

Theorem 10.0.3 (κ-noncollapsing). *For every $T > 0$ there exists $\kappa = \kappa(T) > 0$ such that if X is a smooth manifold and $\{g(t)\}_{t\in[0,T]}$ is a complete Ricci flow of bounded sectional curvature with normalised initial condition on X, then it is κ-noncollapsed on all scales less than or equal to* 1.

The proof relies on the monotonicity of a functional on space-time, Perelman's reduced volume $\widetilde{V}$. The necessary background is reviewed in Section 10.1.

For the long-time study of Ricci flow with bubbling-off (Part III), we shall need a refined version of the κ-noncollapsing result where κ does not depend on t, but depends on the distance to the thick part (cf. Theorem 10.4.1).

This chapter is organised as follows. Section 10.1 is devoted to some preliminaries. We prove Theorem 10.0.3 in Section 10.2. In Section 10.3, we modify this proof to deduce Proposition C. More precisely, we prove a slightly more general version of this proposition, Proposition C′, which will be useful later. In Section 10.4, we prove the long-time κ-noncollapsing result 10.4.1.

10.1 Preliminaries

10.1.1 Basic facts on κ-noncollapsing

Let $v_k^n(\rho)$ denote the volume of a ball of radius ρ in the model space of curvature k and dimension n. When the superscript n is dropped, it is understood that $n = 3$.

Definition 10.1.1. Let $\kappa > 0$. One says that a metric ball $B(x,\rho)$ is κ-noncollapsed if $|\mathrm{Rm}| \leqslant \rho^{-2}$ on $B(x,\rho)$ and if $\operatorname{vol} B(x,\rho) \geqslant \kappa\rho^n$. Similarly, a parabolic ball $P(x,t,\rho,-\rho^{-2})$ is κ-noncollapsed if $|\mathrm{Rm}| \leqslant \rho^{-2}$ on $P(x,t,\rho,-\rho^{-2})$ and if $\operatorname{vol} B(x,t,\rho) \geqslant \kappa\rho^n$.

Lemma 10.1.2. *The following properties always hold:*

(i) *If $B(x,\rho)$ is κ-noncollapsed, then for all $\rho' \in (0,\rho)$, $B(x,\rho')$ is $c_n\kappa$-noncollapsed, where*
$$c_n := \frac{v_0^n(1)}{v_{-1}^n(1)}.$$
The same property holds for $P(x,t,\rho',-\rho'^2) \subset P(x,t,\rho,-\rho^2)$.

(ii) *Let r, δ be surgery parameters and $g(\cdot)$ be a Ricci flow with (r,δ)-bubbling-off. Assume that $P_0 = P(x_0,t_0,\rho_0,-\rho_0^2)$ is a scathed parabolic neighbourhood such that $|\mathrm{Rm}| \leqslant \rho_0^{-2}$ on P_0. Then P_0 is $e^{-12}\kappa_{\mathrm{st}}/2$-noncollapsed.*

Remark 10.1.3. (1) The constant κ_{st} is defined in Proposition 4.3.3. A consequence of (ii) is that in order to prove κ-noncollapsing for a Ricci flow with (r,δ)-bubbling-off, it is sufficient to consider unscathed parabolic neighbourhoods.

(2) If some ball $B(y,\rho)$ is contained in some (ε_0,C_0)-canonical neighbourhood and satisfies $|\mathrm{Rm}| \leqslant \rho^{-2}$, then $B(y,\rho)$ is C_0^{-1}-noncollapsed on the scale ρ by equation (4.1), p. 39.

Proof of Lemma 10.1.2. Let us prove assertion (i): since $\rho^{-2} \leqslant \rho'^{-2}$, we have $|\mathrm{Rm}| \leqslant \rho'^{-2}$ on $B(x,\rho') \subset B(x,\rho)$. We are left to check that $\operatorname{vol} B(x,\rho') \geqslant c_n\kappa\rho'^n$.

The function $v_{-k^2}^n$ satisfies $v_{-k^2}^n(\rho) = k^{-n} v_{-1}^n(k\rho)$ for $k > 0$. Since the sectional curvature is greater than or equal to $-\rho^{-2}$ on $B(x,\rho)$, the Bishop–Gromov inequality (Theorem B.1.1) gives
$$\frac{\operatorname{vol} B(x,\rho')}{\operatorname{vol} B(x,\rho)} \geqslant \frac{v_{-\rho^{-2}}^n(\rho')}{v_{-\rho^{-2}}^n(\rho)} = \frac{v_{-1}^n(\rho'/\rho)}{v_{-1}^n(1)} \geqslant \frac{v_0^n(\rho'/\rho)}{v_{-1}^n(1)} = \frac{v_0^n(1)}{v_{-1}^n(1)}\left(\frac{\rho'}{\rho}\right)^n.$$

Using again the assumption that $B(x,\rho)$ is κ-noncollapsed, we get
$$\operatorname{vol} B(x,\rho') \geqslant \frac{v_0^n(1)}{v_{-1}^n(1)}\kappa\rho'^n.$$

This proves assertion (i).

We now turn to the proof of assertion (ii). Let us recall that we are now working in dimension 3. Let (x,t) be a point in P_0 where $g_+ \neq g$. Assume that t is maximal such that there exists x with this property. Let us consider the restriction of $g(\cdot)$ to $[t, t_0]$, in the sense of Definition 8.1.1. From the definition of Ricci flow with (r,δ)-bubbling-off there exists a δ-almost standard cap U with respect to $g(t)$ such that $x \in U$. The scalar curvature of $g(t)$ at x is then comparable to h^{-2} (where h is the cutoff parameter).

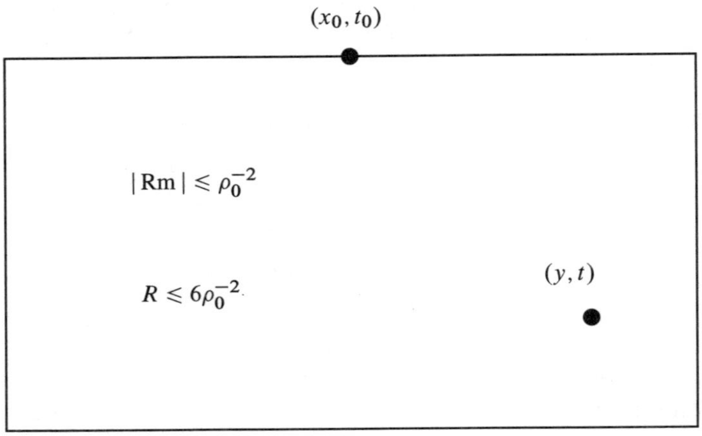

Since $|\mathrm{Rm}| \leqslant \rho_0^{-2}$ on $P_0 \supset B(x_0, t_0, \rho_0) \times [t, t_0]$, we get, taking the limit:

$$\frac{h^{-2}}{2} \leqslant R(x,t) \leqslant 6\rho_0^{-2} \Rightarrow \rho_0 < 4h.$$

Since $B(x_0, t_0, \rho_0) \times [t, t_0]$ is unscathed, we get, using the distance-distortion lemma (Lemma 2.2.7), the following estimate on $B(x_0, t_0, \rho_0)$,

$$e^{-4} g(t) \leqslant g(t_0) \leqslant e^4 g(t),$$

hence

$$B(x_0, t, e^{-2}\rho_0) \subset B(x_0, t_0, \rho_0),$$

and

$$d_t(x_0, x) \leqslant e^2 d_{t_0}(x_0, x) < e^2 \rho_0.$$

It follows from the definition of an almost standard cap that $(U, h^{-2} g(t))$ is $\delta'(\delta)$-close to $B(p_0, 5 + \delta^{-1}) \subset \mathcal{S}_0$. Let p be the tip of U, so that $U \supset B(p, t, h\delta^{-1})$. We have

$$d_t(x_0, p) \leqslant d_t(x_0, x) + d_t(x, p) \leqslant e^2 \rho_0 + 5h \leqslant (4e^2 + 5)h,$$

which implies

$$B(x_0, t, e^{-2}\rho_0) \subset B(p, t, (4e^{-2} + 4e^2 + 5)h) \subset U,$$

since $\delta^{-1} \geqslant 4e^{-2} + 4e^2 + 5$.

By δ'-closeness of $(U, h^{-2}g(t))$ with $B(p_0, 5+\delta^{-1}) \subset \mathcal{S}_0$, which is κ_{st}-noncollapsed on all scales by Proposition 4.3.3, we get:

$$\operatorname{vol}_{g(t)} B(x_0, t, e^{-2}\rho_0) \geqslant \frac{\kappa_{\mathrm{st}}}{2}(e^{-2}\rho_0)^3.$$

Indeed, $\operatorname{vol}_{g(t)} B(x_0, t, e^{-2}\rho_0)(h^3 \operatorname{vol}_{g_0} B(p_0, e^{-2}h^{-1}\rho_0))^{-1}$ is close to 1. Moreover $r := e^{-2}h^{-1}\rho_0 \leqslant 4e^{-2} \leqslant 1$. Hence the curvature condition is satisfied on $B(p_0, r)$, since $|\mathrm{Rm}| \leqslant 1 \leqslant r^{-2}$ on $\mathcal{S}_0$. Let us now estimate the $g(t_0)$-volume of $B(x_0, t_0, \rho_0)$. Since $B(x_0, t, e^{-2}\rho_0) \subset B(x_0, t_0, \rho_0)$, we derive:

$$\begin{aligned}\operatorname{vol}_{g(t_0)} B(x_0, t_0, \rho_0) &\geqslant \operatorname{vol}_{g(t_0)} B(x_0, t, e^{-2}\rho_0)\\ &\geqslant (e^{-2})^3 \operatorname{vol}_{g(t)} B(x_0, t, e^{-2}\rho_0)\\ &\geqslant e^{-12}\frac{\kappa_{\mathrm{st}}}{2}\rho_0^3.\end{aligned}$$

This completes the proof of assertion (ii). □

10.1.2 Perelman's $\mathcal{L}$-length

Let X be a smooth n-manifold, and $g(\cdot)$ be an evolving metric on X.

Definition 10.1.4 ([Per02]). Let τ_0 be a positive number and $\gamma\colon [0, \tau_0] \to X$ be a piecewise smooth curve. The $\mathcal{L}_{\tau_0}$-length of γ is defined by the following formula

$$\mathcal{L}_{\tau_0}(\gamma) := \int_0^{\tau_0} \sqrt{\tau}(R(\gamma(\tau)) + |\dot{\gamma}(\tau)|^2)d\tau,$$

where the scalar curvature as well as the norm are computed using the evolving metric $g(\tau)$.

We now consider a Ricci flow $g(\cdot)$ on X defined on a compact interval $[0, T]$ and set $\tau(t) := T - t$, so that $g(\tau) = g(T-t)$. Let us fix $p \in M$, and $\tau > 0$. We denote by $L(q, \bar{\tau})$ the minimum of the $\mathcal{L}_\tau$-length of curves γ going from $(p, 0)$ to (q, τ), i.e., such that $\gamma(0) = p$ and $\gamma(\tau) = q$.

As in classical Riemannian geometry, there is a notion of minimizing $\mathcal{L}_\tau$-geodesics and hence one can define an $\mathcal{L}_\tau$-exponential map. More precisely, for $v \in T_pM$, $\mathcal{L}_\tau \exp(v) := \gamma(\tau)$, where γ is an $\mathcal{L}_\tau$-geodesic starting at $(p, 0)$ such that $\sqrt{s}\dot{\gamma}(s) \underset{s\to 0}{\longrightarrow} v$. We denote by $J(v, \tau)$ the Jacobian at v of the map $v \to \mathcal{L}_\tau \exp(v)$. We then have the following definitions.

Definition 10.1.5. (i) The *reduced length* is

$$\ell(q, \tau_0) := \frac{1}{2\sqrt{\tau_0}} L(q, \tau_0).$$

(ii) We call *reduced volume* the quantity

$$\tilde{V}(\tau) = \int_M \tau^{-n/2} \exp\big(-\ell(q,\tau)\big) dv_{g(\tau)}(q).$$

We now list some fundamental results about these quantities for later use (see [Ye08a], [Ye08b]).

Theorem 10.1.6 (Monotonicity of reduced length). *For any $\mathcal{L}$-geodesic the map*

$$\tau \to \tau^{-n/2} \exp(-\ell(\gamma(\tau),\tau)) J(\tau)$$

is nonincreasing.

Corollary 10.1.7 (Monotonicity of reduced volume). *The reduced volume is a non-increasing function of τ.*

Here are a few useful estimates (see [KL08], Sections 16, 26).

Lemma 10.1.8. (i) *Assume that* $|\mathrm{Rm}| \leqslant (\rho_0)^{-2}$ *on* $P(p,T,\rho_0,-\rho_0^2)$ *and let* $\gamma \subset P(p,T,\rho_0/2,-\rho_0^2/2)$ *be an $\mathcal{L}$-geodesic issued from p. Then*

$$\left| \frac{d}{d(\tau/\rho_0^2)} \sqrt{\tau}|\dot{\gamma}| \right| \leqslant C(\sqrt{\tau}|\dot{\gamma}| + (\tau/\rho_0^2)^{1/2}), \tag{10.1}$$

for γ an $\mathcal{L}$-geodesic satisfying $\lim_{\tau\to 0} \sqrt{\tau}\dot{\gamma}(\tau) = v$. *The constant $C(n)$ is a generic constant depending only on n.*

(ii) *If* $\lim_{\tau\to 0} \sqrt{\tau}\dot{\gamma}(\tau) = v$, *then*

$$\lim_{\tau\to 0} \tau^{-n/2} \exp(-\ell(v,\tau)) J(v,\tau) = 2^n e^{-|v|^2_{g(t_0)}}. \tag{10.2}$$

The following lemma will also be essential (see [KL08], Lemma 24.3).

Lemma 10.1.9. *For all $\tau \in [0,\tau_0)$, there exists a point $q \in X$ such that*

$$\ell(q,\tau) \leqslant n/2. \tag{10.3}$$

10.2 Proof of Theorem 10.0.3

Fix an n-manifold X, a number $T > 0$ and a complete Ricci flow $g(\cdot)$ of bounded sectional curvature on X defined on $[0,T]$ such that $g(0)$ is normalised. Let (x_0,t_0) be a point in spacetime. Pick $\rho_0 \in (0,1]$ and consider the parabolic ball $P_0 := P(x_0,t_0,\rho_0,-\rho_0^2)$. Suppose that $|\mathrm{Rm}| \leqslant {\rho_0}^{-2}$ on P_0. Set $B_0 := B(x_0,t_0,\rho_0)$ and

$$\kappa := \frac{\mathrm{vol}\, B(x_0,t_0,\rho_0)}{\rho_0^n} = \frac{\mathrm{vol}_{g(t_0)}(B_0)}{\rho_0^n}.$$

Our task is to give a lower bound for κ which depends only on T.

By the doubling time estimate C.1.1, we have $|\mathrm{Rm}| \leqslant 2$ on $[0, 1/16]$. This implies by standard comparison arguments a uniform bound from below for κ on $[0, 1/16]$. (cf. Proposition 5.1.8). Hence we can assume $t_0 > 1/16$.

Set $\tau_0 := \kappa^{1/n}{\rho_0}^2 \leqslant \kappa^{1/n}$ and assume that $\kappa < \left(\frac{1}{16}\right)^n$, which implies $t_0 - \tau_0 > 0$.

By Corollary 10.1.7, we have $\widetilde{V}(t_0) \leqslant \widetilde{V}(\tau_0)$. We shall now bound $\widetilde{V}(\tau_0)$ from above by $C\sqrt{\kappa}$ and $\widetilde{V}(t_0)$ from below by $Ct_0^{-n/2}$. This will give the required lower bound on κ.

Bounding $\widetilde{V}(\tau_0)$ from above. We use the $\mathcal{L}_{\tau_0}$-exponential map to write $\widetilde{V}(\tau_0)$ as an integral on $T_{x_0}X$ endowed with the metric $g(x_0, t_0)$. We then cut $T_{x_0}X$ in two parts.

First step: cutting $T_{x_0}X$. First we show that the vectors $v \in T_{x_0}X$ of norm less than $(10\kappa^{1/2n})^{-1}$ generate $\mathcal{L}$-geodesics which stay in B_0 if κ is small enough. More precisely, Inequality 10.1 gives

$$\left|\frac{d}{d(\tau/\tau_0)}(\kappa^{1/2n}\sqrt{\tau}|\dot{\gamma}|)\right| \leqslant C\kappa^{1/n}(\kappa^{1/2n}\sqrt{\tau}|\dot{\gamma}| + \kappa^{1/n}(\tau/\tau_0)^{1/2}).$$

We limit ourselves to $0 \leqslant \tau/\tau_0 \leqslant 1$ and $\lim_{\tau\to 0}\kappa^{1/2n}\sqrt{\tau}|\dot{\gamma}(\tau)| = \kappa^{1/2n}|v| \leqslant 1/10$. We then set $f(\tau/\tau_0) = \kappa^{1/2n}\sqrt{\tau}|\dot{\gamma}(\tau)|$; the function f defined on $[0, 1]$ satisfies the differential inequality

$$|f'(x)| \leqslant C\kappa^{1/n}(f(x) + \kappa^{1/n}\sqrt{x}) \leqslant C\kappa^{1/n}(f(x) + 1).$$

By Gronwall's lemma, there exists $\kappa(n)$ such that if $\kappa \leqslant \kappa(n)$, then $\kappa^{1/2n}\sqrt{\tau}|\dot{\gamma}(\tau)| \leqslant \frac{1}{9}$ for all $\tau \in [0, \tau_0]$. Thus

$$\int_0^{\tau_0} |\dot{\gamma}(\tau)|_{g(t_0-\tau)}d\tau = \kappa^{-1/2n}\int_0^{\tau_0}\kappa^{1/2n}\sqrt{\tau}|\dot{\gamma}(\tau)|\frac{d\tau}{\sqrt{\tau}} \leqslant \frac{2}{9}\rho_0.$$

On $P(x_0, t_0, \rho_0, -\kappa^{1/n}{\rho_0}^2)$ the curvature bound implies $\left|\frac{\partial g}{\partial t}\right| \leqslant C{\rho_0}^{-2}$, so for all $t_1, t_2 \in [t_0 - \kappa^{1/n}{\rho_0}^2, t_0] = [t_0 - \tau_0, t_0]$, the metrics $g(t_1), g(t_2)$ are $e^{C{\rho_0}^{-2}\kappa^{1/n}{\rho_0}^2} = e^{C\kappa^{1/n}}$-bi-Lipschitz in that neighbourhood. As long as γ stays in this neighbourhood, its $g(t_0)$-length is less than $e^{C\kappa^{1/n}}\frac{2}{9}\rho_0 \leqslant \frac{1}{3}\rho_0$ if κ is small enough. In conclusion, the image by the $\mathcal{L}_{\tau_0}$-exponential map of $\{v \in T_{x_0}X \mid |v|_{g(t_0)} \leqslant \frac{1}{10}\kappa^{-1/2n}\}$ is contained in B_0 for $\kappa \leqslant \kappa(n)$.

Second step: computing the reduced volume on $T_{x_0}X$. From the previous paragraph we deduce

$$I := \int_{|v|_{g(t_0)}\leqslant\frac{1}{10}\kappa^{-1/2n}} \tau_0^{-n/2}e^{-\ell(v,\tau_0)}J(v, \tau_0)dv \leqslant \int_{B_0}\tau_0^{-n/2}e^{-\ell(q,\tau_0)}dq,$$

where dq is the volume form of $g(t_0 - \tau_0)$.

We now bound $\ell(q,\tau_0)$ from below for $q \in B_0$. Let $\gamma\colon [0,\tau_0] \to X$ be an $\mathcal{L}_{\tau_0}$-geodesic such that $\gamma(0) = x_0$ and $\gamma(\tau_0) = q$ and which stays in B_0. Then we have

$$\mathcal{L}_{\tau_0}(\gamma) \geqslant \int_0^{\tau_0} \sqrt{\tau} R(\gamma(\tau)) d\tau \geqslant -n(n-1){\rho_0}^{-2} \int_0^{\tau_0} \sqrt{\tau} d\tau = -\frac{2}{3} n(n-1){\rho_0}^{-2} \tau_0^{3/2},$$

and, if γ is minimising,

$$\ell(q,\tau_0) = \frac{L_{\tau_0}(\gamma)}{2\sqrt{\tau_0}} = \frac{\mathcal{L}_{\tau_0}(\gamma)}{2\sqrt{\tau_0}} \geqslant -\frac{n(n-1)}{3} \tau_0 {\rho_0}^{-2} = -\frac{n(n-1)}{3} \kappa^{1/n}.$$

Thus,

$$I \leqslant e^{\frac{n(n-1)}{3}\kappa^{1/n}} {\tau_0}^{-n/2} \int\limits_{B_0} dq \leqslant e^{\frac{n(n-1)}{3}\kappa^{1/n}} {\tau_0}^{-n/2} \operatorname{vol}_{g(t_0-\tau_0)}(B_0),$$

and

$$I \leqslant e^{\frac{n(n-1)}{3}\kappa^{1/n}} \tau_0^{-n/2} e^{C\kappa^{1/n}} \operatorname{vol}_{g(t_0)}(B_0) \leqslant e^{C\kappa^{1/n}} \sqrt{\kappa}.$$

Define

$$I' := \int_{|v|_{g(t_0)} \geqslant \frac{1}{10}\kappa^{-1/2n}} \tau_0^{-n/2} e^{-\ell(v,\tau_0)} J(v,\tau_0) dv.$$

The integrand is nondecreasing in τ (10.1.6). In particular it is smaller than its limit at $\tau = 0$. Therefore, from (10.2) we deduce

$$I' \leqslant 2^n \int_{|v|_{g(t_0)} \geqslant \frac{1}{10}\kappa^{-1/2n}} e^{-|v|^2_{g(t)}} dv \leqslant e^{-\frac{1}{10}\kappa^{-1/2n}}$$

if $\kappa \leqslant \kappa(n)$.

In conclusion,

$$\widetilde{V}(\tau_0) \leqslant e^{C\kappa^{1/n}} \kappa^{1/2} + e^{-\frac{1}{10}\kappa^{-1/2n}} \leqslant C\kappa^{1/2}, \tag{10.4}$$

for some universal $\kappa \leqslant \kappa(n)$.

Bounding $\widetilde{V}(t_0)$ from below. We need to bound $\ell(q,t_0)$ from above on a ball of controlled volume.

We apply Lemma 10.1.9 to obtain a point $q_0 \in X$ such that Inequality (10.3) holds. For $q \in X$,

$$\ell(q,t_0) = \frac{L(q,t_0)}{2\sqrt{t_0}} = \frac{1}{2\sqrt{t_0}} \inf_\alpha \{\mathcal{L}_{t_0}(\alpha)\}$$

where α is a curve parametrised by τ such that $\alpha(0) = x_0$ and $\alpha(t_0) = q$. For $q \in B(q_0, 0, 1)$, let γ be such a curve, obtained by concatenating a minimising $\mathcal{L}_{t_0-1/16}$-geodesic between x_0 and q_0, and a minimising $g(0)$-geodesic connecting q_0 to q (see

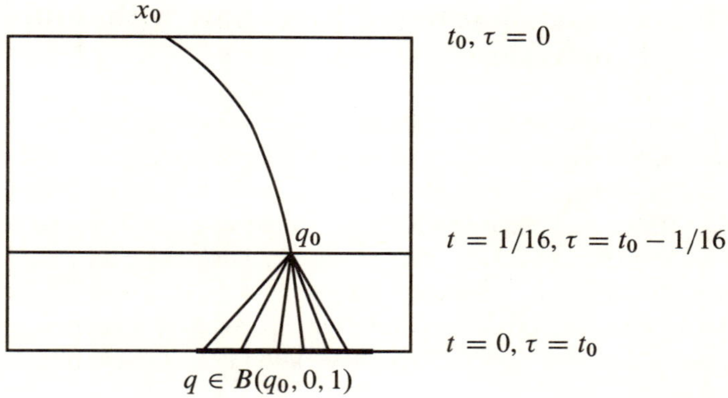

Figure 10.1. Concatenating an $\mathcal{L}$-geodesic with $g(0)$-geodesics.

Figure 10.1). Then

$$\ell(q,t_0) = \frac{L(q,t_0)}{2\sqrt{t_0}} \leqslant \frac{1}{2\sqrt{t_0}} L(q_0, t_0 - 1/16) + \frac{1}{2\sqrt{t_0}} \int_{t_0-1/16}^{t_0} \sqrt{\tau}(R_{g(t_0-\tau)} + |\dot{\gamma}(\tau)|^2_{g(t_0-\tau)})d\tau,$$

which leads to

$$\ell(q,t_0) \leqslant \sqrt{\frac{t_0 - 1/16}{t_0}} \ell(q_0, t_0 - 1/16) + \frac{1}{2}\left(\frac{2n(n-1)}{16} + 16C(n)d_0^2(q,q_0)\right)$$

since for $t \in [0, 1/16]$ (i.e., $\tau \in [t_0 - 1/16, t_0]$) all metrics $g(t)$ satisfy $R \leqslant 2n(n-1)$ and are $C(n)$-Lipschitz equivalent for some universal constant $C(n)$. This comes from the fact that $|\mathrm{Rm}| \leqslant 2$ on this interval. It follows that for all $q \in B(q_0, 0, 1)$ and $t_0 \geqslant 1/16$, we have

$$\ell(q,t_0) \leqslant n/2 + C(n).$$

By integration we get

$$\int_X \tau^{-n/2} e^{-\ell(q,t_0)} dq \geqslant \int_{B(q_0,0,1)} t_0^{-n/2} e^{-\ell(q,t_0)} dq \geqslant t_0^{-n/2} C \operatorname{vol} B(q_0, 0, 1).$$

Recall that $\inf_{x \in X} \operatorname{vol} B(x, 0, 1) \geqslant \frac{1}{2} \upsilon_0(1)$. Thus

$$\widetilde{V}(t_0) = \int_X t_0{}^{-n/2} e^{-\ell(q,t_0)} dq \geqslant \frac{C}{2t_0{}^{n/2}} \upsilon_0(1),$$

and from (10.4) and the monotonicity of the reduced volume we deduce

$$\kappa \geqslant \frac{C}{t_0{}^n} \geqslant \frac{C}{T^n},$$

if $\kappa \leqslant \kappa(n)$. □

10.3 κ-noncollapsing of Ricci flow with bubbling-off: proof of Proposition C

We now turn to the proof of Proposition C, which deals with κ-noncollapsing properties for Ricci flows with (r, δ)-bubbling-off. In fact, we shall prove a stronger result, which applies to Ricci flows with $r(\cdot), \delta(\cdot)$-bubbling-off (see Definition 5.4.6). Indeed, in Part III we shall need another κ-noncollapsing result, where κ is independent of time (see Theorem 11.1.3a)). The proof of this stronger theorem, which is given in Section 10.4 below, uses the same arguments as for Proposition C, with some modifications due to the variation of the parameters. To avoid redundancy, we prove directly Proposition C in the context of Ricci flow with $r(\cdot), \delta(\cdot)$-bubbling-off. More precisely, we have the

Proposition C′. *For any $T \geqslant 0$ there exists $\kappa := \kappa(T)$, such that for any $\hat{r} \in (0, 10^{-3})$, there exists $\bar{\delta}_{C'} := \bar{\delta}_{C'}(T, \hat{r}) \in (0, \delta_0)$ with the following property. Let $g(\cdot)$ be a Ricci flow with $r(\cdot), \delta(\cdot)$-bubbling-off, with normalised initial condition, defined on some interval $[0, b)$. Assume that $\hat{r} \leqslant r(\cdot) \leqslant 10^{-3}$, $0 < \delta(\cdot) \leqslant \bar{\delta}_{C'}$ and $b \leqslant T$. Then $g(\cdot)$ satisfies* $(\mathrm{NC})_\kappa$.

Note that letting $r(\cdot) \equiv r$ and $\delta(\cdot) \equiv \delta$ gives Proposition C.

Before we start the proof of Proposition C′, we prove a lemma, which we state in a slightly more general setting for later reference.

Lemma 10.3.1 (Scathed curves have large $\mathcal{L}$-length). *Let $\hat{r} \in (0, 10^{-3})$ and Δ, Λ be positive numbers. Then there exists $\bar{\delta} = \bar{\delta}(\hat{r}, \Delta, \Lambda) \in (0, \delta_0)$ with the following property. Let $g(\cdot)$ be a $(r(\cdot), \delta(\cdot))$-Ricci flow with bubbling-off defined on an interval $I = [a, a + \Delta]$, where $a \geqslant 0$, $\delta(t) \leqslant \bar{\delta}$ and $r(t) \geqslant \hat{r}$ on I. Let $(x_0, t_0) \in M \times I$ and $\rho_0 \geqslant \hat{r}$ be such that $P_0 := P(x_0, t_0, \rho_0, -\rho_0^2) \subset M \times I$ is unscathed and $|\mathrm{Rm}| \leqslant \rho_0^{-2}$ on P_0.*

Let $t_1 \in [0, t_0]$ and $\gamma \colon [t_1, t_0] \to M$ be such that $\gamma(t_0) = x_0$ and γ is scathed, that is, there exists $t \in [t_1, t_0)$ such that $\gamma(t) \in \Sigma_t$. Then

$$\mathcal{L}_{(t_0 - t_1)}(\gamma) \geqslant \Lambda.$$

Proof. Without loss of generality we assume $\Lambda > 1$. First note that $R \geqslant -6$ by the curvature pinched toward positive assumption. One then has

$$\int_{t_1}^{t_0} \sqrt{t_0 - t}\, R(\gamma(t), t)\, dt \geqslant -4[\tau^{3/2}]_0^{\tau_1} \geqslant -4\Delta^{3/2}.$$

Hence it is sufficient to prove one of the two inequalities

$$\int_{t_1}^{t_0} \sqrt{t_0 - t}\, R(\gamma(t), t)\, dt \geqslant \Lambda, \tag{10.5}$$

$$\int_{t_1}^{t_0} \sqrt{t_0 - t_1}\, |\dot{\gamma}(t)|^2_{g(t)}\, dt \geqslant \Lambda + 4\Delta^{3/2} =: \Lambda'. \tag{10.6}$$

Intuitively, those two conditions mean that a curve has large $\mathcal{L}$-length if it stays for a sufficiently long amount of time in an area of large scalar curvature, or it has large energy (which is the case if it moves very fast or goes a very long way).

By hypothesis, there exists $t \in [t_1, t_0)$ such that $\gamma(t) \in \Sigma_t$. In particular, $\gamma(t)$ lies outside P_0. We shall make a first dichotomy according to whether γ goes out very fast or not.

Set

$$\alpha := \left(\frac{\hat{r}}{4\Lambda'}\right)^2 \in (0, 10^{-6}).$$

Case 1. There exists $t' \in [t_0 - \alpha\rho_0^2, t_0)$ such that $\gamma(t') \notin B_0$.

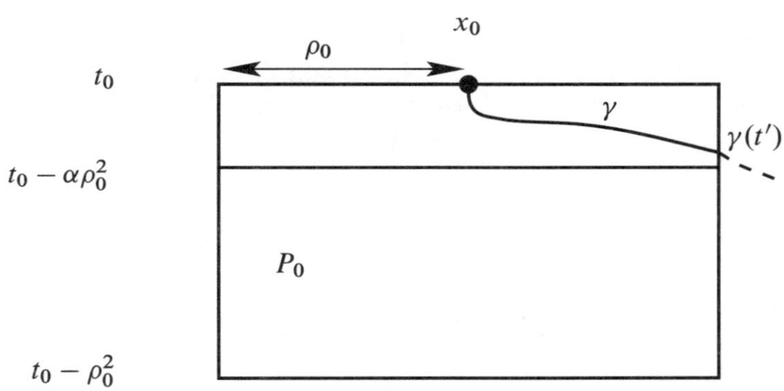

Choose t' maximal with this property. Let us recall that $\tau = t_0 - t$. By abuse of language, we consider γ as a function of τ. We then have

$$\int_0^{t_0-t'} |\dot{\gamma}|\, d\tau \leqslant \left(\int_0^{t_0-t'} \sqrt{\tau}|\dot{\gamma}|^2\, d\tau\right)^{1/2} \left(\int_0^{t_0-t'} \frac{1}{\sqrt{\tau}}\, d\tau\right)^{1/2},$$

so

$$\int_0^{t_0-t_1} \sqrt{\tau}|\dot{\gamma}|^2\, d\tau \geqslant \int_0^{t_0-t'} \sqrt{\tau}|\dot{\gamma}|^2\, d\tau \geqslant \left(\int_0^{t_0-t'} |\dot{\gamma}|\, d\tau\right)^2 \left(\int_0^{t_0-t'} \frac{1}{\sqrt{\tau}}\, d\tau\right)^{-1}.$$

When $t \in (t', t_0]$, we have $\gamma(t) \in B_0$. Since P_0 is unscathed, we have on $B_0 \times (t', t_0]$

$$g(t_0)e^{-4\rho_0^{-2}(t_0-t')} \leqslant g(t) \leqslant g(t_0)e^{4\rho_0^{-2}(t_0-t')},$$

hence

$$\frac{1}{2}g(t_0) \leqslant e^{-4\alpha}g(t_0) \leqslant g(t) \leqslant e^{4\alpha}g(t_0) \leqslant 2g(t_0).$$

Since $\gamma(t') \notin B_0$,

$$\int_0^{t_0-t'} |\dot{\gamma}|_{g(t_0-\tau)} d\tau \geqslant \frac{1}{\sqrt{2}} \int_0^{t_0-t'} |\dot{\gamma}|_{g(t_0)} d\tau \geqslant \frac{\rho_0}{\sqrt{2}},$$

so

$$\int_0^{t_0-t'} |\dot\gamma|^2\sqrt{\tau}\,d\tau \geqslant \frac{\rho_0^2}{2}\big([2\sqrt{\tau}]_0^{t_0-t'}\big)^{-1} \geqslant \frac{\rho_0}{4\sqrt{\alpha}} \geqslant \frac{\hat r}{4\sqrt{\alpha}}.$$

By choice of α, this last quantity is bounded from below by Λ'. This shows that γ satisfies (10.6).

Remark 10.3.2. In this case, there is no constraint on δ.

Case 2. For all $t \in [t_0 - \alpha\rho_0^2, t_0]$, $\gamma(t) \in B_0$.

By assumption there exists $(\bar x, \bar t)$ such that $\gamma(\bar t) = \bar x \in \Sigma_{\bar t}$. Since P_0 is unscathed, we have $\bar t < t_0 - \alpha\rho_0^2$. Assume that $\bar t$ is maximal for this property. We consider the restriction of $g(\cdot)$ to $[\bar t, t_0]$ (in the sense of Definition 8.1.1). Let us choose $\bar x$ such that $R(\bar x, \bar t) \geqslant h^{-2}(t)/2$. Since $h(t) < \delta(t) r(t) \leqslant \bar\delta 10^{-3}$, we may choose $\bar\delta$ small enough (depending on $\hat r$) so that

$$h^{-2}(t)/2 > 10^6\bar\delta^{-2}/2 \geqslant 12\hat r^{-2} \geqslant 12\rho_0^{-2}.$$

We then have $R(\bar x, \bar t) > 12\rho_0^{-2}$. For the sake of simplicity we set $h := h(\bar t)$.

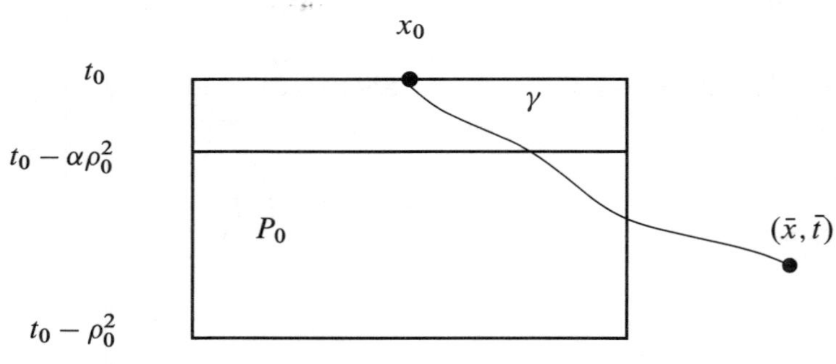

For constants $\theta \in [0,1)$ and $A \gg 1$ to be chosen later, we set

$$P := P(\bar x, \bar t, Ah, \theta h^2)$$

and take $\bar\delta \leqslant \bar\delta_{\mathrm{per}}(A, \theta, \hat r)$, so that the Persistence Theorem 8.1.2 applies. In particular, we have

$$R \geqslant \frac{1}{2}h^{-2} > 6\rho_0^{-2}$$

on P. This implies that $P_0 \cap P = \emptyset$, since on P_0, $|R| \leqslant 6|\mathrm{Rm}| \leqslant 6\rho_0^{-2}$.

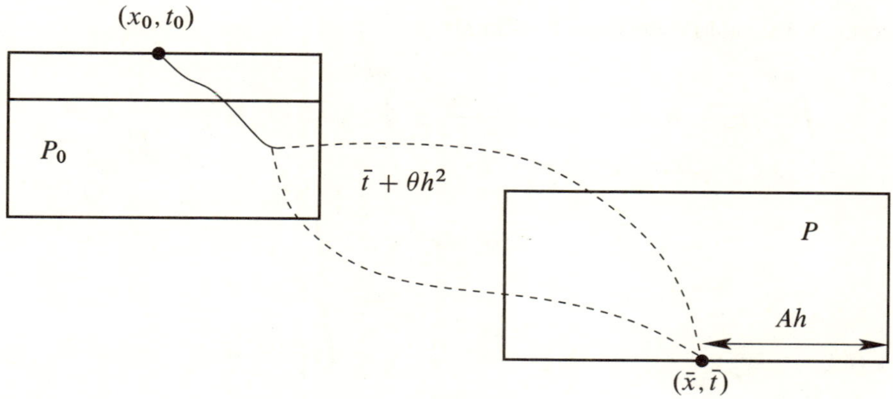

Let $\theta := \theta(\hat{r}, \Delta, \Lambda) \in (0, 1)$ be such that

$$-\mathrm{const}_{\mathrm{st}} \frac{\sqrt{\alpha}\hat{r}}{2} \ln(1-\theta) \geqslant \Lambda',$$

where $\mathrm{const}_{\mathrm{st}}$ is defined in Proposition 4.3.4. We again distinguish two subcases, according to the position of γ as described in the above picture.

Subcase (i). $\gamma([\bar{t}, \bar{t} + \theta h^2]) \subset B(\bar{x}, \bar{t}, Ah)$.

Then by Theorem 8.1.2, P is unscathed. Indeed, otherwise $B(\bar{x}, \bar{t}, Ah) \subset \Sigma_{t'}$ for some $t' \in (\bar{t}, \bar{t} + \theta h^2)$, and in particular $\gamma(t') \in \Sigma_{t'}$, which contradicts our choice of $\bar{t}$.

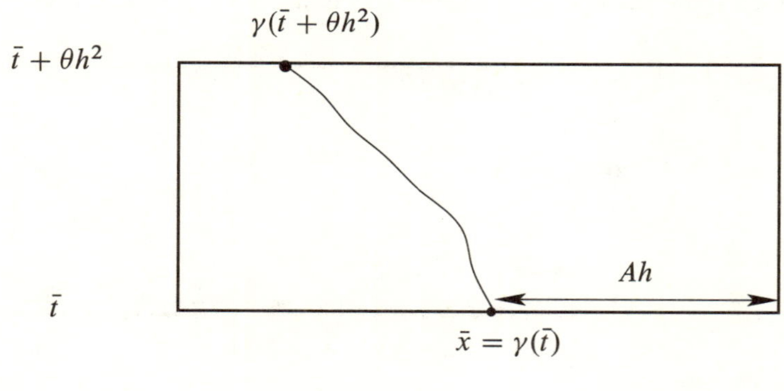

Moreover,

$$\bar{t} + \theta h^2 \leqslant t_0 - \alpha \rho_0^2,$$

which implies that $t_0 - t \geqslant \alpha \rho_0^2$ for all $t \in [\bar{t}, \bar{t} + \theta h^2]$. The closeness part of

Theorem 8.1.2, and Proposition 4.3.4 imply

$$\begin{aligned}
\int_{\bar t}^{\bar t+\theta h^2} \sqrt{t_0-t}\,R(\gamma(t),t)\,dt &\geqslant \frac{\mathrm{const}_{\mathrm{st}}}{2}\int_{\bar t}^{\bar t+\theta h^2} \sqrt{t_0-t}\,\frac{h^{-2}}{1-(t-\bar t)h^{-2}}\,dt \\
&\geqslant \mathrm{const}_{\mathrm{st}}\frac{\sqrt{\alpha\rho_0^2}}{2}\int_{\bar t}^{\bar t+\theta h^2} \frac{h^{-2}}{1-(t-\bar t)h^{-2}}\,dt \\
&= \mathrm{const}_{\mathrm{st}}\frac{\sqrt{\alpha\rho_0^2}}{2}\int_0^{\theta}\frac{1}{1-u}\,du \\
&\geqslant -\mathrm{const}_{\mathrm{st}}\frac{\sqrt{\alpha}\hat r}{2}\ln(1-\theta)\geqslant \Lambda',
\end{aligned}$$

by the choice of θ.

From these inequalities we deduce

$$\begin{aligned}
&\int_{t_1}^{\bar t}\sqrt{t_0-t}\,R\,dt+\int_{t_1}^{t_0}\sqrt{t_0-t}\,R\,dt \\
&\quad=\int_{\bar t}^{\bar t+\theta h^2}\sqrt{t_0-t}\,R\,dt+\int_{\bar t+\theta h^2}^{t_0}\sqrt{t_0-t}\,R\,dt \\
&\quad\geqslant \Lambda'-4\Delta^{3/2}=\Lambda.
\end{aligned}$$

Hence (10.5) holds.

Subcase (ii). There exists $t'\in[\bar t,\bar t+\theta h^2]$ such that $\gamma(t')\notin B(\bar x,\bar t,Ah)$.

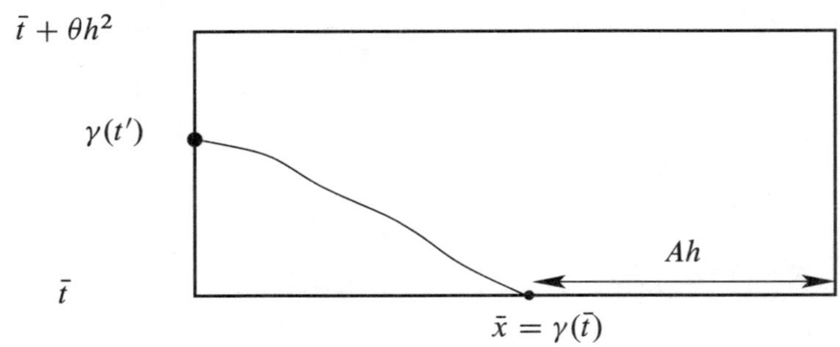

We assume that t' is minimal with this property.

By the same argument as before, $P' := P(\bar x,\bar t,Ah,t')$ is unscathed, and A^{-1}-close to the standard solution by the persistence theorem. As before, this implies that for all $s,s'\in[\bar t,t']$, we have on P',

$$g(s)\leqslant e^{\frac{C}{1-\theta}}g(s'),$$

where C is a universal constant. We thus have

$$\int_{\bar{t}}^{t'} \sqrt{\tau}\, |\dot{\gamma}|^2_{g(t)} dt \geqslant e^{\frac{-C}{1-\theta}} \left(\int_{\bar{t}}^{t'} |\dot{\gamma}|_{g(\bar{t})} dt \right)^2 \left(\int_{t_0-t'}^{t_0-\bar{t}} \frac{1}{\sqrt{\tau}} d\tau \right)^{-1}.$$

Since $t' \leqslant t_0 - \alpha\rho_0^2$ we can bound τ from below by $\alpha\rho_0^2$ on $[\bar{t}, t']$, so

$$\begin{aligned}\int_{\bar{t}}^{t'} \sqrt{\tau}|\dot{\gamma}|^2 dt &\geqslant e^{\frac{-C}{1-\theta}} (Ah)^2 \sqrt{\alpha\rho_0^2} (t' - \bar{t})^{-1} \\ &\geqslant e^{\frac{-C}{1-\theta}} (Ah)^2 \frac{\hat{r}}{10} \sqrt{\alpha} \frac{1}{\theta h^2} = e^{\frac{-C}{1-\theta}} \frac{\hat{r} A^2}{10\theta} \sqrt{\alpha}.\end{aligned}$$

Choosing A large enough, (10.6) holds. □

A consequence of the previous lemma is the following result (for more details, see [KL08], Lemmas 78.3 and 78.6). We shall not give the proof of this statement since in the next section a local version, in the spirit of 8.2 of [Per02], will be sketched.

Lemma 10.3.3. *Let $\hat{r} \in (0, 10^{-3})$ and let Δ, Λ be positive numbers. There exists $\bar{\delta} := \bar{\delta}(\hat{r}, \Delta, \Lambda) \in (0, \delta_0)$ with the following property. Let $g(\cdot)$ be a Ricci flow with $r(\cdot), \delta(\cdot)$-bubbling-off defined on $[a, a + \Delta]$ where $a \geqslant 0$, $\hat{r} \leqslant r(\cdot) < 10^{-3}$ and $\delta(\cdot) \leqslant \bar{\delta}$. Let (x_0, t_0) and $\rho_0 \geqslant \hat{r}$ be such that $P_0 := P(x_0, t_0, \rho_0, -\rho_0^2) \subset M \times [a, a + \Delta]$ is unscathed and $|\mathrm{Rm}| \leqslant \rho_0^{-2}$ on P_0. Then the following statements are true:*

(i) *For all $(q, s) \in M \times [a, a + \Delta]$, if $\ell(q, t_0 - s) < \Lambda$, then there is an unscathed minimising $\mathcal{L}$-geodesic γ connecting (x_0, t_0) to (q, s).*
(ii) *For all $\tau > 0$, $\min_q \ell(q, \tau) \leqslant 3/2$ and is attained.*

We now turn to the proof of Proposition C′. Let us fix $T > 0$, $\hat{r} \in (0, 10^{-3})$, a Ricci flow $g(\cdot)$ with $r(\cdot), \delta(\cdot)$-bubbling-off on $[0, b]$ where $b \leqslant T$, $\hat{r} \leqslant r(\cdot) < 10^{-3}$ and $\delta(\cdot) \leqslant \bar{\delta}$, where $\bar{\delta}$ is a constant to be defined later.

Let (x_0, t_0) be a point in spacetime. Thanks to Lemma 10.1.2 we restrict attention to the maximal scale $\rho_0 \leqslant 1$ such that $|\mathrm{Rm}| \leqslant \rho_0^{-2}$ on $P_0 := P(x_0, t_0, \rho_0, -\rho_0^2)$, and we assume that P_0 is unscathed (see Remark 10.1.3). Hence either $\rho_0 = 1$, or $\rho_0 < 1$ and the upper curvature bound is attained on $\bar{P}_0$. We shall treat the two cases $\rho_0 < \hat{r}$ and $\rho_0 \geqslant \hat{r}$ separately. As before we set $B_0 := B(x_0, t_0, \rho_0)$.

10.3.1 The case $\rho_0 < \hat{r}$

In this case ρ_0 is below the scale of the canonical neighbourhood for all t in the interval. Thus any point (y, t) with scalar curvature greater than ρ_0^{-2} has a canonical neighbourhood. At such a point, the noncollapsing property follows from geometric properties of this neighbourhood. Of course (y, t) need not be equal to (x_0, t_0), so we have to 'propagate' the κ-noncollapsing property using the curvature bound on $\bar{P}_0$. This case follows from the following assertion:

Assertion 10.3.4. $\mathrm{vol}(B_0) \geqslant C_0^{-1} e^{-18} \rho_0^3$.

Let us prove Assertion 10.3.4. Since $\rho_0 < \hat{r} < 1$, there exists $(y,t) \in \bar{P}_0$ such that $|\mathrm{Rm}(y,t)| = \rho_0^{-2} > \hat{r}^{-2}$. By Pinching Lemma 4.4.7, if $R(y,t) \leqslant \hat{r}^{-2}$ then $|\mathrm{Rm}(y,t)| \leqslant \hat{r}^{-2}$. Hence

$$R(y,t) > \hat{r}^{-2} \geqslant r(t)^{-2},$$

and (y,t) is the centre of an (ε_0, C_0)-canonical neighbourhood U.

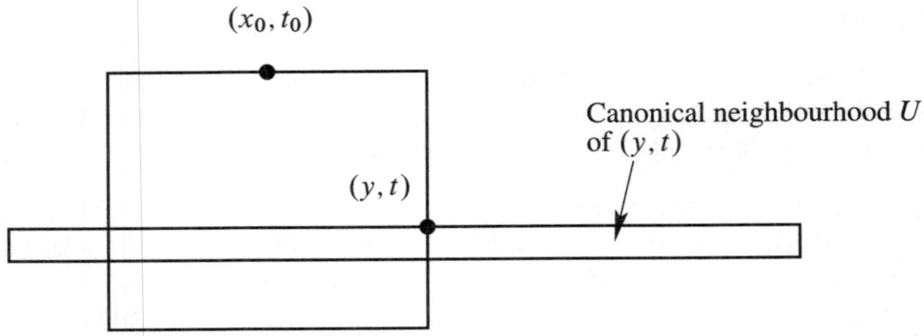

Our next task is to show that $B(x_0, t, e^{-2}\rho_0) \subset U$. By the curvature bounds on $\bar{P}_0$ we have $d_t(x_0, y) \leqslant e^2 \rho_0$ and $B(x_0, t, e^{-2}\rho_0) \subset B(x_0, t_0, \rho_0)$. If U is a ε_0-neck, then $d_t(y, \partial\bar{U}) \geqslant (2\varepsilon_0)^{-1} R(y,t)^{-1/2}$. Since $R(y,t) \leqslant 6\rho_0^{-2}$, we get

$$d_t(y, \partial\bar{U}) \geqslant (2\sqrt{6}\varepsilon_0)^{-1} \rho_0 \geqslant (e^2 + e^{-2})\rho_0,$$

hence

$$d_t(x_0, \partial\bar{U}) \geqslant d_t(y, \partial\bar{U}) - d_t(x_0, y) \geqslant e^{-2}\rho_0,$$

and $B(x_0, t, e^{-2}\rho_0) \subset U$.

If U is an (ε_0, C_0)-cap, then $U = V \cup W$ where V is a core and W is a ε_0-neck. Let $\gamma \colon [0,1] \to \bar{B}_0$ be a minimising $g(t_0)$-geodesic connecting y to x_0. Let us recall that $y \in V$; if $x_0 \notin V$, let $s \in [0,1]$ be maximal such that $\gamma(s) \in \partial V$. Since $\gamma(s) \in B_0$, we have $R(\gamma(s), t) \leqslant 6\rho_0^{-2}$ and we deduce that $d_t(\gamma(s), \partial\bar{U}) \geqslant (\sqrt{6}\varepsilon_0)^{-1}\rho_0$. Since $d_t(\gamma(s), x_0) \leqslant e^2\rho_0$ we get as above, that $B(x_0, t, e^{-2}\rho_0) \subset U$.

By the noncollapsing property of the canonical neighbourhoods ((4.1), p. 39), we get that

$$\mathrm{vol}_{g(t)} B(x_0, t, e^{-2}\rho_0) \geqslant C_0^{-1}(e^{-2}\rho_0)^3.$$

By estimating the distortion of distances and volume as in the proof of Lemma 10.1.2, we conclude that

$$\mathrm{vol}_{g(t_0)}(B_0) \geqslant C_0^{-1} e^{-18} \rho_0^3.$$

10.3.2 The case $\rho_0 \geqslant \hat{r}$

Recall that P_0 is unscathed, and satisfies $|\mathrm{Rm}| \leqslant \rho_0^{-2}$. Let us recall that Σ_t, the singular set at time t, is the closure of $\{x \in M, \quad g(x,t) \neq g_+(x,t)\}$. Let $\gamma\colon [a,b] \to M$ be a curve. Given $v \in T_{x_0}M$ at time t_0, we set, for $\tau = t_0 - t$

$$\gamma_v(\tau) = \mathcal{L}_\tau \exp_{x_0,t_0}(v).$$

It is straightforward to check that the monotonicity of reduced length (see Lemma 10.1.6) is also valid along any minimising $\mathcal{L}$-geodesics which is unscathed. In particular, if $\gamma_v(\tau)$ is minimising and unscathed on $[0,\tau_0]$, then

$$\tau^{-3/2} e^{-\ell(v,\tau)} J(v,\tau)$$

is nonincreasing on $[0,\tau_0]$.

Set

$$Y(\tau) := \{v \in T_{x_0}M \mid \gamma_v \text{ is minimising and unscathed on } [0,\tau)\}.$$

It is easy to check that $\tau \leqslant \tau' \Rightarrow Y(\tau) \supset Y(\tau')$. Then we set

$$\widetilde{V}_{\mathrm{reg}}(\tau) := \int_{Y(\tau)} \tau^{-3/2} e^{-\ell(v,\tau)} J(v,\tau)\, dv.$$

Thus this function is nondecreasing on $[0,t_0]$. We shall adapt the proof of κ-noncollapsing in the smooth case, replacing $\widetilde{V}$ by $\widetilde{V}_{\mathrm{reg}}$. Set

$$\kappa := \frac{\mathrm{vol}_{g(t_0)}(B_0)}{\rho_0^3}, \qquad \tau_0 = \kappa^{1/3}\rho_0^2.$$

Upper bound on $\widetilde{V}_{\mathrm{reg}}(\tau_0)$. As in the smooth case, we get

$$\mathcal{L}_\tau \exp(\{v \in Y(\tau) \mid |v| \leqslant \tfrac{1}{10}\kappa^{-1/6}\}) \subset B_0.$$

Thus

$$I := \int_{\{v \in Y(\tau); |v| \leqslant \frac{1}{10}\kappa^{-1/6}\}} \tau^{-3/2} e^{-\ell(v,\tau)} J(v,\tau) dv \leqslant e^{C\kappa^{1/6}} \sqrt{\kappa}$$

and

$$\begin{aligned} I' &:= \int_{\{v \in Y(\tau); |v| \geqslant \frac{1}{10}\kappa^{-1/6}\}} \tau^{-3/2} e^{-\ell(v,\tau)} J(v,\tau) dv \\ &\leqslant \int_{\{v \in Y(\tau); |v| \geqslant \frac{1}{10}\kappa^{-1/6}\}} \lim_{\tau \to 0} (\tau^{-3/2} e^{-\ell(v,\tau)} J(v,\tau))\, dv \\ &\leqslant e^{-\frac{1}{10}\kappa^{-1/6}}. \end{aligned}$$

In conclusion,

$$I + I' \leqslant e^{C\kappa^{1/3}} \sqrt{\kappa} + e^{-\frac{1}{10}\kappa^{-1/6}} \leqslant C\sqrt{\kappa}$$

for $\kappa \leqslant \kappa(3)$ and some universal C.

Lower bound for $\widetilde{V}_{\mathrm{reg}}(t_0)$. Monotonicity of $\widetilde{V}_{\mathrm{reg}}(\tau_0)$ implies

$$\widetilde{V}_{\mathrm{reg}}(\tau_0) \geqslant \widetilde{V}_{\mathrm{reg}}(t_0).$$

We are going to bound $\widetilde{V}_{\mathrm{reg}}(t_0)$ from below as a function of vol $B(q_0, 0, 1)$, where $q_0 \in M$.

Set $\Lambda = 21$ and apply Lemma 10.3.3 with parameter $\bar{\delta}(\hat{r}, T, \Lambda)$. There exists $q_0 \in M$ such that $\ell(q_0, t_0 - \frac{1}{16}) \leqslant \frac{3}{2}$ and the first part of Lemma 10.3.3 gives us a minimising curve γ connecting x_0 to q_0 realising the minimum $\ell(q_0, t_0 - 1/16)$. Let $q \in B(q_0, 0, 1)$. Consider a curve γ obtained by concatenating some $g(0)$-geodesic from $(q, 0)$ to $(q_0, 1/16)$ with some minimising $\mathcal{L}$-geodesic between $(q_0, 1/16)$ and (x_0, t_0). We have

$$\begin{aligned} \ell(q, t_0) = \frac{L(q, t_0)}{2\sqrt{t_0}} &\leqslant \frac{1}{2\sqrt{t_0}} L(q_0, t_0 - 1/16) \\ &\quad + \frac{1}{2\sqrt{t_0}} \int_{t_0 - 1/16}^{t_0} \sqrt{\tau}(R_{g(t_0-\tau)} + |\dot{\gamma}(\tau)|^2_{g(t_0-\tau)}) d\tau, \end{aligned} \tag{10.7}$$

which leads to

$$\ell(q, t_0) \leqslant \frac{3}{2} + \frac{1}{2} \int_{t_0 - 1/16}^{t_0} (12 + e^{1/2} |\dot{\gamma}(\tau)|^2_{g(0)}) d\tau,$$

since for $s \in [0, 1/16]$ (i.e., $\tau \in [t_0 - 1/16, t_0]$) the metrics $g(s)$ satisfy $\mathrm{Rm} \leqslant 2$, hence $R \leqslant 12$, and are $1/2$-Lipschitz equivalent. We obtain

$$\ell(q, t_0) \leqslant \frac{3}{2} + \frac{1}{2}\Big(\frac{1}{16} 12 + 16 e^{1/2} d^2_{g(0)}(q, q_0)\Big) \leqslant 20.$$

Hence $\ell(q, t_0) < \Lambda$, and Lemma 10.3.3 (i) gives again an unscathed minimising $\mathcal{L}$-geodesic $\tilde{\gamma}$ connecting x_0 to q. Hence $q = \tilde{\gamma}(t_0) = \mathcal{L}_{t_0} \exp(\tilde{v})$ for $\tilde{v} \in Y(t_0)$. This shows that $\mathcal{L}_{t_0} \exp(Y(t_0)) \supset B(q, 0, 1)$.

Moreover, we have $\{\tilde{v} \in Y(t_0) \mid \mathcal{L}_{t_0} \exp(\tilde{v}) \in B(q, 0, 1)\}$, $\ell \leqslant 10$, which implies

$$\begin{aligned} \widetilde{V}_{\mathrm{reg}}(t_0) &= \int_{Y(t_0)} t_0^{-3/2} e^{-\ell(v, t_0)} J(v, t_0)\, dv \\ &\geqslant \int_{B(q_0, 0, 1)} t_0^{-3/2} e^{-20} dv_{g(0)} \\ &\geqslant T^{-3/2} e^{-20} \operatorname{vol}_{g(0)} B(q_0, 0, 1). \end{aligned}$$

This completes the proof of Proposition C$'$.

10.4 κ-noncollapsing at bounded distance of the thick part

In what follows the canonical neighbourhood scale $r(t)$ is assumed to be nonincreasing.

Theorem 10.4.1. *For all $A > 0$, there exists $\kappa = \kappa(A) > 0$ such that for any $t_0 > 0$, there exists $\bar{\delta} := \bar{\delta}(t_0) \in (0, \delta_0)$ decreasing in t_0, with the following property. Let $g(\cdot)$ be a Ricci flow with $r(\cdot), \delta(\cdot)$-bubbling-off on M defined on $[0, +\infty)$ such that $\delta(t) \leqslant \bar{\delta}$ for $t \in [t_0/2, t_0]$; if the parabolic neighbourhood $P(x_0, t_0, \rho_0, -\rho_0^2)$, where $2\rho_0^2 < t_0$, is unscathed and the solution satisfies $|\mathrm{Rm}| \leqslant \rho_0^{-2}$ there, and if*

$$\operatorname{vol} B(x_0, t_0, \rho_0) \geqslant A^{-1}\rho_0^3,$$

then the solution is κ-noncollapsed on the scales less than ρ_0 in the ball $B(x_0, t_0, A\rho_0)$.

Proof. We shall assume that $\rho_0 \geqslant r(t_0)$. To justify this, note that for $\rho_0 < r(t_0)$ we can copy the proof of Proposition C′ in the case where $\rho_0 < \hat{r}$, replacing $\hat{r}$ by $r(t_0)$. (This is where we use that $r(\cdot)$ is nonincreasing.)

Since the noncollapsing property is unsensitive to homothety we can do a parabolic rescaling so that $\rho_0 = 1$ and $t_0 = 1$. More precisely we shall work with the flow,

$$\bar{g}(t) = \frac{1}{\rho_0^2} g(t_0 - \rho_0^2 + t\rho_0^2).$$

From now on, unless otherwise stated, we work with the flow $\bar{g}(\cdot)$. For simplicity we shall not "overline" the geometric quantities related to this metric unless it is necessary.

Let $x \in B(x_0, 1, A)$ and $\rho \in (0, 1]$ be such that $|\mathrm{Rm}| \leqslant \rho^{-2}$ on $P(x, 1, \rho, -\rho^{-2})$. As before, we set

$$\kappa := \frac{\operatorname{vol} B(x, 1, \rho)}{\rho^3},$$

and assume that $\kappa \leqslant \kappa(3)$, where $\kappa(n)$ appears in Section 10.2. We wish to show that there exists $C = C(A) > 0$ such that $\kappa \geqslant C(A)$. We shall consider space-time curves starting from x. As before, we set

$$\widetilde{V}_{\mathrm{reg}}(\tau) := \int_{Y(\tau)} \tau^{-3/2} e^{-\ell(v,\tau)} J(v, \tau)\, dv,$$

for $\tau = 1 - t \in [0, 1]$, and where

$$Y(\tau) := \{v \in T_{x_0}M \mid \mathcal{L}\exp(v) \colon [0, \tau] \to M \text{ is minimising and unscathed}\}.$$

The function $\widetilde{V}_{\mathrm{reg}}$ is nonincreasing in τ. The same computation as before yields the upper bound

$$\widetilde{V}_{\mathrm{reg}}(\kappa^{1/3}\rho^2) \leqslant C\sqrt{\kappa}$$

for some universal $C > 0$. Now the monotonicity of $\widetilde{V}_{\mathrm{reg}}$ gives

$$\widetilde{V}_{\mathrm{reg}}(1) \leqslant \widetilde{V}_{\mathrm{reg}}(\kappa^{1/3}\rho^2) \leqslant C\sqrt{\kappa}.$$

We thus have to estimate $\widetilde{V}_{\mathrm{reg}}(1)$ from below.

10.4.1 A formal computation

By the curvature pinched toward positive assumption we have, before rescaling,

$$R \geqslant -\frac{6}{4t+1} \geqslant -\frac{3}{2t},$$

hence, after rescaling,

$$R \geqslant -\frac{3}{2}\frac{\rho_0^2}{t_0 - \rho_0^2 + t\rho_0^2} \geqslant -\frac{3}{2}\frac{\rho_0^2}{t_0 - \rho_0^2 + \frac{1}{2}\rho_0^2}$$

for $t \in [1/2, 1]$. Recall that $t_0 \geqslant 2\rho_0^2$. Thus

$$R \geqslant -\frac{3}{2}\frac{\rho_0^2}{2\rho_0^2 - \frac{1}{2}\rho_0^2} = -1.$$

Hence, using that $\tau = 1 - t$,

$$\bar{L}(y,\tau) := 2\sqrt{\tau}L(y,\tau) \geqslant -2\sqrt{\tau}\int_0^\tau \sqrt{s}\,ds = -\frac{4}{3}\tau^2$$

(here we have neglected the energy of the curve in the definition of $\mathcal{L}$). We now define $\hat{L}(y,\tau) := \bar{L}(y,\tau) + 2\sqrt{\tau}$. Then, for $\tau \in (0, 1/2]$,

$$\hat{L}(y,\tau) \geqslant -\frac{4}{3}\tau^2 + 2\sqrt{\tau} \geqslant \Big(-\frac{4}{3} + 2\Big)\sqrt{\tau} > 0.$$

Define $h(y,\tau) := \psi(\hat{d}_t(x_0, y))\hat{L}(y,\tau)$, where $\hat{d}_t(x_0, y) = d_t(x_0, y) - A(2t-1)$, and ψ is a cutoff function equal to 1 on $(-\infty, 1/20]$, to $+\infty$ on $[1/10, +\infty)$, nondecreasing and smooth where defined. We now follow the computation in [Per02], 8.2 (p. 21) and use the operator

$$\Box := \frac{\partial}{\partial t} - \Delta_{g(t)};$$

it yields

$$\Box h \geqslant -(\bar{L} + 2\sqrt{\tau})C(A)\psi - 6\psi - \frac{1}{\sqrt{\tau}}\psi = -C(A)h - \Big(6 + \frac{1}{\sqrt{\tau}}\Big)\psi \circ \hat{d}_t, \quad (10.8)$$

where $C(A)$ is a constant depending on A only. At the spatial minimum point, with τ fixed, $\Delta h \geqslant 0$. Let us define $h_0(\tau) = \min h(\cdot, \tau)$. We then have

$$\frac{d}{d\tau}\Big(\log\Big(\frac{h_0(\tau)}{\sqrt{\tau}}\Big)\Big) = h_0^{-1}\frac{dh_0}{d\tau} - \frac{1}{2\tau} \leqslant C(A) + \frac{6\sqrt{\tau}+1}{\sqrt{\tau}\bar{L} + 2\tau} - \frac{1}{2\tau}.$$

As $\bar{L} \geqslant -\frac{4}{3}\tau^2$,

$$\frac{d}{d\tau}\Big(\log\Big(\frac{h_0(\tau)}{\sqrt{\tau}}\Big)\Big) \leqslant C(A) + \frac{50}{\sqrt{\tau}}.$$

By integration we get,

$$\log\left(\frac{h_0(\tau)}{\sqrt{\tau}}\right) - \lim_{s\to 0}\log\left(\frac{h_0(s)}{\sqrt{s}}\right) \geqslant C(A)\tau + 100\sqrt{\tau}.$$

Now for s small,

$$\hat{L} = \bar{L} + 2\sqrt{s} = 2\sqrt{s}L + 2\sqrt{s} > 0 \quad \text{and} \quad h = \psi\hat{L} \geqslant 2,$$

$$2 < \frac{h_0(s)}{\sqrt{s}} = \min\{\psi(2L+2)\} \leqslant 2L(x,s) + 2$$

since $\psi \equiv 1$ near x when s is small. Indeed, $d_t(x_0, x) < A$ and $\psi \equiv 1$ on $B(x_0, 0, \frac{1}{20} + A)$.

Now, using a curve which is constant equal to x, we have that

$$L(x,s) \leqslant \int_0^s \sqrt{s}R\,ds \underset{s\to 0}{\longrightarrow} 0$$

and hence

$$\frac{h_0(s)}{\sqrt{s}} \underset{s\to 0}{\longrightarrow} 2,$$

which gives

$$h_0(\tau) \leqslant 2\sqrt{\tau}\exp(C(A)\tau + 100\sqrt{\tau}).$$

10.4.2 Justification of the formal computations

We now apply Lemma 10.3.1 with the unscaled metric $g(t)$ and with the constant

$$\Lambda = 3\exp(C(A) + 100)\sqrt{\frac{t_0}{2}}$$

and $\hat{r} \leqslant r(t_0)$, $\Delta = t_0/2$. This gives a $\bar{\delta} = \bar{\delta}(A, t_0)$ such that every scathed curve $\gamma(\tau)$ starting at x (in particular, γ exits the set $P(x, t_0, \rho, -\rho^2)$) satisfies the unnormalised inequality

$$\mathcal{L}_g(\gamma) \geqslant \Lambda.$$

This implies that

$$\mathcal{L}_{\bar{g}}(\gamma) \geqslant \Lambda' = \Lambda/\sqrt{t_0/2}.$$

From now on all geometric quantities refer to the normalised metric $\bar{g}(t)$. In particular if

$$L(y,\tau) < 3\exp(C(A)\tau + 100\sqrt{\tau}) \leqslant 3\exp(C(A) + 100) = \Lambda',$$

the $\mathcal{L}$-geodesics joining x to a neighbourhood of y are unscathed. Moreover, near such a point, the formal calculations (10.8) with the function h are valid in the barrier sense.

Let us define $\beta_0(\tau) = \log\left(\frac{h_0(\tau)}{\sqrt{\tau}}\right) - C(A)\tau - 100\sqrt{\tau} - \log 3$. From the preceding argument $\lim_{\tau\to 0}\beta_0(\tau) = \log 2 - \log 3 < 0$. We now follow the proof given in [KL08], Lemma 78.6. The arguments developed there apply to show that β_0 is continuous on $(0, \tau_1)$, where

$$\tau_1 = \sup\{\tau > 0;\ \beta_0(\tau) < 0\}.$$

Similarly the previous computation shows that the upper right derivative of β_0 on $(0, \tau_1)$ is nonpositive. This implies that $\beta_0(\tau_1) < 0$ by the same argument as in [KL08], Lemma 78.6, and hence that $\tau_1 = 1$. This shows that for each $\tau > 0$, there exists y such that $\bar{L}(y, \tau) \leqslant \hat{L}(y, \tau) \leqslant 3\sqrt{\tau}\exp(C(A)\tau + 100\sqrt{\tau})$. Henceforth, for $\tau = 1/2$ there exists $y \in B(x_0, 1/2, 1/10)$ such that

$$L(y, 1/2) < \Lambda' \quad \text{and} \quad \ell(y, 1/2) < \Lambda'/\sqrt{2} = \Lambda/\sqrt{t_0}.$$

Let us come back to the proof of noncollapsing. We now know that there exists $y \in B(x_0, 1/2, 10^{-1})$ such that $\ell(y, 1/2)$ is small (see Figure 10.2 below). By distance comparison and the fact that $|\mathrm{Rm}| \leqslant 1$ on $P(x_0, 0, 1, -1)$ for the normalised metric, we get that on this set,

$$\frac{1}{10} d_1 \leqslant d_0 \leqslant 10 d_1.$$

Hence, $B(x_0, 1, 10^{-2}) \subset B(x_0, 0, 10^{-1}) \subset B(x_0, 1, 1)$ and then

$$\begin{aligned} \mathrm{vol}_{\bar{g}(0)} B(x_0, 0, 10^{-1}) &\geqslant \mathrm{vol}_{\bar{g}(0)} B(x_0, 1, 10^{-2}) \\ &\geqslant 10^{-3/2}\,\mathrm{vol}_{\bar{g}(1)} B(x_0, 1, 10^{-2}) \\ &\geqslant 10^{-3/2}\frac{v_{-1}(10^{-2})}{v_{-1}(1)}\,\mathrm{vol}_{\bar{g}(1)} B(x_0, 1, 1) \\ &\geqslant 10^{-3/2} A^{-1}\frac{v_{-1}(10^{-2})}{v_{-1}(1)}. \end{aligned}$$

The third inequality comes from the Bishop–Gromov inequality and the last one from the hypothesis. Now, as before, in order to estimate $\ell(z, 1)$ for $z \in B(x_0, 0, 1/10)$ we shall concatenate curves from $(x, 1)$ to $(y, 1/2)$ with curves from $(y, 1/2)$ to $(z, 0)$.

From $(x, 1)$ to $(y, 1/2)$ we take a minimising $\mathcal{L}$-geodesic. From $(y, 1/2)$ to $(z, 0)$ we take a minimising $\bar{g}(1)$-geodesic from y to x_0 and a minimising $\bar{g}(1)$-geodesic from x_0 to z; the curve thereby obtained is contained in $B(x_0, 1, 1)$, on which the scalar curvature is bounded from above by 6. The proof given in the previous section now shows that

$$\ell(z, 1) \leqslant C(A) \quad \text{for all } z \in B(x_0, 0, 10^{-1}).$$

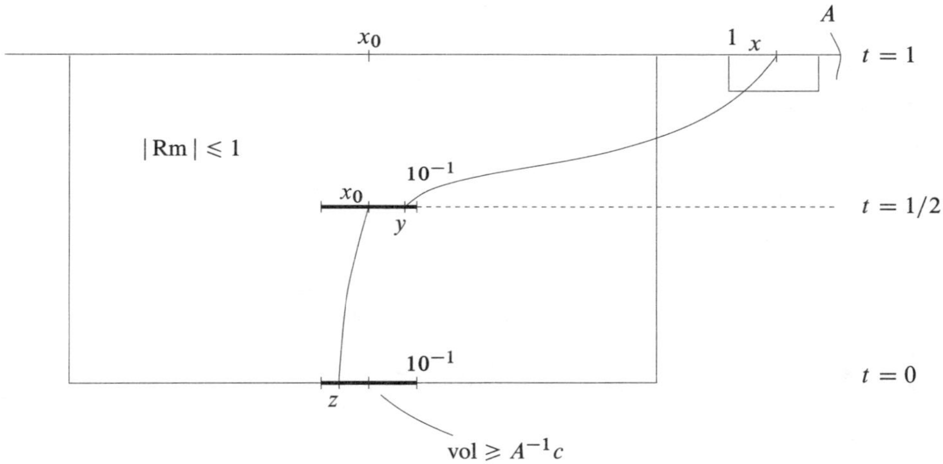

Figure 10.2. Concatenation of $\mathcal{L}$ geodesics with $g(1)$-geodesics.

Finally,

$$\begin{aligned}\widetilde{V}_{\mathrm{reg}}(1) &= \int_{Y(1)} 1^{-3/2} e^{-\ell(z,1)} J(z,1)\,dv \\ &\geqslant \int_{B(x_0,0,10^{-1})} e^{-C(A)}\,dv_{\bar{g}(0)} \\ &\geqslant e^{-C(A)} \operatorname{vol}_{\bar{g}(0)} B(x_0,0,10^{-1}) \\ &\geqslant e^{-C(A)} 10^{-3/2} A^{-1} \frac{v_{-1}(10^{-2})}{v_{-1}(1)}.\end{aligned}$$

We recall that $\tau = 1$ corresponds to $t = 0$ and the parameter in the metric $\bar{g}$ is the time t whereas the parameter in the reduced volume is τ. This is the desired inequality. □

Notes. This chapter corresponds to [Per03b], Lemma 5.2. Our treatment follows closely [KL08], except for the fact that we use the canonical neighbourhood property in the first case. For this, we use the fact that M is not spherical.

For the analytic properties of the $\mathcal{L}$-functional, the papers of Ye [Ye08b], [Ye08a] are a good reference. The reader may also look at Chapters 7 and 8 in the book [CCG$^+$07], where most of the details are treated.

Part III

Long-time behaviour of Ricci flow with bubbling-off

The goal of Part III is to prove the Thin-Thick Decomposition Theorem 1.3.4. To do this, we use the Ricci flow with $r(\cdot)$, $\delta(\cdot)$-bubbling-off provided by the long-time existence theorem (Theorem 5.4.1). We shall need to choose the time-dependent parameters $r(\cdot)$ and $\delta(\cdot)$ carefully in order to have the necessary estimates.

In Chapter 11 we first state the corresponding refined results on the long-time behaviour of Ricci flow with bubbling-off. These results, which correspond to [Per03b], Section 6, in Perelman's work, are the most technical theorems proved in this book. We then prove the thin-thick decomposition theorem assuming these technical results.

Finally, the technical theorems are proved in Chapter 12. One of them gives versions of the κ-noncollapsing, curvature-distance estimates and the canonical neighbourhood property at space-time points where the time multiplied by the curvature is large. The second one essentially shows that certain parabolic neighbourhoods in the thick part are unscathed and with bounded geometry. This is used to show that on the thick part the rescaled metric converges towards the hyperbolic metric.

Chapter 11

The thin-thick decomposition theorem

In this chapter, we fix a closed orientable, irreducible, nonspherical 3*-manifold* M.

Our goal is to prove the Thin-Thick Decomposition Theorem 1.3.4 whose statement is recalled below (Theorem 11.1.1). This theorem asserts the existence of a Ricci flow with bubbling-off on M, defined on $[0, +\infty)$, with the following extra properties. When $t \to \infty$, the rescaled metric $t^{-1}g(t)$, given any sequence of basepoints (x_n, t_n) in the thick part, subconverges in the pointed smooth topology to a complete hyperbolic manifold of finite volume. Moreover, there are local estimates on the curvature tensor and its first covariant derivative, a property we call 'locally controlled curvature in the sense of Perelman'. In Section 11.1 below, we outline the proof of the thin-thick decomposition theorem, splitting it into several technical results. The proof of the theorem, assuming these technical results, is given in Section 11.2. The proof of the technical results is postponed to the next chapter.

11.1 Introduction: main statements

Let (X, g) be a Riemannian manifold and ε be a positive real number. We define the *ε-thin part* of (X, g) to be the subset $X^-(\varepsilon)$ of points $x \in X$ for which there exists $0 < \rho \leqslant 1$ such that the ball $B(x, \rho)$ has sectional curvature $\geqslant -\rho^{-2}$ and volume $< \varepsilon\rho^3$. Its complement is called the *ε-thick part* and denoted by $X^+(\varepsilon)$.

Below we recall the statement of Theorem 1.3.4. For convenience, this statement has been slightly rephrased. In particular, the metric is rescaled by t instead of $4t$, producing 'hyperbolic' limits of sectional curvatures $\frac{-1}{4}$ instead of -1:

Theorem 11.1.1 (Thin-thick decomposition). *Let g_0 be a normalised metric on M. There exists a Ricci flow with bubbling-off $g(\cdot)$ on $[0, +\infty)$ with g_0 as initial condition, and such that the following is true:*

(i) *There exists $C > 0$ such that* $\mathrm{vol}(g(t)) \leqslant Ct^{3/2}$.

(ii) *Let $w > 0$, $x_n \in M$ and $t_n \to +\infty$. If x_n is in the w-thick part of $(M, \frac{1}{t_n}g(t_n))$ for every n, then the sequence of pointed manifolds $(M, \frac{1}{t_n}g(t_n), x_n)$ subconverges in the $\mathcal{C}^\infty$ sense to a complete pointed 3-manifold of finite volume with all sectional curvatures equal to $-1/4$.*

(iii) *For all $w' > 0$ there exists $\bar{r} > 0$, $K_0, K_1 > 0$ and $t_0 < \infty$ depending only on w' such that for all $x \in M$ and $t > t_0$,, for all $r \in (0, \bar{r})$, if the ball*

$B(x,r) \subset (M, t^{-1}g(t))$ *satisfies* $\mathrm{Rm} \geqslant -r^{-2}$ *and* $\mathrm{vol}\, B(x,r) \geqslant w' r^3$, *then* $|\mathrm{Rm}(x)| \leqslant K_0 r^{-2}$ *and* $|\nabla \mathrm{Rm}(x)| \leqslant K_1 r^{-3}$.

Remark 11.1.2. The third assertion in the conclusion implies that if t_n is any sequence tending to infinity, then the sequence $(t_n)^{-1} g(t_n)$ has locally controlled curvature in the sense of Perelman (cf. Definition 13.1.1). Of course, the sequence $(4t_n)^{-1} g(t_n)$ has the same properties for slightly different constants.

Let us outline the proof of this theorem. Since we assume M to be non-spherical, the Long-Time Existence Theorem 5.4.1 yields a nonincreasing positive function $r(\cdot)$, and for any positive function $\bar{\delta}(\cdot)$, a nonincreasing function $\delta(\cdot) \leqslant \bar{\delta}(\cdot)$ such that a Ricci flow with $r(\cdot), \delta(\cdot)$-bubbling-off exists on $[0, +\infty)$ for any normalised initial condition. The given function $r(\cdot)$ is *now fixed forever*. However we have the freedom of choosing $\delta(t)$ going to zero when t goes to infinity as steeply as we need. This we will use to obtain better estimates than those of Part II.

The main difficulty is to obtain a local upper bound for the sectional curvature on the thick part. This bound will be used in order to apply a compactness theorem to prove subconvergence of rescalings of $g(\cdot)$ when the basepoint belongs to the thick part. Then the fact that the time slices of the limit flow have constant sectional curvature will follow from a comparatively easy maximum principle argument. For this argument to work, however, it is essential to consider limits of evolving metrics, although we are ultimately only interested in sequences of metrics.

As a first step toward the required curvature bound, we shall prove the following result, which says that under hypotheses similar to that of belonging to the thick part, we have κ-noncollapsing, canonical neighbourhoods and an upper scalar curvature bound at the appropriate scales:

Theorem 11.1.3. *For all* $A \in (0, \infty)$ *there exist positive numbers* $\kappa, \bar{K}, \bar{K}', \bar{r}$ *and a nonincreasing function* $t \mapsto \bar{\delta}_A(t)$ *depending only on* A *with the following property. Let* $g(\cdot)$ *be a Ricci flow with* $r(\cdot), \delta(\cdot)$*-bubbling-off on* M *with normalised initial condition, and let* (x, t, ρ) *have the following properties. Assume that* $0 < \rho < \sqrt{t/2}$, $0 < \delta(\cdot) < \bar{\delta}_A(t)$ *on* $[t/2, t]$, *and that*

(i) $P(x, t, \rho, -\rho^2)$ *is unscathed,*

(ii) $|\mathrm{Rm}| \leqslant \rho^{-2}$ *there,*

(iii) $\mathrm{vol}\, B(x, t, \rho) \geqslant A^{-1} \rho^3$.

Then the following assertions hold:

(a) *The solution is* κ*-noncollapsed on scales less than* ρ *in* $B(x, t, A\rho)$.

(b) *Every point of* $B(x, t, A\rho)$ *of scalar curvature at least* $\bar{K}\rho^{-2}$ *admits a canonical neighbourhood.*

(c) *If* $\rho \leqslant \bar{r}\sqrt{t}$, *then* $R \leqslant \bar{K}'\rho^{-2}$ *in* $B(x, t, A\rho)$.

An important point in part (a) is that κ does not depend on time. This is crucial since we will use this result to obtain convergence of parabolic rescalings of $g(\cdot)$ at basepoints (x_n, t_n) with t_n going to infinity. Also note that the scale ρ may be much larger than 1.

Note that part (a) has been stated and proved in Section 10.4. Before stating the theorem giving the curvature estimates in the thick part, we define the appropriate function $\bar{\delta}(\cdot)$.

Definition 11.1.4. With the function $\bar{\delta}_A(\cdot)$ as above, set

$$\bar{\delta}(t) := \min\left\{\bar{\delta}_{2t}(2t), \eta(t)\frac{r(4t)}{r(t)}\right\}, \tag{11.1}$$

where $\eta\colon [0,\infty) \to (0,1]$ is some fixed decreasing function such that $\eta(t) \to 0$ as $t \to +\infty$.

Remark 11.1.5. Inequality $\bar{\delta}(t) \leqslant \bar{\delta}_{2t}(2t)$ ensures that any Ricci flow with $r(\cdot), \delta(\cdot)$-bubbling-off with $\delta(\cdot) \leqslant \bar{\delta}(\cdot)$ satisfies the conclusions of Theorem 11.1.3 at time t for all $A \leqslant t$.

We shall then prove a stronger version of Theorem 11.1.3, where the property of having an unscathed parabolic neighbourhood is deduced from the hypotheses rather than included in them.

Theorem 11.1.6 (Curvature estimate in the thick part). *For all $w > 0$, there exist positive numbers $\hat{\theta}, \hat{r}, \widehat{T}, \hat{\tau}, \widehat{K}$ depending only on w with the following property. Assume that $g(\cdot)$ is a Ricci flow with $r(\cdot), \delta(\cdot)$-bubbling-off on $[0,\infty)$, with normalised initial condition and $\delta(\cdot) \leqslant \bar{\delta}(\cdot)$. Let $x \in M$, $t \geqslant \widehat{T}$, $r > 0$ such that*

(i) *$\hat{\theta}^{-1}h \leqslant r \leqslant \hat{r}\sqrt{t}$, where h is the maximal surgery radius on $[t/2, t]$,*

(ii) *$\mathrm{Rm} \geqslant -r^{-2}$ on $B(x,t,r)$,*

(iii) *$\operatorname{vol} B(x,t,r) \geqslant wr^3$.*

Then $P(x,t,r,-\hat{\tau}r^2)$ is unscathed and $|\mathrm{Rm}| < \widehat{K}r^{-2}$ there.

We return to the thin-thick decomposition theorem. Denote $M_n = (M, t_n^{-1}g(t_n))$, and assume that $x_n \in M_n^+(w)$ for all n. Once we have obtained the required curvature estimates, it remains to show that (M_n, x_n) sub-converges to a complete hyperbolic manifold (see Section 11.2.2), proving part (ii) of the thin-thick decomposition theorem. The *locally controlled curvature in the sense of Perelman* property, i.e., part (iii), is obtained from the local curvature estimates and Shi's derivative estimates. (cf. Subsection 11.2.3).

The functions $r(\cdot)$ and $\bar{\delta}(\cdot)$ being fixed as above, Theorem 11.1.1 then reduces to the Long-Time Existence Theorem 5.4.1 and the following result:

Theorem 11.1.7 (Thin-thick decomposition reduced). *Let $g(\cdot)$ be a Ricci flow with $r(\cdot), \delta(\cdot)$-bubbling-off on M defined on $[0,\infty)$, with normalised initial condition, and such that $\delta(\cdot) \leqslant \bar{\delta}(\cdot)$. Then $g(\cdot)$ satisfies assertions* (i)–(iii) *of the conclusion of Theorem* 11.1.1.

This theorem is proved in Section 11.2 below, assuming Theorems 11.1.3 and 11.1.6. The proof of these results occupies Chapter 12.

11.2 Proof of the thin-thick decomposition theorem

In this section, we consider a Ricci flow with $r(\cdot), \delta(\cdot)$-bubbling-off $g(\cdot)$ on $M \times [0,+\infty)$, with normalised initial condition, such that $\delta(\cdot) \leqslant \bar{\delta}(\cdot)$, where $\bar{\delta}(\cdot)$ is the function fixed in Definition 11.1.4. The goal is to prove assertions (i)–(iii) of the conclusion of Theorem 11.1.1.

11.2.1 Rescaled volume is bounded and limits are hyperbolic

The main results of this subsection are Proposition 11.2.1, which immediately implies assertion (i), and Proposition 11.2.3.

We first recall some definitions introduced in Section 1.3.1. We let $\widehat{R}(g) = R_{\min}(g)V(g)^{2/3}$. Recall that our assumption that M is not spherical implies that $R_{\min}(g) \leqslant 0$ for any metric g on M. By Proposition 1.3.1, $t \mapsto \widehat{R}(g(t))$ is then nondecreasing along any Ricci flow with bubbling-off (at singular times, $R_{\min}$ can only go up and V go down).

Proposition 11.2.1. *$V(t)(t+\frac{1}{4})^{-\frac{3}{2}}$ is nonincreasing.*

Proof. From (C.1) of Corollary C.2.2 and (1.3) on p. 10, we get at any regular time t:

$$\frac{V'}{V} \leqslant -R_{\min} \leqslant \frac{6}{4t+1}, \tag{11.2}$$

hence

$$\begin{aligned}(V(t)(4t+1)^{-\frac{3}{2}})' &= V'(4t+1)^{-\frac{3}{2}} - V\,6(4t+1)^{-\frac{5}{2}} \\ &= (4t+1)^{-\frac{3}{2}}\Big(V' - \frac{6V}{4t+1}\Big) \leqslant 0.\end{aligned}$$

□

It follows that $V(t)t^{-\frac{3}{2}}$ is also bounded as $t \to \infty$. This proves part (i) of Theorem 11.1.7.

Next we will show that whenever local limits exist (as $t \to +\infty$), they are hyperbolic. Along the way, we prove a few other useful facts.

Set $\bar{R} := \lim \widehat{R}(t)$ and $\bar{V} := \lim V(t)(t+\frac{1}{4})^{-\frac{3}{2}}$ where the limits are taken when $t \to \infty$.

Lemma 11.2.2. *If $\bar{V} > 0$ then $\bar{R}\bar{V}^{-\frac{2}{3}} = -\frac{3}{2}$ and $R_{\min}(t) \sim -\frac{3}{2t}$.*

Proof. We have

$$\begin{aligned}\bar{R}\bar{V}^{-\frac{2}{3}} &= \lim_{t\to\infty} R_{\min}(t)V(t)^{\frac{2}{3}}\Big(V(t)(t+\tfrac{1}{4})^{-\frac{3}{2}}\Big)^{-\frac{2}{3}} \\ &= \lim_{t\to\infty} R_{\min}(t)(t+\tfrac{1}{4}) \geqslant -\frac{3}{2},\end{aligned} \tag{11.3}$$

by the right inequality of (11.2).

Assume that $\bar{R}\bar{V}^{-\frac{2}{3}} = c > -\frac{3}{2}$. Then $R_{\min}(t)(t + \frac{1}{4}) \geqslant -\frac{3}{2} + \mu$ for some $\mu > 0$ and $t > t_0$. From $V'(t)V(t)^{-1} \leqslant -R_{\min}$,we have

$$\frac{V'(t)}{V(t)} \leqslant \Big(\frac{3}{2} - \mu\Big)\Big(t + \frac{1}{4}\Big)^{-1},$$

hence

$$V(t) \leqslant V(t_0)\Big(t + \frac{1}{4}\Big)^{\frac{3}{2}-\mu}.$$

Then

$$V(t)\Big(t + \frac{1}{4}\Big)^{-\frac{3}{2}} \leqslant V(t_0)\Big(t + \frac{1}{4}\Big)^{-\mu} \to 0,$$

contradicting the assumption $\bar{V} > 0$. Finally, it follows from equality in (11.3) that $R_{\min}(t) \sim -\frac{3}{2t}$. □

Proposition 11.2.3 (Limits are hyperbolic). *Let $0 < r < 1$ and $P(x_k, t_k, r\sqrt{t_k}, -r^2 t_k)$ be a sequence of parabolic neighbourhoods which, after parabolic rescaling by $1/t_k$, converge to a Ricci flow on $P(x_\infty, 1, r, -r^2)$. Then all sectional curvatures of the limit are equal to $\frac{-1}{4t}$, for all $t \in [1 - r^2, 1]$.*

Proof. Let us prove first that under the previous assumptions, $\bar{V} > 0$. Denote by $\bar{g}_k(t)$ the rescaled Ricci flow $t_k{}^{-1} g(t t_k)$. Then

$$V(t_k)(t_k + \frac{1}{4})^{-\frac{3}{2}} \sim \operatorname{vol}(\bar{g}_k(1)) \geqslant \operatorname{vol} B(\bar{x}_k, 1, r) \to \operatorname{vol} B(x_\infty, 1, r) > 0.$$

Lemma 11.2.4. *R is spatially constant on $P(x_\infty, 1, r, -r^2)$.*

Proof. If $R(\cdot, s_0)$ is not constant for some $s_0 \in [1 - r^2, 1]$, then there exist $c > 0$ and $\mu > 0$ such that

$$\forall s \in [s_0 - \mu, s_0 + \mu], \quad \int_{B(x_\infty, s, r)} R_{\min\infty}{}^*(s) - R(x, s)\, dv_{g_\infty} < -c,$$

where $R_{\min\infty}{}^*(s) = \min\{R_\infty(x, s) \mid x \in B(x_\infty, s, r)\}$. Hence for large k,

$$\int_{B(x_k, st_k, r\sqrt{t_k})} R_{\min}{}^*(st_k) - R(x, st_k)\, dv_{g(st_k)} < \frac{-c}{2}\sqrt{t_k},$$

where now $R_{\min}{}^*(st_k) = \min\{R(x, st_k) \mid x \in B(x_k, st_k, r\sqrt{t_k})\} \geqslant R_{\min}(st_k)$. Thus,

$$\int_{B(x_k, st_k, r\sqrt{t_k})} R_{\min}(st_k) - R(x, st_k)\, dv_{g(st_k)} < \frac{-c}{2}\sqrt{t_k}, \tag{11.4}$$

After passing to a subsequence, we can suppose

$$\frac{t_{k+1}}{t_k} > \frac{s_0+\mu}{s_0-\mu},$$

that is $t_{k+1}(s_0-\mu) > t_k(s_0+\mu)$. Then, using Corollary C.2.2 of the Appendix,

$$\begin{aligned}
&\overline{R} - \widehat{R}(0)\\
&\quad= \int_0^\infty \frac{d\widehat{R}}{dt}\,dt \geqslant \sum_{k=0}^\infty \int_{t_k(s_0-\mu)}^{t_k(s_0+\mu)} \frac{d\widehat{R}}{dt}\,dt\\
&\quad\geqslant \sum_{k=0}^\infty \int_{t_k(s_0-\mu)}^{t_k(s_0+\mu)} \tfrac{2}{3}\widehat{R}(t)V^{-1}(t)\int_M (R_{\min}(t)-R(x,t))\,dv_{g(t)}\,dt\\
&\quad= \sum_{k=0}^\infty \int_{s_0-\mu}^{s_0+\mu} \tfrac{2}{3}\widehat{R}(st_k)V^{-1}(st_k)\int_M t_k(R_{\min}(st_k)-R(x,st_k))\,dv_{g(st_k)}\,ds\\
&\quad\geqslant \sum_{k=0}^\infty t_k \int_{s_0-\mu}^{s_0+\mu} \tfrac{2}{3}\widehat{R}(st_k)V^{-1}(st_k)\int_{B(x_k,st_k,r\sqrt{t_k})} (R_{\min}(st_k)-R(x,st_k))\,dv\,ds\\
&\quad\geqslant \frac{2}{3}\sum_{k=0}^\infty t_k \int_{s_0-\mu}^{s_0+\mu} \widehat{R}(st_k)V^{-1}(st_k)\frac{-c}{2}\sqrt{t_k}\,ds\\
&\quad= \frac{-c}{3}\sum_{k=0}^\infty \int_{s_0-\mu}^{s_0+\mu} \widehat{R}(st_k)V^{-1}(st_k)t_k^{\frac{3}{2}}\,ds.
\end{aligned}$$

Using

$$\widehat{R}(st_k) \xrightarrow[k\to\infty]{} \overline{R},$$

and

$$V^{-1}(st_k)t_k^{\frac{3}{2}} = V^{-1}(st_k)(st_k)^{\frac{3}{2}}s^{-\frac{3}{2}} \xrightarrow[k\to\infty]{} \overline{V}^{-1}s^{-\frac{3}{2}},$$

and $-c\,\widehat{R}(st_k) \geqslant 0$, we find that for some large $k_0 \in \mathbb{N}$,

$$\overline{R} - \widehat{R}(0) \geqslant \frac{-c\,\overline{R}\,\overline{V}^{-1}}{4}\sum_{k=k_0}^\infty \int_{s_0-\mu}^{s_0+\mu} s^{-\frac{3}{2}}\,ds = +\infty,$$

which gives the contradiction. □

We continue the proof of Proposition 11.2.3, the scalar curvature on the limit space being denoted by $R(t)$. Since $R_{\min}(t) \sim -\frac{3}{2t}$ in $(M, g(t))$ by Lemma 11.2.2, it follows that $R(t) \geqslant -\frac{3}{2t}$ on $P(x_\infty, 1, r, -r^2)$. Indeed, for $y \in B(x_\infty, 1, r)$,

$$\begin{aligned}
R(t) = R(y,t) &= \lim t_k R(y_k, tt_k) \geqslant \lim R_{\min}(\bar{g}_k(t))\\
&= \lim t_k R_{\min}(g(tt_k)) \underset{k\to\infty}{\sim} t_k\frac{-3}{2t_k t} = -\frac{3}{2t}.
\end{aligned}$$

If there exists $s_0 \in [1 - r^2, 1]$ such that $R(s_0) > \frac{3}{2s_0}$, we can argue as above replacing $R_{\min}(st_k)$ by $\frac{3}{2s_0}$ in (11.4) to get a contradiction.

Hence $R(t) = -\frac{3}{2t}$ on $P(x_\infty, 1, r, -r^2)$. Putting this in (ii) of Proposition C.2.1, we get

$$\frac{3}{2t^2} = 2|\operatorname{Ric}^0|^2 + \frac{2}{3}\Big(-\frac{3}{2t}\Big)^2$$

hence $|\operatorname{Ric}^0|^2 = 0$ on $P(x_\infty, 1, r, -r^2)$. It follows that $\operatorname{Ric} = \frac{R}{3}g = -\frac{1}{2t}g$, finishing the proof of Proposition 11.2.3. □

11.2.2 Hyperbolic limits exist: proof of part (ii)

We continue the proof of Theorem 11.1.7, showing that smoothly convergent subsequences exist for each sequence $x_n \in {M_n}^+(w)$, i.e., $g(\cdot)$ satisfies part (ii). We introduce a new definition:

Definition 11.2.5. If $\xi > 0$, we say that $g(\cdot)$ is ξ-almost hyperbolic at (x, t) if $|2t\operatorname{Ric}(x) + g(x,t)|_{g(t)} < \xi$.

Remark 11.2.6.

$$|2t\operatorname{Ric}(x) + g(x,t)|_{g(t)} = 2\Big|\operatorname{Ric}(x) + \frac{t^{-1}g(x,t)}{2}\Big|_{t^{-1}g(t)},$$

so ξ-almost hyperbolicity implies that the sectional curvatures of $t^{-1}g(t)$ are $\xi/2$-close to $\frac{-1}{4t}$.

Proposition 11.2.7. *For all positive numbers w, r, ξ, A there exist positive numbers Λ, τ and T with the following property. Let $x \in M$ and $t \geqslant T$ be such that*

(i) $\operatorname{vol} B(x,t,r\sqrt{t}) \geqslant w(r\sqrt{t})^3$,
(ii) $\operatorname{Rm} \geqslant -(r\sqrt{t})^{-2}$ *on* $B(x,t,r\sqrt{t})$.

Then

(a) *$P(x,t,Ar\sqrt{t}, -\tau(r\sqrt{t})^2)$ is unscathed and $|\operatorname{Rm}| < \Lambda(r\sqrt{t})^{-2}$ there, and*
(b) *$g(\cdot)$ is ξ-almost hyperbolic at every point of $B(x,t,Ar\sqrt{t})$.*

Proof of Proposition 11.2.7. We start by proving a weaker version of this proposition, Lemma 11.2.8 below. For $w' > 0$, let $\hat\theta' := \hat\theta(w')$, $\hat T' := \hat T(w')$, $\hat K' := \hat K(w')$, $\hat\tau' := \hat\tau(w')$, and $\hat r' := \hat r(w')$ denote the parameters given by Theorem 11.1.6.

Lemma 11.2.8. *For all $w' > 0$, $r \in (0, \hat r')$ and $\xi > 0$, there exists $T' = T'(w', r, \xi)$ with the following property. Let $x \in M$ and $t \geqslant T'$ be such that*

(i) $\operatorname{vol} B(x,t,r\sqrt{t}) \geqslant w'(r\sqrt{t})^3$,
(ii) $\operatorname{Rm} \geqslant -(r\sqrt{t})^{-2}$ *on* $B(x,t,r\sqrt{t})$.

Then

(a) $P(x,t,r\sqrt{t},-\hat{\tau}'(r\sqrt{t})^2)$ *is unscathed and* $|\mathrm{Rm}| < \widehat{K}'(r\sqrt{t})^{-2}$ *there, and*

(b) $g(\cdot)$ *is* ξ*-almost hyperbolic at* (x,t).

Proof of Lemma 11.2.8. We apply Theorem 11.1.6 with $w = w'$. For $r < \hat{r}'$ we have $r\sqrt{t} < \hat{r}'\sqrt{t}$. By taking t sufficiently large, we can ensure that $\hat{\theta}'^{-1}h \leqslant r\sqrt{t}$ (cf. Remark 5.4.8). Part (a) is then given by the conclusion of Theorem 11.1.6.

Assume now that part (b) is false. Then there exists $t_k \to +\infty$ and $x_k \in M$ such that $B(x_k,t_k,r\sqrt{t_k})$ satisfy (i) and (ii) but $(M,g_k(t_k))$ is not ξ-almost hyperbolic at (x_k,t_k). By part a), the neighbourhoods $P(x_k,t_k,r\sqrt{t_k},-\hat{\tau}'(r\sqrt{t_k})^2)$ are unscathed and $|\mathrm{Rm}| \leqslant \widehat{K}'(r\sqrt{t_k})^{-2}$ there for large k. Consider the parabolic rescalings $\bar{g}_k(\cdot) = {t_k}^{-1}g(t_k\cdot)$. The sequence of parabolic neighbourhoods $P(\bar{x}_k,1,r,-\hat{\tau}'r^2)$ then satisfies the assumptions of the Local Compactness Theorem C.3.3. Hence it subconverges towards a Ricci flow $g_\infty(\cdot)$ defined on a neighbourhood $P(x_\infty,1,-r,-\hat{\tau}'r^2)$.

Proposition 11.2.3 applies to a smaller neighbourhood $P(x_k,1,r',-r'^2)$ converging to $P(x_\infty,1,r',-r'^2)$ and shows that $g_\infty(t)$ has constant sectional curvature equal to $-\frac{1}{4t}$ there. We get a contradiction at (x_k,t_k) for k large enough, finishing the proof of Lemma 11.2.8. □

We go back to the proof of Proposition 11.2.7. We begin by extending the conclusion of Lemma 11.2.8 to any radius r (allowing $r > \hat{r}'$). We use Lemma 10.1.2. For $w > 0$, we set $w' = c_3 w$, where c_3 is the constant given by that lemma. For $r > 0$ we define $r' = \min\{r,\hat{r}'\}$. For any $\xi > 0$, Lemma 11.2.8 applies to $B(x,t,r'\sqrt{t})$ if $t \geqslant T'(w',r',\xi) = T_1$. Hence $P(x,t,r'\sqrt{t},-\hat{\tau}'(r'\sqrt{t})^2)$ is unscathed, we have $|\mathrm{Rm}| \leqslant \widehat{K}'(r'\sqrt{t})^{-2}$ there, and $g(\cdot)$ is ξ-almost hyperbolic at (x,t).

Our next goal is to enlarge the ball where the conclusion holds to be $B(x,t,Ar\sqrt{t})$, where $A > 0$ is the given constant. Define

$$B := \min\{\widehat{K}'^{-\frac{1}{2}},\hat{\tau}'^{\frac{1}{2}}\} \ll 1,$$
$$\rho := B\,r'\sqrt{t}.$$

By Lemma 11.2.8, $P(x,t,\rho,-\rho^2)$ is unscathed and $|\mathrm{Rm}| \leqslant \rho^{-2}$ there. Clearly $\rho < \sqrt{t/2}$. Define

$$A' := \max\{3AB^{-1}rr'^{-1},w'^{-1}\} \gg 1,$$

which gives

$$A'\rho \geqslant 3Ar\sqrt{t} \quad\text{and}\quad \mathrm{vol}\,B(x,t,\rho) \geqslant A'^{-1}\rho^3.$$

Recall that by choice of $\bar{\delta}(\cdot)$ (see Remark 11.1.5), Theorem 11.1.3 applies to constant $A' > 0$ for all $t \geqslant A'$. Define $T_2 := \max\{A',T_1\}$ and assume $t \geqslant T_2$.

Setting $\bar{K} := \max\{\bar{K}(A'),10\}$, we obtain by Theorem 11.1.3 (b) that any $(y,t) \in B(x,t,A'\rho)$ where $R(y,t) \geqslant \bar{K}\rho^{-2} =: Q$ is the centre of an (ε_0,C_0)-canonical neighbourhood. Define

$$B' := {C_0}^{-\frac{1}{2}} \bar{K}^{-\frac{1}{2}},$$
$$r'' := \min\{B'Br', \hat{r}({C_0}^{-1})\},$$
$$T_3 := \max\{T'({C_0}^{-1}, r'', 10^{-1}), T_2\},$$

where T' is the parameter defined in Lemma 11.2.8.

Claim. $R < Q$ on $B(x,t,A'\rho)$ if $t \geqslant T_3$.

Proof of the claim. We argue by contradiction: since $R(x,t) \leqslant 6\rho^{-2} < Q$, if the claim does not hold, then there exists $y \in B(x,t,A'\rho)$ such that $R(y,t) = Q$. By definition of Q, (y,t) is the centre of an (ε_0, C_0)-canonical neighbourhood $U \supset B(y,t,C_0^{-1}Q^{-1/2})$, and $R(\cdot,t) \in [C_0^{-1}Q, C_0 Q]$ on U. From Pinching Lemma 4.4.7, $|\mathrm{Rm}| \leqslant C_0 Q$ on U. Define $\rho' := r''\sqrt{t} \leqslant B'\rho$, and note that $\mathrm{Rm} \geqslant -C_0 Q \geqslant -\rho'^{-2}$ on $B(y,t,\rho')$. Moreover, by properties of the canonical neighbourhoods, $\operatorname{vol} B(y,t,\rho') \geqslant {C_0}^{-1}\rho'^3$. As $t \geqslant T_3$, Lemma 11.2.8 applies on $B(y,t,\rho')$ and says that $g(\cdot)$ is 10^{-1}-almost hyperbolic at (y,t). But this is not possible in a canonical neighbourhood. This proves the claim. □

By Pinching Lemma 4.4.7, we have $\mathrm{Rm} \geqslant -Q = -\bar{K}\rho^{-2}$ on $B(x,t,A'\rho)$. Set $\sigma := \bar{K}^{-\frac{1}{2}}\rho$, so that $\mathrm{Rm} \geqslant -\sigma^{-2}$, and pick a point $y \in B(x,t,\frac{A'\rho}{3})$. By the Bishop–Gromov inequality we have

$$\begin{aligned}\frac{\operatorname{vol} B(y,t,\sigma)}{\operatorname{vol} B(y,t,\frac{2A'\rho}{3})} &\geqslant \frac{v_{-\sigma^{-2}}(\sigma)}{v_{-\sigma^{-2}}(\frac{2A'\rho}{3})}\\ &= \frac{v_{-1}(1)}{v_{-1}(\frac{2A'\rho}{3}\sigma^{-1})}\\ &\geqslant \frac{v_{-1}(1)}{v_{-1}(\frac{2A'}{3}\bar{K}^{1/2})} =: E.\end{aligned}$$

Since $B(y,t,\frac{2A'\rho}{3}) \supset B(x,t,\frac{A'\rho}{3}) \supset B(x,t,\rho)$, we deduce

$$\operatorname{vol} B\big(y,t,\tfrac{2A'\rho}{3}\big) \geqslant \operatorname{vol} B(x,t,\rho) \geqslant w'\rho^3,$$

hence

$$\operatorname{vol} B(y,t,\sigma) \geqslant Ew'\rho^3 = Ew'\bar{K}^{\frac{3}{2}}\sigma^3.$$

It follows that for any $\sigma' \in (0,\sigma)$, we have

$$\operatorname{vol} B(y,t,\sigma') \geqslant c_3 E w'\bar{K}^{\frac{3}{2}}\sigma'^3 = w''\sigma'^3.$$

Define

$$
\begin{aligned}
\hat{r}'' &:= \hat{r}(w''),\\
r'' &:= \min\{\bar{K}^{-\frac{1}{2}} Br, \hat{r}''\},\\
T_4 &:= \max\{T'(w'', r'', \xi), T_3\},\\
\widehat{K}'' &:= \widehat{K}(w''),\\
\hat{\tau}'' &:= \hat{\tau}(w'').
\end{aligned}
$$

For $\sigma' = r''\sqrt{t} \leqslant \sigma$, Lemma 11.2.8 can be applied to $B(y,t,r''\sqrt{t})$ and gives that $P(y,t,r''\sqrt{t}, -\hat{\tau}''(r''\sqrt{t})^2)$ is unscathed and $|\mathrm{Rm}| \leqslant \widehat{K}''(r''\sqrt{t})^{-2}$ there, and that $g(\cdot)$ is ξ-almost hyperbolic at (y,t). Thus, if one sets

$$
\tau := \hat{\tau}''\Big(\frac{r''}{r}\Big)^2, \quad \Lambda := \widehat{K}''\Big(\frac{r''}{r}\Big)^{-2}, \quad \text{and} \quad T := T_4,
$$

and one recalls that (y,t) is an arbitrary point of $B\big(x,t,\frac{A'\rho}{3}\big) \supset B(x,t,Ar\sqrt{t})$, it is clear now that the conclusion of Proposition 11.2.7 holds on $P(x,t,Ar\sqrt{t}, -\tau(r\sqrt{t})^2)$, with parameter Λ for $t \geqslant T_4$. □

Lemma 11.2.9 (Lower curvature bound in thick part). *For all $w > 0$ there exists $\bar{\rho} \in (0,1]$ and $\bar{T} \in (0,+\infty)$ such that for any $t \geqslant \bar{T}$ and x in the w-thick part of $(M, t^{-1}g(t))$, then $B(x,\bar{\rho}) \subset (M, t^{-1}g(t))$ satisfies* $\mathrm{Rm} \geqslant -(\bar{\rho})^{-2}$.

Remark 11.2.10. By definition of the w-thick part, this implies $\mathrm{vol}\, B(x,\bar{\rho}) \geqslant w(\bar{\rho})^3$.

Proof of Lemma 11.2.9. We argue by contradiction. Suppose there exists $\rho_k \to 0$, $t_k \to +\infty$ and points x_k in the w-thick part of $(M, {t_k}^{-1}g(t_k))$ such that sectional curvature is not bounded below by $-(\rho_k)^{-2}$ on the ball $B(x_k,\rho_k) \subset (M, {t_k}^{-1}g(t_k))$. Let $\rho'_k \in (0,\rho_k)$ be such that $\inf\{\mathrm{Rm}(y) \mid y \in B(x_k,\rho'_k)\} = -(\rho'_k)^{-2}$, and let $y_k \in \overline{B(x_k,\rho'_k)}$ be a point where the minimum is achieved. We define $r_k := \rho'_k\sqrt{t_k}$, and note that $t_k {r_k}^{-2} = {\rho'_k}^{-2} \to \infty$. It follows from the curvature pinched toward positive assumption, and more precisely from (4.8) p. 44, that (for the unscaled metric):

$$
\frac{R(y_k,t_k)}{{r_k}^{-2}} \geqslant \frac{1}{2}\big(\log((1+t_k){r_k}^{-2}) - 3\big) \to +\infty. \tag{11.5}
$$

We shall apply Theorem 11.1.6, showing first that $r_k \geqslant \theta^{-1}(w)h_k$, where $h_k = h(t_k/2)$ is the maximum of the surgery radii on $[t_k/2, t_k]$. Recall that $h(t_k/2) \leqslant \delta(t_k/2)r(t_k/2) \leqslant \eta(t_k/2)r(t_k)$ by Definition 11.1.4, hence $h(t_k/2) \ll r(t_k)$ for large k. It then suffices to show the following:

Assertion 11.2.11. $r_k \geqslant r(t_k)$ *for large* k.

Proof of Assertion 11.2.11. We argue by contradiction. Up to extracting a subsequence, assume that $r_k < r(t_k)$ for all k. Let us show that

$$\inf\{R(x', t_k){r_k}^2 \mid x' \in B(x_k, t_k, r_k)\} \to +\infty.$$

If this is not true, then let $c \in (1, +\infty)$ and $x'_k \in B(x_k, t_k, r_k)$ be such that $R(x'_k, t_k) < c{r_k}^{-2}$. Consider a point z_k on the segment $[y_k x'_k]$ such that $R(z_k, t_k) = c{r_k}^{-2}$. We apply the Curvature-Distance Theorem 6.1.1 at (z_k, t_k) with $A = 2c^{1/2}$. Since every point of scalar curvature greater than $R(z_k, t_k)$, hence greater than $r(t_k)^{-2}$, has an (ε_0, C_0)-canonical neighbourhood, when $(1 + t_k)R(z_k, t_k) \geqslant Q(A, \varepsilon_0, C_0)$, we get the upper bound $R \leqslant \Lambda(A, \varepsilon_0, C_0)R(z_k, t_k)$ on the ball $B(z_k, t_k, AR(z_k, t_k)^{-1/2}) = B(z_k, t_k, 2r_k)$. In particular $R(y_k, t_k) \leqslant \Lambda c{r_k}^{-2}$ which is a contradiction. Hence $\inf\{R(x', t_k){r_k}^2 \mid x' \in B(x_k, t_k, r_k)\} \to +\infty$.

From that we deduce that for k large enough, the scalar curvature on $B(x_k, t_k, r_k)$ is much greater than ${r_k}^{-2}$. This implies that each point of the ball is the centre of an (ε_0, C_0)-canonical neighbourhood, of size much smaller than r_k. We now recall that x_k is in the w-thick part of $(M, {t_k}^{-1}g(t_k))$ and hence that $\mathrm{vol}\, B(x_k, t_k, r_k) \geqslant w{r_k}^3$ by the very definition of r_k. The sub-ball lemma (Lemma D.1.6) applied to the ball of radius $r_k/2$ yields a ball $B(y_k, t_k, \theta_0(w)r_k/2) \subset B(x_k, t_k, r_k)$ all of whose sub-balls $B(z, r) \subset B(y_k, t_k, \theta_0(w)r_k/2)$ have $(1 - \varepsilon_0)$-almost Euclidean volume. However consider a sub-ball $B(z, r)$ in an ε_0-neck included in $B(y_k, t_k, \theta_0(w)r_k/2)$ with $r = {\varepsilon_0}^{-1}R(z, t_k)$ being the radius of the ε_0-neck. We obtain that $\mathrm{vol}\, B(z, r) \approx {\varepsilon_0}^{-1}2r$ is much less than r^3. This gives a contradiction, proving the assertion. □

We now finish the proof of Lemma 11.2.9. Since $\rho'_k \to 0$, we have

$$\theta^{-1}(w)h_k \leqslant r_k = \rho'_k\sqrt{t_k} \leqslant \bar{r}(w')\sqrt{t_k}.$$

From Theorem 11.1.6, $P(x_k, t_k, r_k, -\hat{\tau}{r_k}^2)$ is unscathed and $|\mathrm{Rm}| \leqslant \widehat{K}{r_k}^{-2}$ there. Hence $R \leqslant 6K{r_k}^{-2}$, contradicting (11.5). □

At last we can prove assertion (ii) of Theorem 11.1.7: fix $w > 0$, a sequence $t_n \to \infty$ and a sequence of points x_n such that for each n, x_n belongs to the w-thick part of $(M, (t_n)^{-1}g(t_n))$. Set r equal to the parameter $\bar{\rho}$ given by Lemma 11.2.9 and choose sequences $\xi_n \to 0$ and $A_n \to \infty$. Applying Proposition 11.2.7 and Hamilton's Compactness Theorem C.3.1, we obtain a subsequence converging to some complete pointed 3-manifold of sectional curvature constant equal to $-1/4$. By the already proved part (i), the limit has finite volume.

11.2.3 Locally controlled curvature: proof of part (iii)

In this subsection we prove that $g(\cdot)$ has *locally controlled curvature in the sense of Perelman*, which is part (iii) of Theorem 11.1.7. Precisely, we check that for every

$w' > 0$ there exists $\bar{r} > 0, K_0, K_1 > 0$ and $t_0 \in (0, +\infty)$ such that for $x \in M$ and $t > t_0$, for all $r \in (0, \bar{r}]$, if $B(x, r) \subset (M, t^{-1}g(t))$ satisfies $\mathrm{Rm} \geqslant -r^{-2}$ and $\mathrm{vol}\, B(x, r) \geqslant w' r^3$, then $|\mathrm{Rm}(x)| \leqslant K_0 r^{-2}$ and $|\nabla \mathrm{Rm}(x)| \leqslant K_1 r^{-3}$.

We shall apply Theorem 11.1.6, with parameters $\hat{r}, \widehat{T}, \hat{\theta}, \widehat{K}$ and $\hat{\tau}$ depending on w'. Set $\bar{r} := \hat{r}$ and consider, for $r \in (0, \bar{r}]$ and $t \geqslant \widehat{T}$, a ball $B(x, r) \subset (M, t^{-1}g(t))$ as above. Denote by h the maximum of the function $h(\cdot)$ on $[t/2, t]$, i.e., $h = h(t/2)$. If $r\sqrt{t} > \hat{\theta}^{-1}h$, then Theorem 11.1.6 applies and says that the neighbourhood (for the unscaled metric) $P(x, t, r\sqrt{t}, -\hat{\tau}(r\sqrt{t})^2)$ is unscathed and $|\mathrm{Rm}| < \widehat{K}(r\sqrt{t})^{-2}$ there. The bound on the covariant derivative is then obtained from Shi's estimate (see [MT07], Theorem 3.29). The conclusion for the rescaled metric $t^{-1}g(t)$ follows.

In the sequel we then argue by contradiction, assuming $r\sqrt{t} \leqslant \theta^{-1}h$. Let $x_k \in M_k$, $t_k \to +\infty$ and $h_k = h(t_k/2)$, $r_k \to 0$ such that $r_k\sqrt{t_k} \leqslant \theta^{-1}h_k$, be sequences such that $B(x_k, r_k) \subset (M, {t_k}^{-1}g(t_k))$ is a counterexample. Up to taking a subsequence there are two cases.

Case 1. $r_k\sqrt{t_k} \leqslant \theta^{-1}h_k$ and $R(x_k, t_k) \geqslant r(t_k)^{-2}$.

We then have that (x_k, t_k) is the centre of a canonical neighbourhood. Property (4.4), p. 40, leads to $|\nabla \mathrm{Rm}| < C_0|\mathrm{Rm}|^{3/2}$ on this neighbourhood, and in particular at (x_k, t_k). Thus it suffices to estimate from above $|\mathrm{Rm}(x_k, t_k)|$ by const. $(r_k\sqrt{t_k})^{-2}$, and for this to show that $R(x_k, t_k)$ is bounded above by $c(r_k\sqrt{t_k})^{-2}$, where c depends on w' only. Pinching Lemma 4.4.7 then allows us to conclude. We argue by contradiction and assume that

$$R(x_k, t_k)(r_k\sqrt{t_k})^2 \to +\infty,$$

possibly after taking a subsequence. Then we can repeat the proof of Lemma 11.2.9, starting at equation (11.5). Indeed, the Distance-Curvature Theorem 6.1.1 implies that the infimum of $R(\cdot, t_k)(r_k\sqrt{t_k})^2$ on $B(x_k, t_k, r_k\sqrt{t_k})$ goes to $+\infty$. As $(r_k\sqrt{t_k})^{-2} \geqslant \theta^2 {h_k}^{-2} \gg r(t_k)^{-2}$ for all k large enough, the ball $B(x_k, t_k, r_k\sqrt{t_k})$ is totally covered by canonical neighbourhoods of very small size compared to r_k. We then take a ball $B(y_k, t_k, \theta_0(w')(r_k\sqrt{t_k})) \subset B(x_k, t_k, r_k\sqrt{t_k})$ given by Sub-Ball Lemma D.1.6, all of whose sub-balls have $(1-\varepsilon_0)$-almost Euclidean volume. Considering an ε_0-neck included in one of these sub-balls, whose volume is not almost Euclidean, gives the contradiction.

Case 2. $r_k\sqrt{t_k} \leqslant \theta^{-1}h_k$ and $R(x_k, t_k) < r(t_k)^{-2}$.

We note (using Definition 11.1.4) that

$$Q_k := |R(x_k, t_k)| + r(t_k)^{-2} < 2r(t_k)^{-2} \ll {h_k}^{-2} \leqslant 2(r_k\sqrt{t_k})^{-2}\theta^{-2}.$$

If $P(x_k, t_k, (2C_0\sqrt{Q_k})^{-1}, -(8C_0Q_k)^{-1})$ is unscathed, then the Curvature-Distance Lemma 6.1.2 shows that $R \leqslant 4Q_k \leqslant (r_k\sqrt{t_k})^{-2}$ on this neighbourhood. Pinching Lemma 4.4.7 implies that $|\mathrm{Rm}| \leqslant (r_k\sqrt{t_k})^{-2}$ on the same neighbourhood. The life-time of the flow being greater than $(16C_0)^{-1}(r_k\sqrt{t_k})^2$, we obtain from Shi's estimate [MT07], Theorem 3.29, the upper bound $|\nabla \mathrm{Rm}| \leqslant c(r_k\sqrt{t_k})^{-3}$ at (x_k, t_k), where c is independent of k, reaching a contradiction.

Now the estimate $R \leqslant 4Q_k \leqslant 8r(t_k)^{-2}$ holds on $B(x_k, t_k, (2C_0\sqrt{Q_k})^{-1}) \times (-\tau, 0]$, if $\tau \in [0, (8C_0 Q_k)^{-1}]$ and if this set is unscathed. As $8r(t_k)^{-2} \ll {h_k}^{-2}$, which is below the curvature scale of surgeries on $[t_k/2, t_k]$, this proves that $P(x_k, t_k, (2C_0\sqrt{Q_k})^{-1}, -(8C_0 Q_k)^{-1})$ are in fact unscathed, and finishes the proof of Theorem 11.1.7.

Notes. Theorem 11.1.3 corresponds to Proposition 6.3 in [Per03b]. Theorem 11.1.6 to Corollary 6.8 of [Per03b].

Arguments of this chapter corresponds to [Per03b], Section 7.1, [KL08], Sections 86–87, Lemma 92.13.

Remark 11.2.12. In Corollary 6.8 of [Per03b], the conclusion is that $R < \widehat{K}r^{-2}$ on $P(x, t, r/4, -\hat{\tau}r^2)$. The extension to $|\mathrm{Rm}|$ follows from Pinching Lemma 4.4.7. The extension to the parabolic ball of radius r follows from part (c) of Theorem 11.1.3. The setting above is made for later convenience.

Chapter 12
Refined estimates for long-time behaviour

In this chapter, we prove the technical Theorems 11.1.3 and 11.1.6, stated and used in the previous chapter to prove the thin-thick decomposition theorem. Section 12.1 is devoted to Theorem 11.1.3. The proof of Theorem 11.1.6 is given in Section 12.2.

We assume throughout that M is a closed, orientable, irreducible, nonspherical 3-manifold.

12.1 Spatial extension of local estimates: proof of Theorem 11.1.3

For convenience we recall the statement of this theorem:

Theorem 12.1.1 (Theorem 11.1.3). *For every positive number A there exist positive numbers κ, $\bar{K}$, $\bar{K}'$, $\bar{r}$ and a nonincreasing function $t \mapsto \bar{\delta}_A(t)$ with the following property. Let $g(\cdot)$ be a Ricci flow with $r(\cdot)$, $\delta(\cdot)$-bubbling-off on M with normalised initial condition, and let (x, t, ρ) have the following properties. Assume that $0 < \rho < \sqrt{t/2}$, that $\delta(\cdot) < \bar{\delta}_A(t)$ on $[t/2, t]$, and that*

(i) *$P(x, t, \rho, -\rho^2)$ is unscathed,*
(ii) *$|\mathrm{Rm}| \leqslant \rho^{-2}$ there,*
(iii) $\mathrm{vol}\, B(x, t, \rho) \geqslant A^{-1}\rho^3$.

Then the following assertions hold true:

(a) *The solution is κ-noncollapsed on scales less than ρ in $B(x, t, A\rho)$.*
(b) *Every point of $B(x, t, A\rho)$ of scalar curvature at least $\bar{K}\rho^{-2}$ admits a canonical neighbourhood.*
(c) *If $\rho \leqslant \bar{r}\sqrt{t}$, then $R \leqslant \bar{K}'\rho^{-2}$ on $B(x, t, A\rho)$.*

Recall that part (a) has been proved in Section 10.4. Parts (b) and (c) are tackled in Sections 12.1.1 and 12.1.2 respectively.

12.1.1 Canonical neighbourhoods: proof of part (b)

As usual we argue by contradiction. As in the proof of Proposition B, the goal is to prove that a sequence of counterexamples subconverges, after rescaling, to a κ-solution for some $\kappa > 0$.

We fix $A > 0$, a sequence $\bar{K}^\alpha \to +\infty$, and $\bar{\delta}^\alpha(\cdot)$ be a sequence of nonincreasing functions going to zero. We assume that $\bar{\delta}^\alpha(\cdot) \leqslant \bar{\delta}_{A'}(\cdot)$ where

$$A' = \max\left\{\frac{\upsilon_{-1}(1)}{\upsilon_{-1}(10^{-2})}10^{-3/2}A, 20A\right\},$$

and $\bar{\delta}_{A'}(\cdot)$ is the function given by part (a) of Theorem 12.1.1. We assume also that

$$\bar{\delta}^\alpha(t) \leqslant \bar{\delta}_{\mathrm{per}}\left(t, 1 - \frac{1}{t}, r(t)\right),$$

which is the parameter given by Persistence Theorem 8.1.2.

For each α, we assume there exists a counterexample. Let $t^\alpha \in [0, +\infty)$ and $g_\alpha(\cdot)$ be a Ricci flow with $r(\cdot), \delta^\alpha(\cdot)$-bubbling-off on M^α, such that $\delta^\alpha(t) \leqslant \bar{\delta}^\alpha(t^\alpha)$ for $t \in [t^\alpha/2, t^\alpha]$, and assume that there exists

- $\rho^\alpha \in (0, \sqrt{t^\alpha/2}]$ and $x^\alpha \in M^\alpha$ such that $P(x^\alpha, t^\alpha, \rho^\alpha, -(\rho^\alpha)^2)$ is unscathed and has curvature $|\mathrm{Rm}| \leqslant (\rho^\alpha)^{-2}$ and $\mathrm{vol}\, B(x^\alpha, t^\alpha, \rho^\alpha) \geqslant A^{-1}(\rho^\alpha)^3$;
- $y^\alpha \in B(x^\alpha, t^\alpha, A\rho^\alpha)$ such that $R(y^\alpha, t^\alpha) \geqslant \bar{K}^\alpha(\rho^\alpha)^{-2}$ and that (y^α, t^α) is not the centre of an (ε_0, C_0)-canonical neighbourhood.

We observe that $R(y^\alpha, t^\alpha) < r(t^\alpha)^{-2}$, and hence that $\bar{K}^\alpha(\rho^\alpha)^{-2} < r(t^\alpha)^{-2}$. It follows that $2\bar{K}^\alpha < t^\alpha r(t^\alpha)^{-2}$. Since $t\, r(t)^{-2}$ is bounded on finite intervals, $t^\alpha \to +\infty$.

We first refine the choice of 'bad' points (x^α, t^α) using a standard point picking argument.

In the sequel, for the sake of simplicity, we shall suppress the indices α.

Lemma 12.1.2 (Point picking). *For all α large enough, there exist $\bar{t} \in [t - \rho/2, t]$ and $\bar{x} \in B(x, \bar{t}, 2A\rho)$ such that $\bar{Q} := R(\bar{x}, \bar{t}) \geqslant \bar{K}\rho^{-2}$, $(\bar{x}, \bar{t})$ is not the centre of an (ε_0, C_0)-canonical neighbourhood but every (x', t') such that*

$$\bar{t} - \frac{1}{4}\bar{K}\bar{Q}^{-1} \leqslant t' \leqslant \bar{t}, \quad d_{t'}(x', x) < d_{\bar{t}}(x, \bar{x}) + \bar{K}^{1/2}\bar{Q}^{-1/2},$$

and $R(x', t') \geqslant 2\bar{Q}$, is centre of an (ε, C_0)-canonical neighbourhood.

Proof of Lemma 12.1.2. Starting with (y, t), we inductively define (y_i, t_i) as follows. We set $Q_i = R(y_i, t_i)$. If there exists (y', t') such that $t_i - \frac{1}{4}\bar{K}{Q_i}^{-1} \leqslant t' \leqslant t_i$, $d_{t'}(y', x) < d_{t_i}(x, y_i) + \bar{K}^{1/2}{Q_i}^{-1/2}$, $R(y', t') \geqslant 2Q_i$, and (y', t') is not centre of an (ε_0, C_0)-canonical neighbourhood, then we set $(y_{i+1}, t_{i+1}) := (y', t')$. Otherwise, $(\bar{x}, \bar{t}) := (y_i, t_i)$ satisfies the requirement. We observe that as long as the sequence is not stationary, we have

$$Q_i = R(y_i, t_i) \geqslant 2Q_{i-1} \geqslant 2^i Q_0 \geqslant 2^i \bar{K}\rho^{-2},$$

and $Q_i < r(t_i)^{-2} \leqslant r(t)^{-2} < +\infty$ (α is fixed here). The sequence thus becomes stationary for i large enough. Since $t_{i+1} \in [t_i - \frac{1}{4}\bar{K}{Q_i}^{-1}, t_i]$, we have $t_i - t_{i+1} \leqslant$

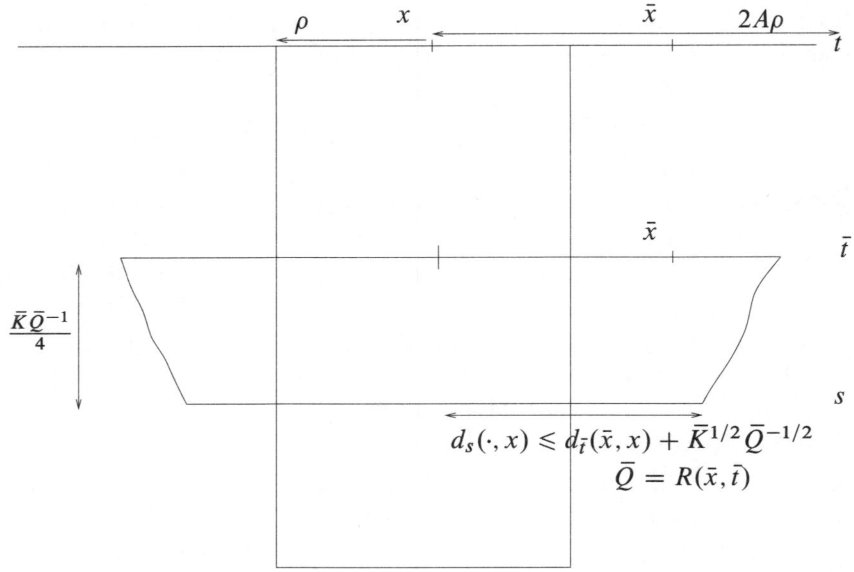

$\frac{1}{4}\bar{K}Q_i{}^{-1} \leqslant \frac{1}{4}2^{-i}\rho^2$, from which we deduce

$$t - t_{i+1} \leqslant \sum_{j=0}^{i} t_j - t_{j+1} \leqslant \frac{1}{4}\sum_{j=0}^{\infty} 2^{-j}\rho^2 = \rho^2/2.$$

On the other hand

$$\begin{aligned} d_{t_{i+1}}(y_{i+1}, x) &< d_{t_i}(y_i, x) + \bar{K}^{1/2}Q_i{}^{-1/2} \\ &< A\rho + \sum_{j=0}^{i} \bar{K}^{1/2}Q_j{}^{-1/2} < A\rho + \sum_{j=0}^{\infty}(\sqrt{2})^{-j}\rho < 2A\rho \end{aligned}$$

since one may assume, without loss of generality, that $A > \sum_{j=0}^{\infty}(\sqrt{2})^{-j}$. □

Lemma 12.1.3. *There exists $\kappa' = \kappa'(A) > 0$ such that for all α large enough, $g(\cdot)$ is κ'-noncollapsed on scales less than ρ on $B(x, \bar{t}, 2A\rho)$.*

Proof. We shall apply part (a) of Theorem 12.1.1 to the ball $B(x, \bar{t}, 10^{-1}\rho)$, with the constant A' which satisfies $A'10^{-1}\rho \geqslant 2A\rho$. Let us check the assumptions of the theorem at $(x, \bar{t})$. We first remark that $10^{-1}\rho \leqslant \sqrt{\bar{t}/2}$ as $\bar{t} \geqslant t - \rho^2/2 \geqslant t/2 \geqslant \rho^2$.

Since $|\mathrm{Rm}| \leqslant \rho^{-2}$ on $P(x, t, \rho, -\rho^2)$ and $\bar{t} \geqslant t - \rho^2/2$, it follows from the distance-distortion lemma (Lemma 2.2.7) that

$$10^{-1} \leqslant \frac{g(t)}{g(\bar{t})} \leqslant e^{4\rho^{-2}(\rho^2/2)} = e^2 \leqslant 10,$$

and

$$B(x,t,10^{-2}\rho) \subset B(x,\bar{t},10^{-1}\rho) \subset B(x,t,\rho).$$

The neighbourhood $P(x,\bar{t},10^{-1}\rho,-10^{-2}\rho^2)$ is thus unscathed and satisfies the required curvature bound. Moreover, using the Bishop–Gromov inequality:

$$\begin{aligned}
\mathrm{vol}_{g(\bar{t})} B(x,\bar{t},10^{-1}\rho) &\geqslant 10^{-3/2}\,\mathrm{vol}_{g(t)} B(x,\bar{t},10^{-1}\rho)\\
&\geqslant 10^{-3/2}\,\mathrm{vol}_{g(t)} B(x,t,10^{-2}\rho)\\
&\geqslant 10^{-3/2}\frac{v_{-\rho^{-2}}(10^{-2}\rho)}{v_{-\rho^{-2}}(\rho)}\,\mathrm{vol}_{g(t)} B(x,t,\rho)\\
&\geqslant 10^{-3/2}\frac{v_{-1}(10^{-2})}{v_{-1}(1)}A^{-1}\rho^3\\
&\geqslant A'^{-1}(10^{-1}\rho)^3.
\end{aligned}$$

Recall that $\bar{\delta}^\alpha(\cdot) < \bar{\delta}_{A'}(\cdot)$. Part (a) of Theorem 12.1.1 yields a constant $\kappa = \kappa(A')$ such that $g(\cdot)$ is κ-noncollapsed on $B(x,\bar{t},A'10^{-1}\rho)$ on scales less than $10^{-1}\rho$. We deduce that $g(\cdot)$ is $10^{-3}\kappa'$-noncollapsed at scales less than ρ, on $B(x,\bar{t},A'10^{-1}\rho) \supset B(x,\bar{t},2A\rho)$. Thus Lemma 12.1.3 is proved by setting $\kappa' := 10^{-3}\kappa$. □

We now consider the parabolic rescalings

$$\bar{g}(\cdot) = R(\bar{x},\bar{t})g\Big(\bar{t} + \frac{\cdot}{R(\bar{x},\bar{t})}\Big),$$

and we argue as in the proof of Proposition B in order to extract a subsequence converging towards a κ-solution. Using Theorem 4.2.11 and Definition 5.1.1, this will give a contradiction for α large enough. Unless otherwise stated we work with these rescaled metrics. We note that $(\bar{x},0)$ does not have an (ε_0,C_0)-canonical neighbourhood, but all points (x',t') such that $t' \in [-\bar{K}/4,0]$, $d_{t'}(x',x) < d_0(\bar{x},x) + \bar{K}^{1/2}$ and $R(x',t') \geqslant 2$ do have such neighbourhoods. In the sequel, we set[1]

$$\widetilde{P} := \{(x',t') \mid t' \in [-\bar{K}/4,0], d_{t'}(x',x) < d_0(\bar{x},x) + \bar{K}^{1/2}\}.$$

Since $\delta^\alpha(\cdot)$ goes to 0, one can check that Lemma 9.2.1 is still valid, although $\delta^\alpha(\cdot)$ is now a function. For convenience we give this modified statement:

Lemma 12.1.4. *For all positive numbers $K, \bar{A}, \tau$, there exists $\alpha_0 = \alpha_0(K,\bar{A},\tau) \in \mathbb{N}$ such that for all $\alpha \geqslant \alpha_0$, if $|\mathrm{Rm}| \leqslant K$ on $B(\bar{x},0,\bar{A}) \times (-\tau,0]$, then $P(\bar{x},0,\bar{A},-\tau)$ is unscathed.*

Step 1. The sequence $(M,\bar{g}(\cdot),\bar{x})_\alpha$ subconverges towards a partial Ricci flow $(M_\infty,g_\infty(\cdot),x_\infty)$, where $g_\infty(0)$ is complete and $g_\infty(t)$ has nonnegative curvature operator.

[1] Note that this is not a parabolic neighbourhood. This entails some complications.

Proof of Step 1. Let us check that arbitrarily large (in space) parabolic neighbourhoods are unscathed with bounded curvature. From the Curvature-Distance Theorem 6.1.1, for all $\bar{A} > 0$, there exist $Q = Q(\bar{A}) > 0$ and $\Lambda = \Lambda(\bar{A}) > 0$ such that if (x', t') satisfies

(i) $(1 + t')R(x', t') \geqslant Q$,

(ii) every $y \in B(x', t', \bar{A}R(x', t')^{-1/2})$ such that $R(y, t') \geqslant 2R(x', t')$ has an (ε_0, C_0)-canonical neighbourhood,

then $R(z, t') \leqslant \Lambda R(x', t')$ for all $z \in B(x', t', \bar{A}R(x', t')^{-1/2})$.

Assumption (ii) is satisfied at $(\bar{x}, \bar{t})$ by definition, as soon as $\bar{K}^{1/2} \geqslant \bar{A}$ since then $B(x', t', \bar{A}R(x', t')^{-1/2}) \subset \widetilde{P}$. Assumption (i) is also verified for α large enough, since

$$(1 + \bar{t})R(\bar{x}, \bar{t}) \geqslant \bar{t}\bar{K}\rho^{-2} \geqslant 2\bar{K}.$$

We conclude that the evolving metrics $\bar{g}(\cdot)$ satisfy $R(\cdot, 0) \leqslant \Lambda(\bar{A})$ on $B(\bar{x}, 0, \bar{A})$ for all α large enough. In order to be able to use the local compactness theorem for flows, Theorem C.3.3, let us show that there exists $\tau(\bar{A}) > 0$ such that $P(\bar{x}, 0, \bar{A}, -\tau(\bar{A}))$ is unscathed and has bounded curvature, for all α large enough. Precisely, we now intend to exhibit an honest parabolic neighbourhood contained in $\widetilde{P}$. We set $K = K(\bar{A}) := 2(\Lambda(\bar{A}) + 2)$. We may assume this number to be large enough so that, by Pinching Lemma 4.4.7, if $R \leqslant K$ then $|\operatorname{Rm}| \leqslant K$. We define $\alpha_1(\bar{A}) \geqslant \alpha_0(K, \bar{A}, (C_0K)^{-1})$, where α_0 is the parameter given by Lemma 12.1.4, large enough such that for $\alpha \geqslant \alpha_1$, $\bar{K}^{1/2} > 100\bar{A}$.

Lemma 12.1.5. *For all $\alpha \geqslant \alpha_1$, $P(\bar{x}, 0, \bar{A}, -(C_0K)^{-1})$ is contained in $\widetilde{P}$, is unscathed and satisfies $|\operatorname{Rm}| \leqslant K(\bar{A})$.*

Proof. Let us consider $s \in [-(C_0K)^{-1}, 0]$, and minimal (for α fixed) such that $B(\bar{x}, 0, \bar{A}) \times (s, 0] \subset \widetilde{P}$ and is unscathed.

As $R(\cdot, 0) \leqslant K/2$ on $B(\bar{x}, 0, \bar{A})$ and every point in $B(\bar{x}, 0, \bar{A}) \times (s, 0]$ with scalar curvature not smaller than 2 is the centre of an (ε_0, C_0)-canonical neighbourhood, the curvature-time lemma (Lemma 6.1.3) shows that $R \leqslant K$ on this set. It follows that $|\operatorname{Rm}| \leqslant K$ and hence by Lemma 12.1.4, $P(\bar{x}, 0, \bar{A}, s)$ is unscathed. We now claim that if $s > -(C_0K)^{-1}$, then there exists $\sigma > 0$ such that $P(\bar{x}, 0, \bar{A}, s - \sigma) \subset \widetilde{P}$. Recall that $(x', t') \in \widetilde{P}$ if (for the rescaled evolving metric $\bar{g}(\cdot)$) $d_{t'}(x', x) < d_0(x, \bar{x}) + \bar{K}^{1/2}$. We shall show that if $x' \in B(\bar{x}, 0, \bar{A})$, then $d_s(x', x) < d_0(x, \bar{x}) + \bar{K}^{1/2}/2$, which suffices to prove the claim, by continuity. Let $x' \in B(\bar{x}, 0, \bar{A})$, then we have $d_s(x', x) \leqslant d_s(x', \bar{x}) + d_s(\bar{x}, x)$. As $|\operatorname{Rm}| \leqslant K$ on $P(\bar{x}, 0, \bar{A}, s)$, by the usual distance distortion estimate (Lemma 2.2.7), for $t' \in [s, 0]$,

$$e^{-4C_0^{-1}} \leqslant \frac{\bar{g}(0)}{\bar{g}(s)} \leqslant e^{4C_0^{-1}}.$$

In particular,

$$d_s(x', \bar{x}) \leqslant 10 d_0(x', \bar{x}) < 10\bar{A}. \tag{12.1}$$

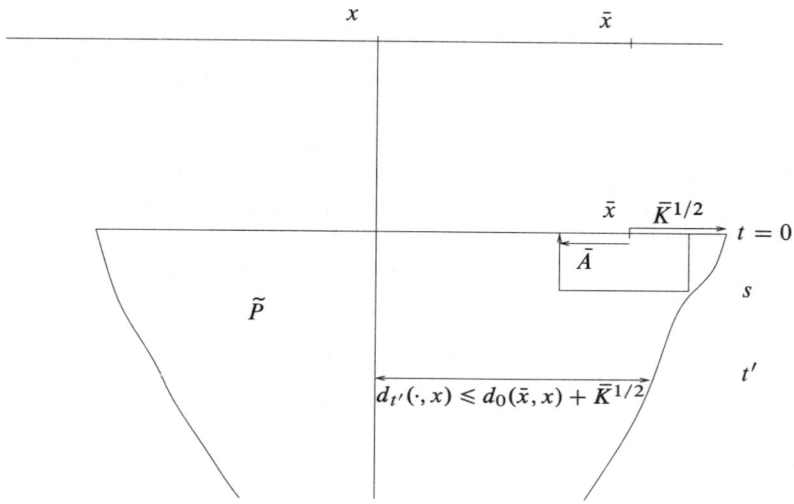

In order to estimate $d_s(\bar{x}, x)$ we shall use the additive distance-distortion, Lemma 2.2.10, which gives a better estimate for large distances. Set $r_1 := K^{-1/2} \ll 1$ (hence $r_1^{-2} = K$). For $t' \in [s, 0]$, we have $|\mathrm{Rm}| \leqslant K$ on $B(x, t', r_1) \cup B(\bar{x}, t', r_1)$. Indeed, we have $|\mathrm{Rm}| \leqslant \bar{Q}^{-1}\rho^{-2} \leqslant \bar{K}^{-1} \ll 1$ on $B(x, t', \bar{Q}^{1/2}\rho/10) \supset B(x, t', \bar{K}^{1/2}/10) \supset B(x, t', r_1)$, on the one hand. Similarly, since $B(\bar{x}, t', 10^{-1}\bar{A}) \subset B(\bar{x}, 0, \bar{A})$ by the above estimate for all $t' \in [s, 0]$, we get $|\mathrm{Rm}| \leqslant K$ on $B(\bar{x}, t', r_1)$. Lemma 2.2.10 gives

$$\frac{d}{dt'} d_{t'}(x, \bar{x}) \geqslant -4(\tfrac{2}{3} K r_1 + r_1^{-1}) \geqslant -10 r_1^{-1}.$$

Integrating between $s \geqslant -(C_0 K)^{-1}$ and $t = 0$ we obtain

$$d_0(x, \bar{x}) - d_s(x, \bar{x}) \geqslant -10 C_0^{-1} r_1 \geqslant -1.$$

Thus

$$d_s(x, \bar{x}) \leqslant d_0(x, \bar{x}) + 1.$$

Combining this with (12.1) and the definition of α_1, we obtain

$$d_s(x', x) \leqslant 10\bar{A} + d_0(x, \bar{x}) + 1 < d_0(x, \bar{x}) + \bar{K}^{1/2}/2.$$

This finishes the proof of Lemma 12.1.5 □

It follows from Lemma 12.1.3 that $\bar{g}(\cdot)$ is κ'-noncollapsed at $(\bar{x}, 0)$ on scales less than $\bar{Q}^{1/2}\rho$, hence on scales less than $\bar{K}^{1/2}$. Combined with the curvature bound given by Lemma 12.1.5, Theorem B.1.2 shows that the injectivity radius of the metrics $\bar{g}(0)$ is bounded from below independently of α. By Theorem C.3.3, the sequence $(M, \bar{g}(\cdot), \bar{x})_\alpha$ subconverges towards a partial Ricci flow $(M_\infty, g_\infty(\cdot), x_\infty)$ where $g_\infty(0)$ is complete. Moreover, the curvature operator of $g_\infty(t)$ is nonnegative by Proposition 6.1.6. This finishes the proof of Step 1. □

Step 2. The manifold $(M_\infty, g_\infty(0))$ has bounded curvature.

The proof of this step is identical to the corresponding one in the proof of Proposition B. Since the points in $P(\bar{x}, 0, \bar{A}, -\tau(A))$ with scalar curvature larger than 2 are centres of (ε_0, C_0)-canonical neighbourhoods for all α large enough, we get by taking limits that each point of $(M_\infty, g_\infty(0))$ with scalar curvature larger than 3 is the centre of a $(2\varepsilon_0, 2C_0)$-weak canonical neighbourhood. We conclude in the same way that the scalar curvature of $g_\infty(0)$ is bounded above by some constant Q_0.

From that we deduce that the constant $K(\bar{A})$ used in Lemma 12.1.5 can be chosen independently of $\bar{A}$. From this we see that

$$\tau_0 := \sup\{\tau > 0 \mid \exists K(\tau) > 0, \forall \bar{A} > 0, \exists \alpha(\bar{A}, \tau), \forall \alpha \geqslant \alpha(\bar{A}, \tau),\\ P(\bar{x}, 0, \bar{A}, -\tau) \subset \widetilde{P}, \text{ is unscathed and } |\mathrm{Rm}| \leqslant K(\tau) \text{ there}\}$$

is positive.

Step 3. There exists $Q > 0$ such that the scalar curvature of $g_\infty(t)$ is bounded above by Q, for all $t \in (-\tau_0, 0]$.

The proof is identical to the one done for Proposition B, using the nonnegativity of the curvature and the canonical neighbourhoods property.

Step 4. $\tau_0 = +\infty$.

We argue as in the proof of Proposition B. We deduce from Step 3 that for all $\tau \in (0, \tau_0)$, for all $\bar{A} > 0$, for all α large enough $P(\bar{x}, 0, \bar{A}, -\tau) \subset \widetilde{P}$, is unscathed and with sectional curvature bounded by a constant K independent of $\bar{A}$ and of α. Let us assume that $\tau_0 < +\infty$. We now need to extend the above property beyond $-\tau_0$. Let us consider τ close enough to τ_0 so that $\tau - \tau_0 < \frac{1}{2}(2C_0(K+2))^{-1}$. Let $s \in [-\tau_0 - \frac{1}{2}(2C_0(K+2))^{-1}, -\tau]$ be minimal (depending on α) such that $B(\bar{x}, 0, \bar{A}) \times (s, 0]$ is contained in $\widetilde{P}$ and is unscathed. We can also use the Curvature-Time Lemma 6.1.3 in order to show that $B(\bar{x}, 0, \bar{A}) \times (s, 0]$ has scalar curvature (hence sectional curvatures) bounded by $2(K+2)$. Lemma 12.1.4 then shows that $P(\bar{x}, 0, \bar{A}, s)$ is unscathed for α large enough depending only on $\bar{A}$ and K. Let us show that $P(\bar{x}, 0, \bar{A}, s - \sigma) \subset \widetilde{P}$ for $\sigma > 0$. We do the same computations as in Step 1.

Let $x' \in B(\bar{x}, 0, \bar{A})$. We set $K' := 2(K+2)$. On $B(\bar{x}, 0, \bar{A})$, we have

$$e^{-4K'|s|} \leqslant \frac{\bar{g}(0)}{\bar{g}(s)} \leqslant e^{4K'|s|} \leqslant c_1^2,$$

for some constant $c_1 > 0$ independent from $\bar{A}$ and α. Thus $d_s(x', \bar{x}) \leqslant c_1 \bar{A}$.

As before, set $r_1 := K'^{-1/2}$. Integrating inequality (2.2) given by Lemma 2.2.10 between s and 0 we find

$$d_0(x, \bar{x}) - d_s(x, \bar{x}) \geqslant -10 r_1^{-1} s \geqslant -c_2,$$

for some constant $c_2 > 0$ independent of $\bar{A}$ and α. Hence

$$d_s(x', x) \leqslant d_0(x, \bar{x}) + c_1\bar{A} + c_2 < d_0(x, \bar{x}) + \bar{K}^{1/2}/2,$$

for α large enough depending only on $\bar{A}$ and τ_0. By continuity, we have $d_{s-\sigma}(x', x) \leqslant d_0(x, \bar{x}) + \bar{K}^{1/2}$ for all $x' \in B(\bar{x}, 0, \bar{A})$, for $\sigma > 0$ small enough. It follows that $P(\bar{x}, 0, \bar{A}, -\tau_0 - \frac{1}{2}(2C_0(K+2))^{-1}) \subset \bar{P}$ is unscathed and with curvature bounded by K' for α large enough depending on $\bar{A}$ and on τ_0. This contradicts the definition of τ_0.

It follows that $\tau_0 = +\infty$, and we can conclude, as in proof of Proposition B that a subsequence of $(M, \bar{g}(\cdot), \bar{x})$ converges to a κ'-solution. As explained before, this finishes the proof of part (b).

12.1.2 Curvature-distance estimates: proof of part (c)

Fix $A > 0$. Applying parts (a) and (b) of Theorem 12.1.1 with A replaced by $2A$, we obtain a constant $\bar{K} = \bar{K}(2A) \gg 1$ and a function $\bar{\delta}(t) = \bar{\delta}_{2A}(t)$. Let $Q(\cdot, \cdot, \cdot)$ and $\Lambda(\cdot, \cdot, \cdot)$ be given by Theorem 6.1.1. Define

$$\begin{aligned} A' &:= A\bar{K}^{1/2}, \\ Q &:= Q(A', \varepsilon_0, C_0), \\ \Lambda &:= \Lambda(A', \varepsilon_0, C_0), \\ \bar{K}' &:= \bar{K}\Lambda, \\ \bar{r} &:= \bar{K}^{\frac{1}{2}} Q^{-\frac{1}{2}}. \end{aligned}$$

Let us fix $t > 0$ and $\rho \in (0, \bar{r}\sqrt{t}]$. We note that for all $y \in B(x, t, 2A\rho)$, if $R(y, t) \geqslant \bar{K}\rho^{-2}$ then (y, t) has an (ε_0, C_0)-canonical neighbourhood.

Let us show that assertion (c) holds with $\bar{K}' := \bar{K}\Lambda$, that is $R(\cdot, t) \leqslant \bar{K}\Lambda\rho^{-2}$ on $B(x, t, A\rho)$. If not, there exists $x' \in B(x, t, A\rho)$ such that $R(x', t) > \bar{K}\Lambda\rho^{-2}$. Since $R(x, t) \leqslant 6\rho^{-2}$, we can find $y \in [xx']$ such that $R(y, t) = \bar{K}\rho^{-2}$. Then

$$(1+t)R(y, t) \geqslant t\rho^{-2}\bar{K} \geqslant \bar{r}^{-2}\bar{K} = Q,$$

and every point of

$$B(y, t, A'R(y, t)^{-1/2}) = B(y, t, A\rho) \subset B(x, t, 2A\rho)$$

with scalar curvature not less than $R(y, t)$ is centre of an (ε_0, C_0)-canonical neighbourhood.

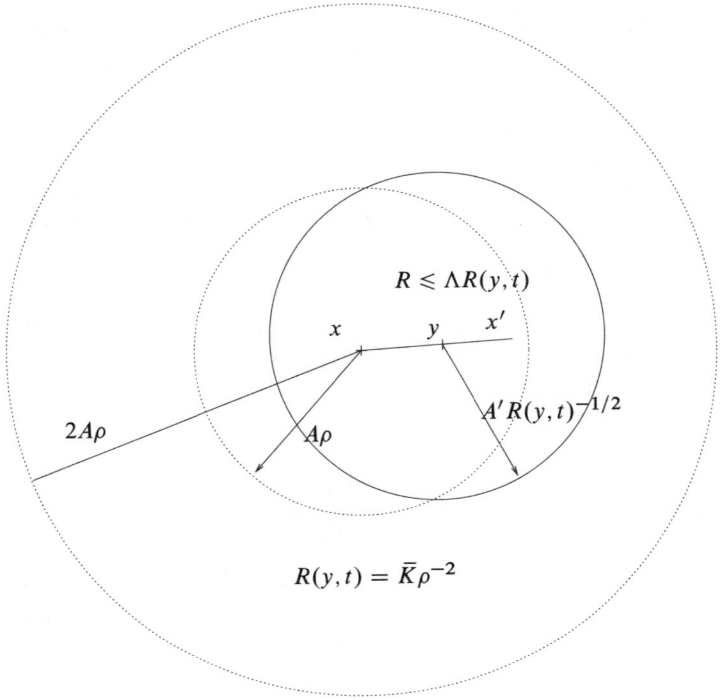

Theorem 6.1.1 thus shows that

$$R(\cdot\,,t) \leqslant \Lambda R(y,t) = \Lambda \bar{K}\rho^{-2}$$

on $B(y,t,A'R(y,t)^{-1/2}) = B(y,t,A\rho)$. It is in particular true at x' since $d_t(x',y) < A\rho$.

This ends the proof of Theorem 12.1.1.

To conclude this section, we give a corollary to Theorem 11.1.3 which will be useful to prove Theorem 12.2.1.

Corollary 12.1.6. *For all $A, w > 0$ there exist positive numbers $\bar{r}_1 \in (0, 10^{-3})$, $\bar{\theta}_1$, $\bar{K}_1$, $\bar{\tau}_1$ and $\overline{T}_1$ with the following property. Let $g(\cdot)$ be a Ricci flow with $r(\cdot), \delta(\cdot)$-bubbling-off on $[0,\infty[$, with normalised initial condition and $\delta(\cdot) \leqslant \bar{\delta}(\cdot)$. Let $x \in M$, $t \geqslant \overline{T}_1$, $r > 0$ such that*

(i) $\bar{\theta}_1^{-1} h \leqslant r \leqslant \bar{r}_1\sqrt{t}$,
(ii) *$P(x,t,r,-r^2)$ is unscathed,*
(iii) *$|\mathrm{Rm}| \leqslant r^{-2}$ there,*
(iv) *$\operatorname{vol} B(x,t,r) \geqslant wr^3$.*

Then $P(x,t,Ar,-\bar{\tau}_1 r^2)$ is unscathed and $|\mathrm{Rm}| < \bar{K}_1 r^{-2}$ there.

Proof. Define

$$\begin{aligned}
A' &:= \max\left\{4A, \frac{e^{3/2}}{8}\frac{\upsilon_{-1}(1)}{\upsilon_{-1}(1/4)}w\right\},\\
\bar{T}_1 &:= 2A'\\
\bar{r}_1 &:= \bar{r}(A'),\\
\bar{K}_1 &:= 4\bar{K}'(A'),\\
\bar{\theta}_1 &:= (2K_1)^{-1/2},\\
\bar{\tau}_1 &:= (4K_1)^{-1},
\end{aligned}$$

where $\bar{r}$ and $\bar{K}'$ are parameters given by Theorem 12.1.1.

Let (x,t,r) be a triple that satisfies the assumptions. Set $r' := \frac{r}{2}$. By assumptions (ii) and (iii) and the multiplicative distortion lemma (Lemma 2.2.7),

$$4^{-1} \leqslant e^{-1} \leqslant \frac{g(t)}{g(s)} \leqslant e \leqslant 4, \tag{12.2}$$

for every $s \in [t - r'^2, t] = [t - \frac{r^2}{4}, t]$, hence

$$B\left(x,t,\frac{r'}{2}\right) \subset B(x,s,r') \subset B(x,t,r)$$

for all such s. In particular $P(x,s,r',-r'^2) \subset P(x,t,r,-r^2)$. It follows that $P(x,s,r',-r'^2)$ is unscathed and $|\mathrm{Rm}| \leqslant r^{-2} \leqslant r'^{-2}$, satisfying assumptions (i) and (ii) of Theorem 11.1.3. Moreover, by (12.2) and the above inclusion,

$$\begin{aligned}
\mathrm{vol}_{g(s)} B(x,s,r') &\geqslant \mathrm{vol}_{g(s)} B\left(x,t,\frac{r'}{2}\right)\\
&\geqslant e^{-3/2}\,\mathrm{vol}_{g(t)} B\left(x,t,\frac{r}{4}\right)\\
&\geqslant e^{-3/2}\frac{\upsilon_{-1}(1/4)}{\upsilon_{-1}(1)}\,\mathrm{vol}_{g(t)} B(x,t,r)\\
&\geqslant e^{-3/2}\frac{\upsilon_{-1}(1/4)}{\upsilon_{-1}(1)}wr^3\\
&= 8e^{-3/2}\frac{\upsilon_{-1}(1/4)}{\upsilon_{-1}(1)}wr'^3\\
&\geqslant A'^{-1}r'^3.
\end{aligned}$$

Hence assumption (iii) of Theorem 11.1.3 is satisfied at (x,s,r') for all $s \in [t-r'^2,t]$. As $r' < r < \bar{r}_1\sqrt{t} \leqslant 10^{-3}\sqrt{t}$, we have $r'^2 \leqslant 10^{-6}t$ hence $s \geqslant t-r'^2 \geqslant (1-10^{-6})t \geqslant t/4$, so $r' \leqslant 10^{-3}\sqrt{t} \leqslant 10^{-3}2\sqrt{s} \leqslant \sqrt{s/2}$. In addition,

$$r' = \frac{r}{2} \leqslant \frac{\bar{r}_1\sqrt{t}}{2} \leqslant \bar{r}_1\sqrt{s},$$

as required in part (c) of Theorem 11.1.3. Finally, $s \geqslant t/2 \geqslant A'$. Our choice or $\bar{\delta}(\cdot)$ in Definition 11.1.4 implies that Theorem 11.1.3 applies at time s for the constant A' (cf. Remark 11.1.5).

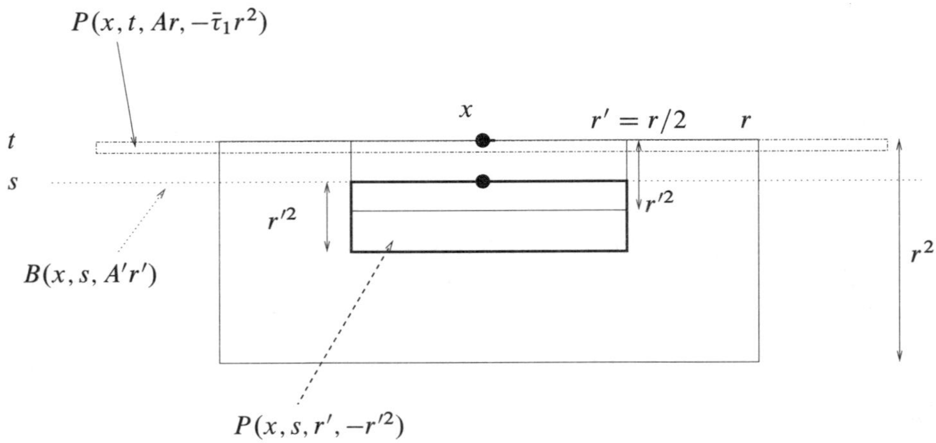

Then $R < \bar{K}'(A')r'^{-2} = \bar{K}_1 r^{-2}$ on $B(x,s,A'r') \supset B(x,s,2Ar)$. From $\bar{\theta}_1^{-1} h \leqslant r$, we then have

$$R < \bar{K}_1 \bar{\theta}_1^2 h^{-2} \leqslant \frac{h^{-2}}{2} \leqslant \frac{h^{-2}(s)}{2},$$

hence the solution is unscathed on $\bigcup_{s\in[t-r'^2,t]} B(x,s,2Ar)$.

Moreover, by Lemma 4.4.7, $|\mathrm{Rm}| < \bar{K}_1 r^{-2}$ on this set. Then, using Lemma 2.2.7 as usual, we deduce that $B(x,t,Ar) \subset B(x,s,2Ar)$ for all $s \in [t - \bar{\tau}_1 r^2, t]$. Hence

$$P(x,t,Ar,-\bar{\tau}_1 r^2) \subset \bigcup_{s\in[t-\bar{\tau}_1 r^2,t]} B(x,s,2Ar).$$

This completes the proof of Corollary 12.1.6. □

12.2 Curvature estimates in the thick part: proof of Theorem 11.1.6

Let us first recall the statement of Theorem 11.1.6:

Theorem 12.2.1. *For all $w > 0$, there exist positive numbers $\hat{\theta}$, $\hat{r}$, $\widehat{T}$, $\hat{\tau}$, $\widehat{K}$ with the following property. Assume that $g(\cdot)$ is a Ricci flow with $r(\cdot),\delta(\cdot)$-bubbling-off on M defined on $[0,\infty)$, with normalised initial condition and $\delta(\cdot) \leqslant \bar{\delta}(\cdot)$. Let $x \in M$, $t \geqslant \widehat{T}$, $r > 0$ be such that*

(i) $\hat{\theta}^{-1}h \leqslant r \leqslant \hat{r}\sqrt{t}$, *where h is the maximal surgery radius on* $[t/2, t]$,

(ii) $\mathrm{Rm} \geqslant -r^{-2}$ *on* $B(x, t, r)$,

(iii) $\mathrm{vol}\, B(x, t, r) \geqslant wr^3$.

Then $P(x, t, r, -\hat{\tau}r^2)$ is unscathed and $|\mathrm{Rm}| < \hat{K}r^{-2}$ there.

The proof of Theorem 12.2.1 is quite involved. We shall deduce it from a weaker statement, Theorem 12.2.3, and from Corollary 12.1.6. In the same way, the proof of Theorem 12.2.3 is reduced to another result, Proposition 12.2.4, and to Corollary 12.1.6. Finally, Proposition 12.2.4 is deduced from a core lemma (Lemma 12.2.6), whose proof ends the section.

We begin the reduction by replacing assumption (iii) of Theorem 12.2.1, which depends on w, by a stronger requirement. We need a definition:

Definition 12.2.2 (cf. Lemma D.1.6). We say that a metric ball $B(x, r)$ has $(1 - \varepsilon_0)$-*almost Euclidian volume* if $\mathrm{vol}\, B(x, r) \geqslant (1 - \varepsilon_0)v_0(r)$.

Theorem 12.2.3. *There exist positive numbers $\hat{\theta}_2, \hat{r}_2, \hat{\tau}_2, \hat{K}_2$ with the following property. Assume that $g(\cdot)$ is a Ricci flow with $r(\cdot), \delta(\cdot)$-bubbling-off on M defined on $[0, \infty)$, with normalised initial condition and $\delta(\cdot) \leqslant \bar{\delta}(\cdot)$. Let (x, t, r) satisfy*

(i) $\hat{\theta}_2^{-1}h \leqslant r \leqslant \hat{r}_2\sqrt{t}$, *where h is the maximal surgery radius on* $[t/2, t]$,

(ii) $\mathrm{Rm} \geqslant -r^{-2}$ *on* $B(x, t, r)$,

(iii) *every sub-ball of $B(x, t, r)$ has $(1 - \varepsilon_0)$-almost Euclidean volume.*

Then $P(x, t, r, -\hat{\tau}_2 r^2)$ is unscathed and $|\mathrm{Rm}| < \hat{K}_2 r^{-2}$ there.

The only difference to Theorem 12.2.1 is the stronger assumption on the volume of $B(x, t, r)$. We now deduce Theorem 12.2.1 from Theorem 12.2.3, postponing the proof of 12.2.3.

Proof of Theorem 12.2.1 *assuming Theorem* 12.2.3. The idea is the following: Sub-ball Lemma D.1.6 provides a sub-ball of $B(x, t, r)$ all of whose sub-balls have $(1 - \varepsilon_0)$-almost Euclidean volume, on which we apply Theorem 12.2.3. Then we obtain curvature bounds on the larger parabolic ball by Corollary 12.1.6.

The details are as follows. Fix $w > 0$ and let $\theta_0 = \theta_0(w)$ be the parameter given by Lemma D.1.6. Set

$$B := \min\{\hat{K}_2^{-\frac{1}{2}}, \hat{\tau}_2^{\frac{1}{2}}\}\theta_0.$$

Applying Corollary 12.1.6 with $A = 2/B$ and $w = (1 - \varepsilon_0)v_0(1)$, we obtain numbers $\bar{r}_1, \bar{\theta}_1, \bar{\tau}_1, \bar{T}_1, \bar{K}_1$. Define

$$\hat{\theta} := \min\{B\bar{\theta}_1, \theta_0\hat{\theta}_2\},$$
$$\hat{r} := \min\{\bar{r}_1, \hat{r}_2\},$$
$$\hat{T} := \max\{\bar{T}_1, \hat{T}_2\},$$

$$\hat{\tau} := \bar{\tau}_1 B^2,$$
$$\widehat{K} := \bar{K}_1 B^{-2}.$$

Pick (x,t,r) satisfying the assumptions:

$$t \geqslant \widehat{T}, \quad \hat{\theta}^{-1}h \leqslant r \leqslant \hat{r}\sqrt{t}, \quad \mathrm{Rm} \geqslant -r^{-2} \text{ on } B(x,t,r), \quad \mathrm{vol}\, B(x,t,r) \geqslant wr^3.$$

Set $r' := \theta_0 r$. By Lemma D.1.6 there exists $B(x',t,r') \subset B(x,t,r)$ such that every sub-ball of $B(x',t,r')$ has $(1-\varepsilon_0)$-almost Euclidean volume. Clearly $\mathrm{Rm} \geqslant -r'^{-2}$on $B(x',t,r')$. Moreover, $t \geqslant \widehat{T}_2$ and

$$\hat{\theta}_2^{-1}h \leqslant \hat{\theta}^{-1}\theta_0 h \leqslant \theta_0 r = r' \leqslant r \leqslant \hat{r}\sqrt{t} \leqslant \hat{r}_2\sqrt{t},$$

hence Theorem 12.2.3 applies to (x',t,r'). Therefore $P(x',t,r',-\hat{\tau}_2 r'^2)$ is unscathed and satisfies $|\mathrm{Rm}| < \widehat{K}_2 r'^{-2}$. Set

$$r'' := Br = \min\{\widehat{K}_2^{-\frac{1}{2}}, \hat{\tau}_2^{\frac{1}{2}}\}r'.$$

This is precisely the choice such that $\widehat{K}_2 r'^{-2} \leqslant r''^{-2}$ and $\hat{\tau}_2 r'^2 \geqslant r''^2$. Then $P(x',t,r'',-r''^2)$ is unscathed and $|\mathrm{Rm}| \leqslant r''^{-2}$ there (see Figure 12.1).

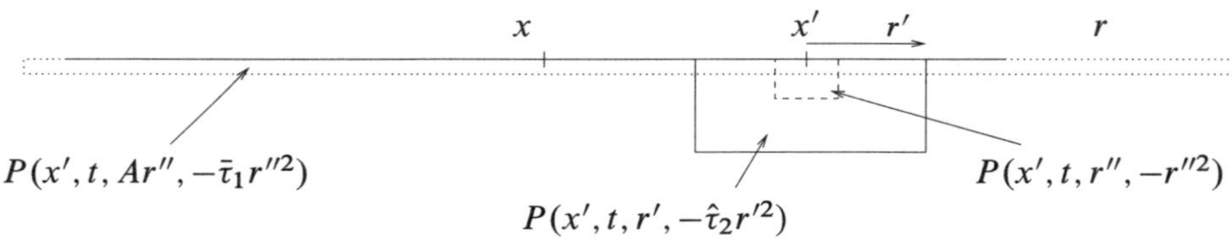

Figure 12.1. Extending curvature bounds from a smaller ball.

We have $t \geqslant \overline{T}_1$ and

$$\bar{\theta}_1^{-1}h \leqslant \hat{\theta}^{-1}Bh \leqslant Br = r'' \leqslant r \leqslant \hat{r}\sqrt{t} \leqslant \bar{r}_1\sqrt{t}.$$

Moreover, $B(x',t,r'') \subset B(x',t,r')$, hence

$$\mathrm{vol}\, B(x',t,r'') \geqslant (1-\varepsilon_0)v_0(r'') = (1-\varepsilon_0)v_0(1)r''^3.$$

Therefore Corollary 12.1.6 applies to the triple (x',t,r'') with parameters A and $(1-\varepsilon_0)v_0(1)$. It follows that $P(x',t,Ar'',-\bar{\tau}_1 r''^2)$ is unscathed and $|\mathrm{Rm}| < \bar{K}_1 r''^{-2} = \bar{K}_1 B^{-2} r^{-2} = \widehat{K} r^{-2}$. As $Ar'' = 2r$ and $\bar{\tau}_1 r''^2 = \bar{\tau}_1 B^2 r^2 \geqslant \hat{\tau} r^2$, it follows that $P(x,t,r,-\hat{\tau}r^2)$ is unscathed and $|\mathrm{Rm}| < \widehat{K} r^{-2}$ there. □

We now deduce Theorem 12.2.3 from Corollary 12.1.6 and the following proposition, whose proof is postponed.

Proposition 12.2.4. *There exist positive numbers $\hat{\theta}_3$, $\hat{r}_3$, $\widehat{T}_3$, $\hat{\tau}_3$, $\widehat{K}_3$ with the following property. Assume that $g(\cdot)$ is a Ricci flow with $r(\cdot)$, $\delta(\cdot)$-bubbling-off on $[0,\infty[$, with normalised initial condition and $\delta(\cdot) \leqslant \bar{\delta}(\cdot)$. Let $x \in M$, $t \geqslant \widehat{T}_3$, and $r > 0$ satisfy*

(i) *$\hat{\theta}_3^{-1} h \leqslant r \leqslant \hat{r}_3 \sqrt{t}$, where h is the maximal surgery radius on $[t/2, t]$,*
(ii) *$\mathrm{Rm} \geqslant -r^{-2}$ on $B(x,t,r)$,*
(iii) *any sub-ball of $B(x,t,r)$ has $(1-\varepsilon_0)$ almost Euclidean volume.*

Then $\bigcup_{[t-\hat{\tau}_3 r^2, t]} B(x,s,r/2)$ is unscathed and $|\mathrm{Rm}| < \widehat{K}_3 r^{-2}$ there.

Remark 12.2.5. The proposition is weaker in the sense that it involves a union of balls of radius $r/2$ rather than a parabolic ball of radius r.

Proof of Theorem 12.2.3 *assuming Proposition* 12.2.4. Set

$$\tau := \min\left\{(4\widehat{K}_3)^{-\frac{1}{2}}, \hat{\tau}_3^{\frac{1}{2}}\right\},$$
$$B := \min\left\{\tau^{-\frac{1}{2}}, \widehat{K}_3^{-\frac{1}{2}}, \tfrac{1}{8}\right\}.$$

Let $\bar{r}_1, \bar{\theta}_1, \bar{\tau}_1, \overline{T}_1, \overline{K}_1$ be the parameters obtained by applying Corollary 12.1.6 with $A = 1/B$ and $w = (1-\varepsilon_0)\upsilon_0(1)$. Define

$$\begin{aligned}
\hat{\theta}_2 &:= \min\{B\bar{\theta}_1, \hat{\theta}_3\},\\
\hat{r}_2 &:= \min\{\bar{r}_1, \hat{r}_3\},\\
T_2 &:= \max\{\overline{T}_1, \widehat{T}_3\},\\
\hat{\tau}_2 &:= \bar{\tau}_1 B^2,\\
\widehat{K}_2 &:= \overline{K}_1 B^{-2}.
\end{aligned}$$

Choose (x,t,r) satisfying the assumptions of Theorem 12.2.3. Since $t \geqslant \widehat{T}_2 \geqslant \widehat{T}_3$ and

$$\hat{\theta}_3^{-1} h \leqslant \hat{\theta}_2^{-1} h \leqslant h \leqslant \hat{r}_2\sqrt{t} \leqslant \hat{r}_3\sqrt{t},$$

we deduce that t and r satisfy assumption (i) of Proposition 12.2.4. Assumptions (ii), (iii) are also satisfied, thus $\bigcup_{[t-\hat{\tau}_3 r^2, t]} B(x,s,r/2)$ is unscathed and $|\mathrm{Rm}| < \widehat{K}_3 r^{-2}$ there. By the Multiplicative Distance-Distortion Lemma 2.2.7,

$$4^{-1} \leqslant \frac{g(t)}{g(s)} \leqslant 4,$$

on $B(x,t,r/4)$ if $s \in [t-\tau r^2, t]$. Hence $P(x,t,r/4,-\tau r^2) \subset \bigcup_{[t-\hat{\tau}_3 r^2, t]} B(x,s,r/2)$ is unscathed and $|\mathrm{Rm}| < \widehat{K}_3 r^{-2}$ there. Set $r' := Br$. We have $P(x,t,r',-r'^2) \subset P(x,t,r/4,-\tau r^2)$ and $|\mathrm{Rm}| \leqslant r'^{-2}$. On the other hand, $t \geqslant \overline{T}_1$ and

$$\bar{\theta}_1^{-1} h \leqslant B\hat{\theta}_2^{-1} h \leqslant Br = r' \leqslant \hat{r}_2\sqrt{t} \leqslant \bar{r}_1\sqrt{t}.$$

Moreover, vol $B(x,t,r') \geqslant (1-\varepsilon_0)\upsilon_0(r')$ by assumption (iii). Hence the assumptions of Corollary 12.1.6 are satisfied for the constants A and $(1-\varepsilon_0)\upsilon_0(1)$. Then

$$P(x,t,Ar',-\bar{\tau}_1 r'^2) \supset P(x,t,r,-\hat{\tau}_2 r^2)$$

is unscathed and $|\mathrm{Rm}| < \bar{K}_1 r'^{-2} = \hat{K}_2 r^{-2}$ there. □

We now turn to the proof of Proposition 12.2.4. The core of the argument relies on the following lemma, whose proof is postponed again.

Lemma 12.2.6 (Core lemma). *There exist positive numbers τ_*, K_* such that for any Ricci flow $g(\cdot)$ on an open subset $\Omega \subset M$ (not necessarily complete), $\tau \in [0,\tau_*]$, and $x \in \Omega$ the following holds. Assume that*

(i) *for all $t \in [-\tau,0]$, $B(x,t,1) \subset \Omega$ is relatively compact,*
(ii) *$g(\cdot)$ is smooth on $\bigcup_{[-\tau,0]} B(x,t,1)$ and $\mathrm{Rm} \geqslant -1$ on this set,*
(iii) vol $B(x,0,1) \geqslant (1-\varepsilon_0)\upsilon_0(1)$.

Then

(a) *$R \leqslant K_*\tau^{-1}$ on $\bigcup_{[-\tau/2,0]} B(x,t,1/2)$,*
(b) vol $B(x,-\tau,1/4) \geqslant \frac{1}{10}\upsilon_0(1/4)$.

Proof of Proposition 12.2.4 *assuming Lemma* 12.2.6. The proof will be by contradiction. We shall make another reduction, considering that the radius r is smaller or greater than the scale $r(t)$ of the canonical neighbourhoods. The case where $r \leqslant r(t)$, which is easier, will be tackled by the following lemma.

Lemma 12.2.7 (Small radius lemma). *There exist positive numbers $\hat{\theta}_4$, $\hat{T}_4$, $\hat{\tau}_4$, $\hat{K}_4$ with the following property. Assume that $g(\cdot)$ is a Ricci flow with $r(\cdot), \delta(\cdot)$-bubbling-off on M defined on $[0,\infty)$, with normalised initial condition and $\delta(\cdot) \leqslant \bar{\delta}(\cdot)$. Let $x \in M, t \geqslant \hat{T}_4, r > 0$ satisfy*

(i) $\hat{\theta}_4^{-1} h \leqslant r \leqslant r(t)$,
(ii) $\mathrm{Rm} \geqslant -r^{-2}$ *on $B(x,t,r)$,*
(iii) *any sub-ball of $B(x,t,r)$ has $(1-\varepsilon_0)$-almost Euclidean volume.*

Then $P(x,t,r,-\hat{\tau}_4 r^2)$ is unscathed and $|\mathrm{Rm}| < \hat{K}_4 r^{-2}$ there.

Proof. Define

$$K_0 := 32C_0^2,$$
$$\tau_0 := (32C_0^3)^{-1},$$
$$B := \min\{K_0^{-\frac{1}{2}}, \tau_0^{\frac{1}{2}}, \tfrac{1}{4}\}.$$

Let $\bar{r}_1, \bar{\theta}_1, \bar{\tau}_1, \overline{T}_1, \overline{K}_1$ be the parameters given by Corollary 12.1.6 applied with $A = 1/B$ and $w = (1-\varepsilon_0)\upsilon_0(1)$. Then define

$$\begin{aligned}
\hat{\theta}_4 &:= \min\{B\bar{\theta}_1, (8C_0)^{-1}\},\\
\hat{\tau}_4 &:= \bar{\tau}_1 B^2,\\
\widehat{K}_4 &:= \overline{K}_1 B^{-2},\\
\widehat{T}_4 &\geqslant \overline{T}_1 \ \text{ such that } \ r(t) \leqslant \bar{r}_1\sqrt{t} \ \text{ if } \ t \geqslant \widehat{T}_4.
\end{aligned}$$

Let (x,t,r) satisfy the assumptions of the small radius lemma. In particular we have $8C_0 h \leqslant r \leqslant r(t)$.

Assertion 12.2.8. *$P(x,t,r/4,-\tau_0 r^2)$ is unscathed and $R < K_0 r^{-2}$ there.*

Proof of Assertion 12.2.8. Let us prove first that $R < 16C_0^2 r^{-2}$ on $B(x,t,r/4)$. Assume that it is not true and pick $x' \in B(x,t,r/4)$ such that

$$R(x',t) \geqslant 16C_0^2 r^{-2} \geqslant 16C_0^2 r(t)^{-2} \geqslant r(t)^{-2}.$$

Then (x',t) has an (ε_0, C_0)-canonical neighbourhood U, and $B(x',t,r') \subset U \subset B(x,t_0,2r')$, where $r' \in (C_0^{-1}R(x',t)^{-1/2}, C_0 R(x',t)^{-1/2})$. In particular,

$$r' \leqslant C_0 R(x',t)^{-1/2} \leqslant r/4.$$

Hence $2r' \leqslant r/2$. We deduce that $U \subset B(x',t,r/2) \subset B(x,t,r)$. This contradicts the assumption of almost Euclidean volume for sub-balls. Indeed, there exists $y \in U$ and $s \approx R(y,t)^{-1/2}\varepsilon_0^{-1} \leqslant r$ such that $B(y,t,s)$ is close to an ε_0-neck included in U, hence has volume

$$\approx R(y,t)^{-3/2}\frac{2}{\varepsilon_0}\operatorname{vol}(S^2) \approx (s\varepsilon_0)^3\frac{2}{\varepsilon_0}\operatorname{vol}(S^2) = 2\varepsilon_0^2\operatorname{vol}(S^2)s^3,$$

which is much less than $(1-\varepsilon_0)\upsilon_0(1)s^3$. Hence $R < 16C_0^2 r^{-2}$ on $B(x,t,r/4)$.

Now consider $s \in [t-\tau_0 r^2, t]$ minimal such that $B(x,t,r/4)\times(s,t]$ is unscathed. By the same argument as in the proof of Lemma 6.1.3, $R < 32C_0^2 r^{-2} = K_0 r^{-2}$ on this set. On the other hand $8C_0 h \leqslant r$ hence

$$R < 32C_0^2(64)^{-1}C_0^{-2}h^{-2} = \frac{h^{-2}}{2}.$$

It follows that $P(x,t,r/4,s-t)$ is unscathed and $s = t-\tau_0 r^2$, proving Assertion 12.2.8. □

By Assertion 12.2.8, $P(x,t,r/4,-\tau_0 r^2)$ is unscathed and $R < K_0 r^{-2}$ there. Set $r' := Br$. We have $r' \leqslant r/4$ and $r'^2 = B^2 r^2 \leqslant \tau_0 r^2$, hence $P(x,t,r',-r'^2) \subset P(x,t,r/4,-\tau_0 r^2)$. Moreover, we have

$$\bar{\theta}_1^{-1}h \leqslant B\hat{\theta}_4^{-1}h \leqslant Br = r' \leqslant r \leqslant r(t) \leqslant \bar{r}_1\sqrt{t}.$$

In particular, Lemma 4.4.7 applies, giving $|\mathrm{Rm}| < K_0 r^{-2} \leqslant r'^{-2}$ on $P(x,t,r',-r'^2)$. Recall that $t \geqslant \bar{T}_1$ and $\mathrm{vol}\, B(x,t,r') \geqslant (1-\varepsilon_0)\upsilon_0(r')$. Hence the hypotheses of Corollary 12.1.6 are satisfied for the constants A and $(1-\varepsilon_0)\upsilon_0(1)$. Therefore $P(x,t,Ar',-\bar{\tau}_1 r'^2)$ is unscathed and

$$|\mathrm{Rm}| < \bar{K}_1 r'^{-2} = \bar{K}_1 B^{-2} r^{-2} = \widehat{K}_4 r^{-2}$$

there. Since $P(x,t,r,-\hat{\tau}_4 r^2) \subset P(x,t,Ar',-\bar{\tau}_1 r'^2)$, this proves Lemma 12.2.7 □

We continue the proof of Proposition 12.2.4. We use Lemmas 12.2.6 and 12.2.7 to define the following parameters and constants:

$$\begin{aligned}
\hat{\theta}_3 &:= \hat{\theta}_4,\\
\widehat{T}_3 &:= \widehat{T}_4,\\
\hat{\tau}_3 &:= \min\left\{\frac{\tau_*}{4}, (4\widehat{K}_4)^{-1}, \hat{\tau}_4\right\},\\
\widehat{K}_3 &:= \max\left\{\frac{K_*}{\tau_*}, \widehat{K}_4\right\},\\
B &:= \frac{1}{4}\theta_0\left(\frac{\upsilon_0(1)}{10}\right),\\
B' &:= \min\left\{\frac{\hat{\tau}_3^{-\frac{1}{2}}}{\sqrt{2}}, \left(\frac{K_*}{\hat{\tau}_3}\right)^{-\frac{1}{2}}, \frac{1}{40}\right\},\\
A &:= 4(BB')^{-1}.
\end{aligned}$$

Here $\theta_0(\cdot)$ is the parameter given by Lemma D.1.6. Let $\bar{r}_1$, $\bar{\theta}_1$, $\bar{\tau}_1$, $\bar{T}_1$ and $\bar{K}_1$ be the parameters given by Corollary 12.1.6 applied to the above-defined A and $w = (1-\varepsilon_0)\upsilon_0(1)$. Set

$$\begin{aligned}
K &:= \bar{K}_1 (BB')^{-2},\\
\sigma &:= \min\left\{(4K)^{-1}, \bar{\tau}_1 B^2 B'^2\right\}.
\end{aligned}$$

Let us make an observation. For any triple (x,t,r) satisfying the assumptions of Proposition 12.2.4, and moreover $r \leqslant r(t)$, the conclusion of the proposition is true. Indeed, in this case the small radius lemma (Lemma 12.2.7) applies, hence $P(x,t,r,-\hat{\tau}_4 r^2)$ is unscathed and $|\mathrm{Rm}| < \widehat{K}_4 r^{-2}$ there. By the multiplicative distance-distortion lemma (Lemma 2.2.7), we get

$$B(x,s,\tfrac{r}{2}) \subset B(x,t,r),$$

for all $s \in [t - \min\{(4\widehat{K}_4)^{-1}, \hat{\tau}_4\} r^2, t]$. By definition of $\hat{\tau}_3$, $\widehat{K}_3$, this proves that the conclusion of Proposition 12.2.4 is satisfied for such (x,t,r).

Thus, arguing by contradiction, all counterexamples to Proposition 12.2.4 satisfy $r \geqslant r(t)$. We hence assume that there exist a sequence $\hat{r}^\alpha \to 0$, Ricci flows with

$r(\cdot), \delta^\alpha(\cdot)$-bubbling-off $g_\alpha(\cdot)$ on M_α, defined on $[0,+\infty)$, with normalised initial condition and $\delta^\alpha(\cdot) \leqslant \bar\delta(\cdot)$, and counterexamples $(x^\alpha, t^\alpha, r^\alpha)$, which satisfy $t^\alpha \geqslant \widehat{T}_3$ and

(i) $\hat\theta_3^{-1} h^\alpha \leqslant r^\alpha \leqslant \hat r^\alpha \sqrt{t^\alpha}$, where h^α is the maximal surgery radius on $[t^\alpha/2, t^\alpha]$,

(i′) $r^\alpha \geqslant r(t^\alpha)$,

(ii) $\mathrm{Rm} \geqslant -(r^\alpha)^{-2}$ on $B(x^\alpha, t^\alpha, r^\alpha)$,

(iii) any sub-ball of $B(x^\alpha, t^\alpha, r^\alpha)$ has $(1-\varepsilon_0)$-almost Euclidean volume,

but $\bigcup_{[t^\alpha - \hat\tau_3(r\alpha)^2, t^\alpha]} B(x^\alpha, t, r^\alpha/2)$ is scathed or $|\mathrm{Rm}| < \widehat{K}_3 (r^\alpha)^{-2}$ is not true there.

The idea of the proof is the following. We want to consider the counterexamples $B(x,t,r)$ of "smallest" radius. Then we shall apply the conclusion of Proposition 12.2.4 to a sub-ball of much smaller, but controlled, radius, and then extend to the previous one by Corollary 12.1.6, getting a contradiction.

For each α, define

$$r_0^\alpha := \inf\{r > 0 \mid \text{ a counterexample } (x,t,r) \text{ exists with } t \leqslant t^\alpha\}.$$

Note that by the fact that $r(\cdot)$ is nonincreasing, $r \geqslant r(t) \geqslant r(t^\alpha)$. Hence $r_0^\alpha \geqslant r(t^\alpha) > 0$. The infimum need not be attained, but clearly we can redefine $(x^\alpha, t^\alpha, r^\alpha)$ in such a way that

$$\frac{r^\alpha}{r_0^\alpha} \xrightarrow[\alpha\to+\infty]{} 1.$$

In particular no counterexample (x,t,r) now exists with $r \leqslant r^\alpha/2$ and $t \leqslant t^\alpha$. Let us remark that $t^\alpha \to +\infty$. Indeed, from $r(t^\alpha) \leqslant r^\alpha \leqslant \hat r^\alpha \sqrt{t^\alpha}$, we get

$$t^\alpha r(t^\alpha)^{-2} \geqslant (\hat r^\alpha)^{-2} \to +\infty,$$

and the remark follows from the fact that $t\, r(t)^{-2}$ is bounded on any finite interval $[0,T]$. Define

$$\tau^\alpha := \sup\Big\{\tau \geqslant 0 \mid \bigcup_{[t^\alpha - \tau (r^\alpha)^2, t^\alpha]} B(x^\alpha, t, r^\alpha) \text{ is unscathed and } \mathrm{Rm} \geqslant -(r^\alpha)^{-2} \text{ there}\Big\}.$$

Remark 12.2.9. $B(x^\alpha, t^\alpha, r^\alpha)$ is unscathed by definition of our surgery, and due to the discreteness of the surgery times, $\bigcup_{t\in[t^\alpha - \tau(r^\alpha)^2, t^\alpha]} B(x^\alpha, t, r^\alpha)$ is unscathed for some $\tau > 0$. However, maybe $\tau^\alpha = 0$ if the curvature control is barely realised.

Lemma 12.2.10. *For large α, $\tau^\alpha > \tau_*$.*

Proof of Lemma 12.2.10. By contradiction. Let us assume that $\tau^\alpha \leqslant \tau_*$ for a subsequence, which we consider thereafter as the sequence. Define $\eta := \sigma(2\tau_*)^{-1}$, where σ has been defined on p. 162. Set $\tau'^\alpha := (1-\eta)\tau^\alpha$. Note that

$$\bigcup_{t\in[t^\alpha - \tau'^\alpha (r^\alpha)^2, t^\alpha]} B(x^\alpha, t, r^\alpha)$$

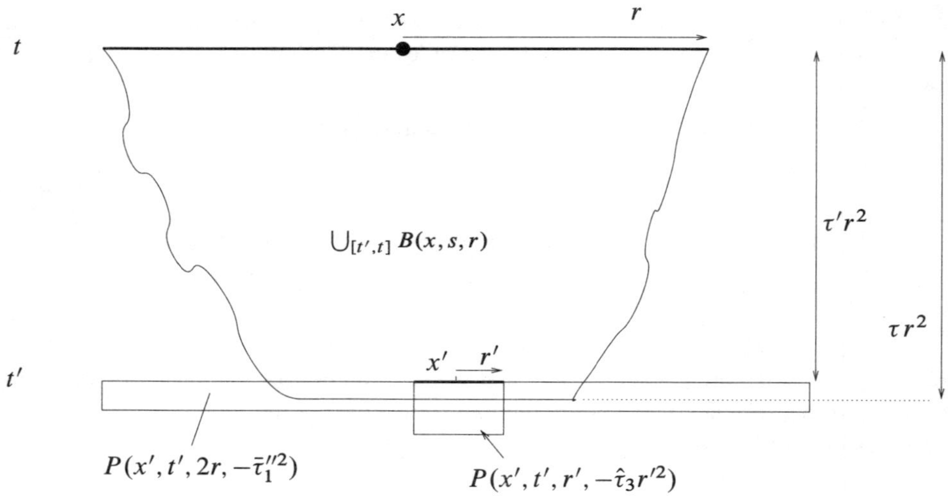

Figure 12.2. Finding a sub-ball of almost Euclidian volume.

is unscathed and $\mathrm{Rm} \geqslant -(r^\alpha)^2$ there.

From now on we sometimes drop the indices α. Define

$$t' := t - \tau' r^2 \quad \text{and} \quad r' := \frac{1}{4}\theta_0\left(\frac{v_0(1)}{10}\right) r = Br < r/2.$$

We claim that there exists $B(x',t',r') \subset B(x,t',r/4)$, all of whose sub-balls have $(1-\varepsilon_0)$-almost Euclidean volume.

Indeed, if $t = t'$ we can take $x' = x$. If not, observe that $\tau' \leqslant \tau_*$, and that (up to rescaling the metric by r^{-2}), the hypotheses of Lemma 12.2.6 are satisfied on

$$\bigcup_{s\in[t-\tau' r^2,t]} B(x,s,r).$$

Then by Lemma 12.2.6 (b) we have

$$\operatorname{vol} B\left(x,t',\frac{r}{4}\right) \geqslant \frac{1}{10} v_0\left(\frac{r}{4}\right) = \frac{1}{10} v_0(1)\left(\frac{r}{4}\right)^3.$$

Then we apply Lemma D.1.6 to $B(x,t',r/4)$ (after rescaling by $(\frac{r}{4})^{-2}$) to get a ball

$$B\left(x',t',\theta_0\left(\frac{v_0(1)}{10}\right)\frac{r}{4}\right) \subset B\left(x,t',\frac{r}{4}\right),$$

satisfying the required assertion. Let us now check that the triple (x',t',r') satisfies the assumptions of Proposition 12.2.4 for large α (recall that we omit some mentions to α). In particular we must have

$$\hat{\theta}_3^{-1} h' \leqslant r' \leqslant \hat{r}^{\alpha} \sqrt{t'},$$

where now $h' = \max_{[t'/2,t']} h^{\alpha}(\cdot)$.

Recall that $h(\cdot)$ is nonincreasing and $h(\cdot) \leqslant \delta(\cdot) r(\cdot)$. Clearly, $[t'/2, t'] \subset [t/4, t]$, as

$$t' = t - \tau' r^2 \geqslant t - \tau' \hat{r}^2 t \geqslant (1 - 10^{-6}) t.$$

On the one hand, using the fact that $\eta(t^{\alpha}) \to 0$ in Definition 11.1.4 of $\bar{\delta}(\cdot)$, for large α we have

$$r' = Br \geqslant Br(t) \geqslant \hat{\theta}_3^{-1} r(\tfrac{t}{4}) \bar{\delta}(\tfrac{t}{4}) \geqslant \hat{\theta}_3^{-1} r(\tfrac{t}{4}) \delta(\tfrac{t}{4}) \geqslant \hat{\theta}_3^{-1} h(\tfrac{t}{4}) \geqslant \hat{\theta}_3^{-1} h(\tfrac{t'}{2}) \geqslant \hat{\theta}_3^{-1} h'.$$

On the other hand,

$$r' = Br \leqslant B\hat{r}\sqrt{t} \leqslant \hat{r}\sqrt{t'} B \sqrt{\tfrac{t}{t'}} \leqslant \hat{r}\sqrt{t'}. \tag{12.3}$$

Finally recall that $t^{\alpha} \to \infty$ hence $t'^{\alpha} \geqslant t\alpha/2 \geqslant \widehat{T}_3$ for large α. Assumption (ii) of Proposition 12.2.4 is clearly true as $r' < r$ and (iii) holds by definition of $B(x', t', r')$. Finally, as $r'^{\alpha} < \frac{r^{\alpha}}{2}$ and $t'^{\alpha} \leqslant t^{\alpha}$, the conclusion of Proposition 12.2.4 holds true by definition of r^{α} and $r_0{}^{\alpha}$. Then

$$\bigcup_{s \in [t' - \hat{\tau}_3 r'^2, t']} B(x', s, r'/2)$$

is unscathed and $|\mathrm{Rm}| < \widehat{K}_3 r'^{-2}$ there. We now need to extend the curvature control to $B(x, s, r)$ using Corollary 12.1.6. Before doing this we need better curvature estimates. Note that on the set above, by the curvature pinched toward positive property, we have (see Section 4.4)

$$\mathrm{Rm} \geqslant -\phi_s(R) \geqslant -\phi_s(\widehat{K}_3 r'^{-2}) = -\frac{\phi_0((1+s)\widehat{K}_3 r'^{-2})}{(1+s)\widehat{K}_3 r'^{-2}} \widehat{K}_3 \cdot r'^{-2} \geqslant -r'^{-2}$$

for large α. Indeed, using (12.3) we get $t' - \hat{\tau}_3 r'^2 \geqslant t' - \hat{\tau}_3 \hat{r}^2 t' \geqslant t'/2$, hence

$$(1+s)\widehat{K}_3 r'^{-2} \geqslant \frac{t'}{2} r'^{-2} \geqslant \frac{\hat{r}^{-2}}{2} \to \infty,$$

as α goes to $+\infty$. Doing the parabolic rescaling by $4r'^{-2}$ at (x', t'), one finds that

$$\bigcup_{[-4\hat{\tau}_3, 0]} B(x', s, 1)$$

is unscathed and $\mathrm{Rm} \geqslant -1$ there. Applying Lemma 12.2.6 (a) (note that $4\hat{\tau}_3 \leqslant \tau_*$), we obtain $R < \frac{K_*}{4\hat{\tau}_3}$ on

$$\bigcup_{[-2\hat{\tau}_3, 0]} B(x', s, 1/2),$$

hence, unscaling and using Pinching Lemma 4.4.7, we get $|\mathrm{Rm}| < \frac{K_*}{\hat{\tau}_3} r'^{-2}$ on

$$\bigcup_{[t' - \frac{\hat{\tau}_3}{2} r'^2, t']} B(x', s, r'/2).$$

Set $r'' := B'r'$, where B' has been defined on p. 162. Then $|\mathrm{Rm}| < r''^{-2}$ on the set above. Using the multiplicative distance-distortion lemma (Lemma 2.2.7),

$$B(x', t', r'') \subset B(x', s, 10r'') \subset B(x', s, \tfrac{r'}{4}),$$

for $s \in [t' - r''^2, t']$. Then

$$P(x', t', r'', -r''^2) \subset \bigcup_{[t' - \frac{\hat{\tau}_3}{2} r'^2, t']} B(x', s, \tfrac{r'}{4})$$

and hence it is unscathed. Let us check that the assumptions of Corollary 12.1.6 hold true at (x', t', r'') for the constants $A = 4(BB')^{-1}$ and $(1 - \varepsilon_0)\upsilon_0(1)$. As before, for large α we have

$$r'' = BB'r \geqslant BB'r(t) \geqslant \bar{\theta}_1^{-1} r(\tfrac{t}{4})\bar{\delta}(\tfrac{t}{4}) \geqslant \bar{\theta}_1^{-1} h(\tfrac{t}{4}) \geqslant \bar{\theta}_1^{-1} h'.$$

On the other hand, since $t' \geqslant (1 - 10^{-6})t$, we have

$$r'' = BB'r \leqslant \hat{r}^{\alpha} \sqrt{t} BB' \leqslant \bar{r}_1 \sqrt{t'}.$$

Finally, $t' \geqslant (1 - 10^{-6})t \geqslant \bar{T}_1$ for large α. Then $P(x', t', Ar'', -\bar{\tau}_1 r''^2)$ is unscathed and $|\mathrm{Rm}| < \bar{K}_1 r''^{-2} = \bar{K}_1 (BB') r^{-2} = K r^{-2}$ there by Corollary 12.1.6. Since $Ar'' = 4r$, we also have $|\mathrm{Rm}| < K r^{-2}$ on $P(x, t', 2r, -\bar{\tau}_1 (BB')^2 r^2)$ (recall that $d_{t'}(x, x') < r/4$). Using the distance-distortion lemma, we deduce that

$$\bigcup_{[t' - \sigma r^2, t']} B(x, s, r) \subset P(x, t', 2r, -\bar{\tau}_1 (BB')^2 r^2),$$

where σ has been defined at the beginning of the proof. By the same pinching argument as before, we deduce that $\mathrm{Rm} \geqslant -r^{-2}$ on this set. We thus have obtained that, for large α (see Figure 12.2),

$$\bigcup_{[t - (\tau' + \sigma) r^2, t]} B(x, s, r)$$

is unscathed and $\mathrm{Rm} \geqslant -r^{-2}$ there. But

$$\tau' + \sigma = (1 - \eta)\tau + 2\eta\tau_* \geqslant \tau - \eta\tau_* + 2\eta\tau_* \geqslant \tau + \eta\tau_*$$

contradicting the definition of τ^{α}. This proves Lemma 12.2.10. □

We now conclude the proof of Proposition 12.2.4. Having obtained that $\tau^\alpha > \tau_*$ for large α, we get by definition of τ^α that

$$\bigcup_{[t-\tau_* r^2, t]} B(x, s, r)$$

is unscathed and $\mathrm{Rm} \geqslant -r^{-2}$ there. Recall that the sub-balls of $B(x,t,r)$ have $(1-\varepsilon_0)$-almost Euclidean volume, and in particular $B(x,t,r)$ itself. Hence we can apply (after rescaling) Lemma 12.2.6 a) at (x,t,r) to get that $R < \frac{K_*}{\tau_*} r^{-2} \leqslant \widehat{K}_3 r^{-2}$ on

$$\bigcup_{[t-\frac{\tau_*}{2} r^2, t]} B(x, s, \tfrac{r}{2}) \supset \bigcup_{[t-\tau_2' r^2, t]} B(x, s, \tfrac{r}{2}).$$

This proves Proposition 12.2.4. □

It remains to prove Lemma 12.2.6.

Proof of Lemma 12.2.6. We begin with a time-dependent estimate, which will be sufficient to prove part (a) of Lemma 12.2.6.

Sublemma 12.2.11. *There exists $L > 0$, $\tau_\sharp > 0$ such that under the assumptions* (i) *and* (ii) *of Lemma* 12.2.6, *if $\tau \in [0, \tau_\sharp]$ then the scalar curvature satisfies*

$$R < L(t+\tau)^{-1} \quad \text{on} \quad \bigcup_{t\in[-\tau,0]} B(x, t, \tfrac{1}{2}).$$

Proof of Sublemma 12.2.11. We argue by contradiction. If the sublemma is not true there exist $\tau_k \to 0$, $L_k \to +\infty$, $g_k(\cdot)$ Ricci flows on Ω_k, smooth on $\mathcal{B}_k = \bigcup_{[-\tau_k,0]} B(x_k, t, \frac{1}{2})$ where $x_k \in \Omega_k$. The sectional curvature is bounded below by -1 on $\mathcal{B}_k$ and $\operatorname{vol} B(x_k, 0, 1) \geqslant (1-\varepsilon_0)v_0(1)$. Finally there exist points (x_k'', t_k'') such that $x_k'' \in \bigcup_{[-\tau_k,0]} B(x_k, t, \frac{1}{2})$ and $t_k'' \in [-\tau_k, 0]$, contradicting the conclusion, that is

$$R(x_k'', t_k'') \geqslant L_k(t_k'' + \tau_k)^{-1}.$$

Define

$$C := \frac{v_{-1}(1/4)}{4v_{-1}(1)} \in (0, 10^{-1}), \tag{12.4}$$

$$t_k := \inf\{t \in [-\tau_k, 0] \mid \operatorname{vol} B(x_k, t', 1) \geqslant C(1-\varepsilon_0)v_0(1) \text{ for all } t' \in [t, 0]\}. \tag{12.5}$$

By continuity $t_k < 0$, and we can assume moreover that $-1 \leqslant -\tau_k \leqslant t_k$.

The first step of the proof is to obtain on $[t_k, 0]$ a curvature bound as in Sublemma 12.2.11, $-\tau_k$ being replaced by t_k. Then we have to prove that in fact, $t_k = -\tau_k$, and this a contradiction. We begin with the curvature bound:

Assertion 12.2.12. *There exists L' such that, for k large enough,*

$$R < L'(t-t_k)^{-1} \quad \textit{on} \bigcup_{[t_k,0]} B(x_k,t,\tfrac{1}{2}).$$

Proof of Assertion 12.2.12. If it is not true, then up to taking a subsequence, there exists $L'_k \to +\infty$ and $(x'_k,t'_k) \in \bigcup_{[t_k,0]} B(x_k,t,1/2) \times [t_k,0]$ with

$$R(x'_k,t'_k) \geqslant L'_k(t'_k-t_k)^{-1}.$$

Now (x'_k,t'_k) may not be the "best" point with this property; we have to choose a point with almost maximal curvature. This is done in the following point-picking claim, whose proof is postponed at the end of the section.

Let $A_k := \frac{1}{25}\sqrt{L'_k}$ and $H_k := \frac{A_k}{\max(\frac{1}{5},9C(3))}$ where $C(n)$ is the constant given by Lemma 2.2.11. We may assume $A_k \geqslant 2$.

Claim (Point-picking claim). There exists $(\bar{x}_k,\bar{t}_k)$ with $\bar{t}_k \in [t_k,0]$, such that $\bar{Q}_k = R(\bar{x}_k,\bar{t}_k) \geqslant L'_k(\bar{t}_k-t_k)^{-1}$ and $\bar{t}_k - H_k\bar{Q}_k^{-1} > t_k$, and such that $B(\bar{x}_k,\bar{t}_k,A_k\bar{Q}_k^{-1/2}) \subset B(x_k,\bar{t}_k,2/3)$ and $R < 2\bar{Q}_k$ on

$$\bigcup_{[\bar{t}_k-H_k\bar{Q}_k^{-1},\bar{t}_k]} B(\bar{x}_k,t,A_k\bar{Q}_k^{-1/2}) \subset \bigcup_{(t_k,0]} B(x_k,t,1)$$

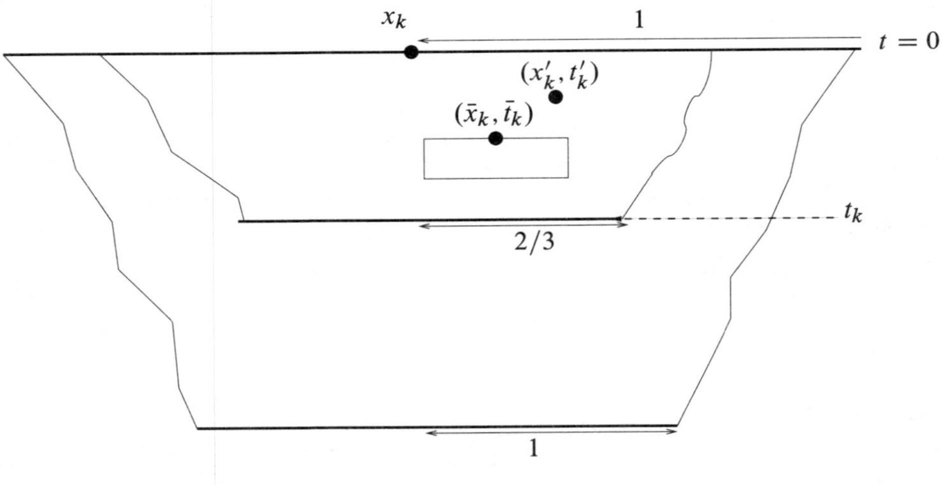

Figure 12.3. Point-picking: finding $(\bar{x}_k,\bar{t}_k)$ with local curvature estimates.

Using the point-picking claim, we continue the proof of Assertion 12.2.12. After rescaling by $\bar{Q}_k$ at $(\bar{x}_k,\bar{t}_k)$ we get a sequence $\bar{g}_k(\cdot)$ satisfying

$$R_{\bar{g}_k} < 2 \quad \text{on} \bigcup_{[H_k,0]} B(\bar{x}_k,t,A_k),$$

and $\mathrm{Rm}(\bar{g}_k(\cdot)) \geqslant -\bar{Q}_k^{-1} \to 0$ as $k \to \infty$. Since $\bar{t}_k \in (t_k, 0]$ we have a lower bound on the volume of $B(\bar{x}_k, \bar{t}_k, A_k \bar{Q}_k^{-1/2})$. Indeed, on a Riemannian manifold whose sectional curvature is not smaller than -1, any ball $B(x,r) \subset B(x_0, 2/3)$ satisfies

$$\frac{\mathrm{vol}\, B(x,r)}{r^n} \geqslant c_n \,\mathrm{vol}\, B(x_0, 1),$$

for a universal constant c_n, by Bishop–Gromov's comparison inequality. Applying this to the unscaled metrics, we get

$$\begin{aligned}\mathrm{vol}\, B(\bar{x}_k, \bar{t}_k, A_k \bar{Q}_k^{-1/2}) &\geqslant c_3 \,\mathrm{vol}\, B(\bar{x}_k, \bar{t}_k, 1)(A_k \bar{Q}_k^{-1/2})^3 \\ &\geqslant c_3 C(1-\varepsilon_0) v_0(1)(A_k \bar{Q}_k^{-1/2})^3,\end{aligned}$$

and, after rescaling,

$$\mathrm{vol}\, B(\bar{x}_k, 0, A_k) \geqslant c_3 C(1-\varepsilon_0) v_0(1) A_k^3.$$

Since $A_k \to \infty$ and $R(x,t) \leqslant 2$ on this last ball, we have a positive lower bound on the injectivity radius of $\bar{g}_k(0)$ at $\bar{x}_k$ by Theorem B.1.2. This allows us to apply the local version of Hamilton's Compactness Theorem C.3.3 to get a complete limit Ricci flow $\bar{g}_\infty(\cdot)$ defined on $\Omega_\infty \times (-\infty, 0]$, which has nonnegative curvature operator, bounded sectional curvature and is ancient (recall that $H_k \to \infty$) and non flat.

Let us denote by $\mathcal{V}(t)$ the asymptotic volume of $\bar{g}_\infty(t)$. Since $A_k \to \infty$, the above volume estimate shows that $\mathcal{V}(0) > 0$. By Theorem 4.2.2, $\bar{g}_\infty(\cdot)$ cannot be a κ-solution.

This means that for each $\kappa > 0$ there exist x_κ, t_κ and r_κ such that $\mathrm{Rm}_\infty \leqslant r_\kappa^{-2}$ on $B(x_\kappa, t_\kappa, r_\kappa)$ and $\mathrm{vol}\, B(x_\kappa, t_\kappa, r_\kappa) < \kappa r_\kappa^3$. Recall that in nonnegative curvature, $r^{-n} \,\mathrm{vol}\, B(x,r)$ is nonincreasing. It follows that $\mathcal{V}(t_\kappa) < \kappa$. On the other hand, since the curvature is nonnegative, $\mathcal{V}(t)$ is also nonincreasing in t. Indeed, for $U \subset \Omega_\infty$

$$\frac{d\,\mathrm{vol}(U)}{dt} = -\int_U R\, dv_{g_\infty} \leqslant 0.$$

Moreover, the additive distance-distortion lemma (Lemma 2.2.10) shows that $d_t \leqslant d_0 + c|t|$ for $t < 0$, where c denotes a generic constant. This implies that $B(x_\kappa, 0, r) \subset B(x_\kappa, t_\kappa, r + c|t_\kappa|)$. Consequently

$$\begin{aligned}\frac{\mathrm{vol}\, B(x_\kappa, 0, r)}{r^3} &\leqslant \frac{\mathrm{vol}_{g_\infty(0)}\, B(x_\kappa, t_\kappa, r + c|t_\kappa|)}{(r + c|t_\kappa|)^3}\Big(\frac{r + c|t_\kappa|}{r}\Big)^3 \\ &\leqslant \frac{\mathrm{vol}_{g_\infty(t_\kappa)}\, B(x_\kappa, t_\kappa, r + c|t_\kappa|)}{(r + c|t_\kappa|)^3}\Big(\frac{r + c|t_\kappa|}{r}\Big)^3,\end{aligned}$$

which gives, by letting r go to infinity, $\mathcal{V}(0) \leqslant \mathcal{V}(t_\kappa) < \kappa$. Hence $\mathcal{V}(0) \leqslant \kappa$ for all $\kappa > 0$, a contradiction. This ends the proof of Assertion 12.2.12. □

To conclude the proof of Sublemma 12.2.11, we need a last distance comparison estimate. We recall that $C(3)$ is the constant of Lemma 2.2.11.

Assertion 12.2.13. *For any $L \geqslant 1$ and $a \in (0, 1/2)$, there exists $\tau(L, a) > 0$ with the following property. Assume that for some $\tau \in [0, \tau(L, a)]$,*

$$\mathrm{Rm} \geqslant -1 \quad \text{and} \quad R < L(t+\tau)^{-1}, \quad \text{on} \bigcup_{[-\tau,0]} B(x, t, 1/2),$$

then

$$d_t(x, \cdot) \leqslant d_0(x, \cdot) + 4C(3)(L|t|)^{\frac{1}{2}} \quad \text{on } B(x, 0, a) \times [-\tau, 0].$$

Proof of Assertion 12.2.13. Define

$$\eta := \frac{1}{3}\left(\frac{1}{2} - a\right) > 0,$$
$$\tau(L, a) := \left(\frac{\eta}{4C(3)}\right)^2 L^{-1}.$$

Let $\tau \in [0, \tau(L, a)]$ and $t' \in [-\tau, 0]$ be minimal such that $B(x, 0, a) \subset B(x, t, \frac{1}{2} - \eta)$ for all $t \in [t', 0]$. Consider $t \in [t', 0]$ and $x' \in B(x, 0, a)$. Then $B(x', t, \eta) \subset B(x, t, 1/2)$. Indeed, if $y \in B(x', t, \eta)$ then

$$d_t(x, t) \leqslant d_t(x, x') + d_t(x', y) < 1/2 - \eta + \eta = 1/2.$$

By assumption $\mathrm{Rm} \geqslant -1$ on $B(x, t, 1/2)$, hence denoting $\nu \leqslant \mu \leqslant \lambda$ its eigenvalues,

$$\mathrm{Rm} \leqslant \lambda = \frac{R}{2} - \mu - \nu \leqslant \frac{L(t+\tau)^{-1}}{2} + 2 \leqslant L(t+\tau)^{-1},$$

on this ball. Applying Lemma 2.2.11 at time t with constant $K = 2L(t+\tau)^{-1}$, we get

$$\frac{d}{dt} d_t(x, x') \geqslant -C(3)(2L)^{1/2}(t+\tau)^{-1/2}.$$

Integrating between t and 0 we get

$$d_0(x, x') - d_t(x, x') \geqslant -2C(3)(2L)^{1/2}((t+\tau)^{1/2} - \tau^{1/2}) \geqslant -2C(3)(2L)^{1/2}|t|^{1/2}.$$

In particular,

$$d_{t'}(x, x') < d_0(x, x') + 4C(3)(L|t'|)^{\frac{1}{2}} < d_0(x, x') + 4C(3)(L\tau)^{\frac{1}{2}} < a + \eta \leqslant \frac{1}{2} - 2\eta.$$

It follows that $B(x, t', a) \subset B(x, 0, \frac{1}{2} - 2\eta)$, hence that $t' = -\tau$. This proves Assertion 12.2.13. □

An immediate consequence of Assertion 12.2.13 is that

$$B(x,0,a) \subset B(x,t,a+4C(3)(L|t|)^{\frac{1}{2}}) \quad \text{if } t \in [-\tau(L,a),0].$$

Recall that by Assertion 12.2.12, we have for large k,

$$R < L'(t-t_k)^{-1} \quad \text{on} \bigcup_{[t_k,0]} B(x_k,t,1/2).$$

Hence the assumptions of Assertion 12.2.13, with the constant L' and $a = \frac{1}{4}$, are satisfied on $[t_k,0]$ when k is large enough so that $t_k \in [-\tau(L',\frac{1}{4}),0]$. Since $t_k \to 0$, Assertion 12.2.13 implies

$$B\Big(x_k,0,\frac{1}{4}\Big) \subset B\Big(x_k,t_k,\frac{1}{4}+4C(3)(L'|t_k|)^{\frac{1}{2}}\Big) \subset B(x_k,t_k,1),$$

for k large enough. Then, using again the Bishop–Gromov inequality when $\mathrm{Rm} \geqslant -1$, we have for k large enough,

$$\begin{aligned}
\mathrm{vol}_{g(t_k)} B(x_k,t_k,1) &\geqslant \mathrm{vol}_{g(t_k)} B(x_k,0,\tfrac{1}{4}) \\
&\geqslant e^{6t_k}\,\mathrm{vol}_{g(0)} B(x_k,0,\tfrac{1}{4}) \\
&\geqslant e^{6t_k}\frac{v_{-1}(\frac{1}{4})}{v_{-1}(1)}\,\mathrm{vol}_{g(0)} B(x_k,0,1) \\
&\geqslant e^{6t_k}\frac{v_{-1}(\frac{1}{4})}{v_{-1}(1)}(1-\varepsilon_0)v_0(1) \\
&\geqslant 2C(1-\varepsilon_0)v_0(1),
\end{aligned}$$

the last inequality using (12.4) on p. 167. If $t_k > -\tau_k$ the above estimate contradicts the definition (12.5) of t_k. Indeed by continuity $\mathrm{vol}_{g(t_k)} B(x_k,t,1) \geqslant C(1-\varepsilon_0)v_0(1)$ on $[t_k-\sigma,0]$ for some $\sigma > 0$. We thus have proved that $t_k = -\tau_k$. Then Assertion 12.2.12 contradicts the assumptions made at the beginning of the proof of Sublemma 12.2.11 that we have no such scalar curvature bound on $\mathcal{B}_k$. This finishes the proof of Sublemma 12.2.11. □

We set $K_* := 2L$, where L is given by Sublemma 12.2.11, and

$$\tau_* := \min\Big\{\tau_\sharp, \tau\Big(L,\frac{1}{4}\Big), \Big(\frac{10^{-3}}{4C(3)}\Big)^2 L^{-1}\Big\},$$

where $\tau_\sharp$ is given by Sublemma 12.2.11, and $\tau(L,\frac{1}{4})$ is the parameter of Assertion 12.2.13.

Then part (a) of Lemma 12.2.6 follows from Assertion 12.2.12, by noticing that

$$R < L(t+\tau)^{-1} < \frac{2L}{\tau}, \quad \text{if } t \in [-\tau/2,0].$$

Let us check that part (b) of Lemma 12.2.6 holds with the above choices. By Sublemma 12.2.11 and Assertion 12.2.13, if $\tau \in [0, \tau_*]$, then

$$B(x, 0, 1/4 - 4C(3)(L\tau)^{\frac{1}{2}}) \subset B(x, -\tau, 1/4).$$

Set $\zeta := 4C(3)(L\tau)^{\frac{1}{2}} \leqslant 4C(3)(L\tau_\sharp)^{\frac{1}{2}} \leqslant 10^{-3}$. Then using as in the computation above that $\mathrm{Rm} \geqslant -1$, we get

$$\begin{aligned}
\mathrm{vol}_{g(-\tau)} B(x, -\tau, \tfrac{1}{4}) &\geqslant \mathrm{vol}_{g(-\tau)} B(x, 0, (\tfrac{1}{4} - \zeta)) \\
&\geqslant (e^{-4\tau})^{3/2} \mathrm{vol}_{g(0)} B(x, 0, (\tfrac{1}{4} - \zeta)) \\
&\geqslant (e^{-4\tau_*})^{3/2} \frac{\upsilon_{-1}(\frac{1}{4} - \zeta)}{\upsilon_{-1}(1)} \mathrm{vol}_{g(0)} B(x, 0, 1) \\
&\geqslant e^{-6\tau_*} \frac{\upsilon_{-1}(\frac{1}{4} - \zeta)}{\upsilon_{-1}(1)} (1 - \varepsilon_0)\upsilon_0(1) \\
&\geqslant e^{-6\tau_*} \frac{\upsilon_{-1}(\frac{1}{4} - \zeta)}{\upsilon_{-1}(1)} 4^3 (1 - \varepsilon_0)\upsilon_0(\tfrac{1}{4}) \\
&\geqslant \frac{1}{10} \upsilon_0(\tfrac{1}{4})
\end{aligned}$$

as $\frac{\upsilon_{-1}(1/5)}{\upsilon_{-1}(1)} 4^3 \geqslant 0.42$.

This finishes the proof of Lemma 12.2.6. □

We end this chapter by proving the point-picking claim of p. 168.

Proof of the Point-Picking Claim. Since we are working with a fixed Riemannian manifold we may drop the index k. Recall that we assume that there exists

$$(x', t') \in \bigcup_{s \in [t,0]} B(x, s, \tfrac{1}{2}) \times [t, 0]$$

such that $R(x', t') \geqslant L'(t' - t)^{-1}$. We are looking for a 'better' point satisfying this property. More precisely, the claim is that there exists $(\bar{x}, \bar{t})$ with $\bar{t} \in [t, 0]$, such that $\bar{Q} = R(\bar{x}, \bar{t}) \geqslant L'(\bar{t} - t)^{-1}$ and $\bar{t} - H\bar{Q}^{-1} > t$, and such that

$$B(\bar{x}, \bar{t}, A\bar{Q}^{-1/2}) \subset B(x, \bar{t}, 2/3),$$

and $R < 2\bar{Q}$ on

$$\bigcup_{[\bar{t} - H\bar{Q}^{-1}, \bar{t}]} B(\bar{x}, s, A\bar{Q}^{-1/2}) \subset \bigcup_{(t,0]} B(x, s, 1).$$

We recall that $A := \frac{1}{25}\sqrt{L'}$ and $H := \frac{A}{\max(\frac{1}{5}, 9C(3))}$.

We prove first that there exists $(\bar{x}, \bar{t})$ such that

(i) $R(y,s) < 2\bar{Q} = R(\bar{x}, \bar{t})$ for all (y,s) such that $s \in [\bar{t} - H\bar{Q}^{-1}, \bar{t}]$, $d_s(y,x) < d_{\bar{t}}(\bar{x}, x) + A\bar{Q}^{-1/2}$,

(ii) $B(\bar{x}, \bar{t}, A\bar{Q}^{-1/2}) \subset B(x, \bar{t}, 2/3)$.

Let us start with a space-time point (x_1, t_1) such that $x_1 \in B(x, t_1, 1/2)$, $t_1 \in [t, 0]$ and $Q_1 = R(x_1, t_1) \geqslant L'(t_1 - t)^{-1}$. The point (x_1, t_1) may satisfy the curvature requirement, in which case we take $(\bar{x}, \bar{t}) = (x_1, t_1)$. In that case

$$HQ_1^{-1} \leqslant HL'^{-1}(t_1 - t) \leqslant \frac{1}{4}(t_1 - t)$$

(we may assume $L' \geqslant 1$). If this is not the case, there exists (x_2, t_2) with $t_2 \in [t_1 - HQ_1^{-1}, t_1] \subset [t_1 - \frac{1}{4}(t_1 - t), t_1] \subset [t, 0]$, such that

$$d_{t_2}(x_2, x) < d_{t_1}(x_1, x) + AQ_1^{-1/2} < \frac{1}{2} + AQ_1^{-1/2},$$

and $Q_2 = R(x_2, t_2) \geqslant 2R(x_1, t_1) = 2Q_1$. Let us notice that with our choices

$$t_2 - HQ_2^{-1} \geqslant t_2 - \frac{H}{2}Q_1^{-1} \geqslant t_2 - \frac{1}{8}(t_1 - t) \geqslant t_1 - \frac{1}{4}(t_1 - t) - \frac{1}{8}(t_1 - t).$$

Either (x_2, t_2) is the point that we are looking for or we iterate the procedure. It has to stop after finitely many iterations. Indeed, if not one finds a sequence (x_n, t_n) such that $t_{n+1} \in [t_n - HQ_n^{-1}, t_n]$,

$$d_{t_{n+1}}(x_{n+1}, x) < d_{t_n}(x_n, x) + AQ_n^{-1/2}$$

and $Q_{n+1} = R(x_{n+1}, t_{n+1}) \geqslant 2Q_n$. We thus get

$$\begin{aligned}
(t_1 - t_n) &\leqslant (t_1 - t_2) + \cdots + (t_{n-1} - t_n) \\
&\leqslant \sum_{n\geqslant 1} HQ_n^{-1} \\
&\leqslant HQ_1^{-1} \sum_{n\geqslant 0} 2^{-n} \\
&= 2HQ_1^{-1} \leqslant 2\frac{H}{L'}(t_1 - t) \leqslant \frac{1}{2}(t_1 - t).
\end{aligned}$$

Similarly

$$\begin{aligned}
d_{t_n}(x_n, x) &< d_{t_{n-1}}(x_{n-1}, x) + AQ_{n-1}^{-1/2} \\
&< d_{t_1}(x_1, x) + A\sum_{k=1}^{n-1} Q_k^{-1/2} \\
&< d_{t_1}(x_1, x) + AQ_1^{-1/2} \sum_{n\geqslant 0} (\sqrt{2})^{-n}
\end{aligned}$$

$$
\begin{aligned}
&< d_{t_1}(x_1, x) + 4AQ_1^{-1/2} \\
&< \frac{1}{2} + 4\frac{A}{\sqrt{L'}}(t_1 - t)^{1/2} \\
&< \frac{1}{2} + 4\frac{A}{\sqrt{L'}}.
\end{aligned}
$$

(Recall we assume that $t \leqslant 1$.) Now with our choice of A we get $1/2 + 4A/\sqrt{L'} < 2/3$, that is

$$B(x_n, t_n, AQ_n^{-1/2}) \subset B(x, t, 2/3).$$

This yields the contradiction if the above sequence is not finite since then $R(x_n, t_n) \to +\infty$ and we are in a compact (part of) a smooth Ricci flow.

We now show that

$$B\left(\bar{x}, s, \frac{A}{2}\bar{Q}^{-1/2}\right) \subset B(x, s, d_{\bar{t}}(\bar{x}, x) + A\bar{Q}^{-1/2}). \tag{12.6}$$

Let $s' \in [\bar{t} - H\bar{Q}^{-1}, \bar{t}]$ be minimal such that (12.6) holds on $[s', \bar{t}]$. We will show that for $t \in [t', \bar{t}]$,

$$B\left(\bar{x}, t, \frac{A}{2}\bar{Q}^{-1/2}\right) \subset B\left(x, t, d_{\bar{t}}(\bar{x}, x) + \frac{3A}{4}\bar{Q}^{-1/2}\right),$$

proving that $s' = \bar{t} - H\bar{Q}^{-1}$. We have to show that

$$d_s(y, \bar{x}) < \frac{A}{2}\bar{Q}^{-1/2} \quad \text{implies} \quad d_s(y, x) < d_{\bar{t}}(\bar{x}, x) + \frac{3A}{4}\bar{Q}^{-1/2}.$$

By the triangle inequality we have for such y and s,

$$d_s(y, x) < d_s(y, \bar{x}) + d_s(\bar{x}, x) < \frac{A}{2}\bar{Q}^{-1/2} + d_s(\bar{x}, x).$$

Hence it suffices to show that

$$d_s(\bar{x}, x) < d_{\bar{t}}(\bar{x}, x) + \frac{A}{4}\bar{Q}^{-1/2}.$$

This comes as before from Lemma 2.2.11,

$$\frac{d}{ds} d_s(\bar{x}, x) \geqslant -C(3)K^{1/2},$$

where $K > 0$ bounds from above the Ricci curvature. As $\mathrm{Rm} \geqslant -1$ and $R < 2\bar{Q}$ we get $\mathrm{Rm} \leqslant \bar{Q} + 2 \leqslant 2\bar{Q}$ on $B(x, s, d_{\bar{t}}(\bar{x}, x) + A\bar{Q}^{-1/2})$. Setting $K = 4\bar{Q}$ we get

$$\frac{d}{ds} d_s(\bar{x}, x) \geqslant -2C(3)\bar{Q}^{\frac{1}{2}}.$$

Integrating for $s \in [s', t]$ we obtain, as $s' \in [\bar{t} - H\bar{Q}^{-1}, \bar{t}]$,

$$d_{\bar{t}} \geqslant d_{s'} - 2C(3)H\bar{Q}^{-\frac{1}{2}}.$$

This yields

$$d_{\bar{t}}(\bar{x}, x) \geqslant d_s(\bar{x}, x) - 2C(3)H\bar{Q}^{-\frac{1}{2}} > d_{s'}(\bar{x}, x) - \frac{A}{4}\bar{Q}^{-\frac{1}{2}}.$$

This proves (12.6) for all $s \in [\bar{t} - H\bar{Q}^{-1}, \bar{t}]$.

Now for each $s \in [\bar{t} - H\bar{Q}^{-1}, \bar{t}]$, we have

$$B(\bar{x}, s, A\bar{Q}^{-1/2}) \subset B\left(x, s, d_{\bar{t}}(\bar{x}, x) + \frac{3A}{2}\bar{Q}^{-1/2}\right) \subset B(x, s, 1)$$

as $d_{\bar{t}}(\bar{x}, x) < \frac{2}{3}$ and $A\bar{Q}^{-1/2} \leqslant AL'^{-1/2} = \frac{1}{25}$. □

Notes. Theorem 12.1.1 is similar to [Per03b], Proposition 6.3. Our proof follows [Per03b], [KL08].

Theorem 12.2.1, Theorem 12.2.3 and Lemma 12.2.6 are similar to [Per03b], Corollary 6.8, [Per03b], Proposition 6.4, and Lemma [Per03b], Lemma 6.5, respectively. Our proof of Theorem 12.2.3 uses arguments of [Per03b], [KL08], but slightly rearranged. Precisely we prove first the weaker version 12.2.4.

Part IV

Weak collapsing and hyperbolisation

Chapter 13
Collapsing, simplicial volume and strategy of proof

13.1 Collapsing and weak collapsing

The phenomenon of *collapsing* has received much attention in all dimensions from Riemannian geometers: a sequence of Riemannian metrics on a fixed manifold *collapses with bounded curvature* if all the sectional curvatures remain bounded while the injectivity radius goes uniformly everywhere to zero. For example, any flat torus T^n admits such sequences, obtained by choosing a fixed metric g_0 and rescaling the metric on some of the S^1 factors. More generally, any product of a closed manifold by a closed flat manifold has a sequence which collapses with bounded curvature.

J. Cheeger and M. Gromov [CG86], [CG90] show that a manifold M admits a sequence of collapsing metrics with bounded curvature if and only if it carries a so-called *F-structure*, which is a kind of generalised torus action. If M is a closed, orientable 3-dimensional manifold, then the existence of an F-structure on M implies that of a partition into orbits which are circles and tori, such that each orbit has a saturated neighbourhood. It follows that 3-manifolds supporting an F-structure are precisely the so-called *graph manifolds*. A compact, orientable 3-manifold is called a *graph manifold* if it is a union of circle bundles glued together along their boundaries. This important class of 3-manifolds first appeared in D. Mumford's seminal work on the topology of normal singularities of algebraic surfaces in 1961 [Mum61]. Shortly thereafter, F. Waldhausen [Wal67] gave the general definition of graph manifolds and classified them.

Another, more geometric, description of the class of graph manifolds is as follows: M is a graph manifold if and only if all the summands in its Kneser decomposition are graph manifolds; moreover, if M is irreducible, then it is a graph manifold if and only if all pieces in its JSJ decomposition are Seifert fibred. Hence this class includes the class of compact *Seifert manifolds* discussed in Chapter 1.

R. Hamilton, based on Theorem 1.2.2, suggested that Cheeger–Gromov theory might be used in his Ricci flow program for proving the Geometrisation Conjecture. However, it is still unknown as of this writing whether the sequences of metrics which can be extracted from the long-time analysis of Ricci flow with bubbling-off have bounded sectional curvature (compare Theorem 1.3.4). In order to circumvent this difficulty, Perelman introduced a different notion of collapsing, which fits better with the results he obtained about the long-time behaviour of Ricci flow and where no global upper curvature bound is imposed.

Let us give some definitions, which are due to Perelman up to slight modifications. Let (M, g) be a Riemannian 3-manifold and $\varepsilon > 0$ be a real number. Recall from Chapter 1 that a point $x \in M$ is *ε-thin with respect to g* if there exists $0 < \rho \leqslant 1$ such that on the ball $B(x,\rho)$, the sectional curvature is greater than or equal to $-\rho^{-2}$ and the volume of this ball is less than $\varepsilon\,\rho^3$.

Definition 13.1.1. Let M be a 3-manifold and g_n be a sequence of Riemannian metrics on M.

- We say that g_n has *locally controlled curvature in the sense of Perelman* if it has the following property: for all $\varepsilon > 0$ there exist $\bar{r}(\varepsilon) > 0$, $K_0(\varepsilon) > 0$, $K_1(\varepsilon) > 0$, such that for n large enough, if $0 < r \leqslant \bar{r}(\varepsilon)$, $x \in (M, g_n)$ satisfies $\operatorname{vol} B(x,r)/r^3 \geqslant \varepsilon$, and if the sectional curvature on $B(x,r)$ is greater than or equal to $-r^{-2}$, then $|\operatorname{Rm}(x)| < K_0 r^{-2}$ and $|\nabla \operatorname{Rm}(x)| < K_1 r^{-3}$.
- We say that g_n *collapses in the sense of Perelman* if it has locally controlled curvature in the sense of Perelman and there exists a sequence of positive numbers ε_n going to zero such that for every n, each point of M is ε_n-thin with respect to g_n.

In [Per03b], Perelman stated a collapsing result without proof, a special case of which is the following:

Theorem 13.1.2 (Perelman's collapsing theorem, closed case, see also [MT08], [CG09], [SY05], [KL10]). *Let M be a closed, orientable* 3*-manifold. If M admits a sequence of Riemannian metrics that collapses in the sense of Perelman, then M is a graph manifold.*

The main result of Part IV, Theorem 13.1.3 below, can be viewed as a *weak collapsing* result in the sense that we allow the thick part to be nonempty, but require a control on its volume. For sequences of metrics coming from the Ricci flow, the thick part is not compatible with a global collapse [Ham99], but we do not use this in our approach. In addition, Theorem 13.1.3 applies also to other sequences of metrics.

Recall the definition of the topological invariant $V_0(M)$: a *link* in M is a (possibly empty, possibly disconnected) closed 1-submanifold of M. A link is *hyperbolic* if its complement is a hyperbolic 3-manifold (necessarily of finite volume, as M is closed). The invariant $V_0(M)$ is defined as the infimum of the volumes of all hyperbolic links in M. When M is closed, this quantity is finite because any closed 3-manifold contains a hyperbolic link [Mye82]. Since the set of volumes of hyperbolic 3-manifolds is well-ordered, this infimum is always realised by some hyperbolic 3-manifold H_0; in particular, it is positive.

The following is a restatement of Theorem 1.3.13:

Theorem 13.1.3 (Weak collapsing). *Let M be a non-simply connected, closed, orientable, irreducible* 3*-manifold. Suppose that there exists a sequence g_n of Riemannian metrics on M satisfying the following:*

(1) *The sequence* $\operatorname{vol}(g_n)$ *is bounded.*
(2) *For all $\varepsilon > 0$ and for every sequence $x_n \in M$ so that x_n is ε-thick with respect to g_n for each n, the sequence of pointed manifolds (M, g_n, x_n) subconverges*

in the pointed $\mathcal{C}^2$-topology towards a hyperbolic pointed manifold with volume strictly less than $V_0(M)$.

(3) *The sequence g_n has locally controlled curvature in the sense of Perelman.*

Then M is a Seifert fibred manifold or contains an incompressible torus.

Remark 13.1.4. Since any irreducible graph manifold is Seifert or contains an incompressible torus, the conclusion of Theorem 13.1.3 is logically equivalent to saying that M is a graph manifold or contains an incompressible torus. In the proof, we will in fact assume that M does not contain any incompressible torus, and show that M is a graph manifold.

Remark 13.1.5. A special case of Theorem 13.1.3 is when the sequence g_n collapses in the sense of Perelman. In a sense, this amounts to Condition (2) being vacuously true. In this case, Hypothesis (1) is not used, so that we recover Perelman's collapsing theorem for closed manifolds (Theorem 13.1.2) under some extra (in fact unnecessary) topological hypotheses on M. This special case is sufficient to complete the proof of the Geometrisation Conjecture provided one is willing to apply some minimal surface arguments of Hamilton's (cf. [Ham99]) to show that in the case of sequences of metrics coming from Ricci flow with bubbling-off, the tori appearing in the boundary of the thick part are incompressible. Namely, when the irreducible manifold has infinite fundamental group, the flow runs for infinite time and there are two possibilities: either there is a thick part and Hamilton's argument [Ham99] provides a hyperbolic piece in the geometric decomposition of the manifold, or the manifold collapses and (since it is not simply connected) the collapsing case in Theorem 13.1.3 applies. We will not use [Ham99] here, preferring to deal with the incompressibility issue by an argument inspired by M. Anderson's approach to geometrisation [And02]. See also the comments in Section 13.4.

13.2 Simplicial volume

The purpose of this section is to review the notion of *simplicial volume,* sometimes called Gromov's norm, introduced by M. Gromov in [Gro82]. This is a key tool in the proof of Theorem 13.1.3. In particular, we need an additivity result for the simplicial volume under gluing along incompressible tori (see [Gro82], [Kue03], [Som81]) which implies that the simplicial volume of a 3-manifold admitting a geometric decomposition is proportional to the sum of the volumes of the hyperbolic pieces. In particular, such a manifold has zero simplicial volume if and only if it is a graph manifold. We also use in an essential way Gromov's vanishing theorem, recalled below as Theorem 13.2.6.

The main references for this section are [Gro82], [Iva85].

13.2.1 Definition and first examples

The *simplicial volume* of a compact, orientable n-manifold M (possibly with boundary) is defined as

$$\|M\| := \inf \left\{ \sum_{i=1}^{k} |\lambda_i| \;\middle|\; \begin{array}{l} \sum_{i=1}^{k} \lambda_i \sigma_i \text{ is a cycle representing a fundamental} \\ \text{class in } H_n(M, \partial M; \mathbb{R}), \text{ where } \sigma_i : \Delta^n \to M \\ \text{is a singular simplex and } \lambda_i \in \mathbb{R},\ i = 1, \dots, k. \end{array} \right\}$$

It follows from the definition that if $f : (M, \partial M) \to (N, \partial N)$ is a proper map, then

$$\|M\| \geqslant |\deg(f)|\, \|N\|.$$

In particular, if a manifold admits a self-map of degree d with $|d| > 1$, then its simplicial volume is zero. Hence, for the n-sphere and the n-torus, $\|S^n\| = \|T^n\| = 0$. The same formula implies that the simplicial volume of spherical or Euclidean manifolds is zero. In general, the simplicial volume of a Seifert fibred manifold vanishes, by the following result of [Yan82]:

Proposition 13.2.1 (Yano). *The simplicial volume of a compact, connected, smooth manifold M which admits a nontrivial smooth S^1-action is zero.*

The following is a non-vanishing result [Gro82], [Thu80]. We let v_n denote the volume of the regular ideal hyperbolic n-simplex.

Theorem 13.2.2 (Gromov–Thurston). *If M is a hyperbolic n-manifold with torus or empty boundary, then*

$$\|M\| = \frac{1}{v_n} \operatorname{vol}(M).$$

13.2.2 Simplicial volume and geometric decompositions

The following result is useful for computing the simplicial volume of a 3-manifold from its geometric decomposition [Gro82], [Kue03], [Som81], provided it exists.

Theorem 13.2.3 (Gromov). *The simplicial volume of* 3*-manifolds is additive for connected sums and gluing along incompressible tori.*

Combining 13.2.1, 13.2.2 and 13.2.3, we get:

Corollary 13.2.4. *If M is a closed, orientable* 3*-manifold admitting a geometric decomposition, and $H_1, \dots, H_k$ are the hyperbolic pieces of this decomposition, then*

$$\|M\| = \frac{1}{v_3} \sum_{i=1}^{k} \operatorname{vol}(H_i).$$

Remark 13.2.5. By applying this corollary, once we know that M has a geometric decomposition, then M is a (possibly reducible) graph manifold iff $\|M\| = 0$.

We will also use a vanishing theorem of Gromov. We say that a covering $\{U_i\}_i$ by open sets of a topological space X has *dimension at most k* if every point of X belongs to at most $k + 1$ open sets of the covering. The dimension of the covering is defined by taking a minimum of such k; it equals the dimension of the nerve of the covering.

A connected subset $U \subset M$ is said to be *amenable* if the image of its fundamental group $\pi_1(U) \to \pi_1(M)$ is amenable. This definition extends to disconnected subsets by requiring that all connected components should be amenable.

Theorem 13.2.6 (Gromov's vanishing theorem). *Let M be a closed and orientable n-manifold. If M admits a covering of dimension at most $n-1$ by amenable open sets, then $\|M\| = 0$.*

This theorem is proved in [Gro82], [Iva85] using bounded cohomology, a notion dual to simplicial volume. Namely, [Gro82], §3.1, implies that the image of the bounded cohomology of an n-dimensional manifold M in the usual cohomology vanishes in dimension n. Thus the dual to the fundamental class in $H^n(M; \mathbb{R})$ is not a bounded cohomology class. By duality this is equivalent to the vanishing of the simplicial volume, cf. [Gro82], §1.1.

13.2.3 Simplicial volume and collapsing

Philosophically, the connection between *collapsing* and simplicial volume is that any manifold which admits a collapsing sequence in any reasonable sense should have zero simplicial volume. The model for such a result is Gromov's isolation theorem [Gro82], p. 14. His argument extends to collapsing in the sense of Perelman. Although we will not need this fact, we give a quick sketch, since it provides an example of a *covering argument*, a technique essential to our proof of Theorem 13.1.3.

Let M be a closed, orientable 3-manifold and g_n be a sequence of Riemannian metrics on M which collapses in the sense of Perelman. For n large enough, thanks to the local control on the curvature, each point $x \in M$ has a neighbourhood U_x in (M, g_n) which is close to a metric ball in some manifold of nonnegative sectional curvature, whose volume is small compared to the cube of the radius (cf. Proposition 14.2.1). These neighbourhoods will be called *local models* throughout Part IV. By Corollary B.2.6, they have virtually abelian, hence amenable fundamental groups.

Adapting Gromov's arguments, one can extract from this open cover a finite cover $U_1, \dots, U_p$, of dimension at most 2. Since these open sets are essentially small metrics balls, it is not difficult to obtain a 3-dimensional covering. The crucial part in the argument is to go down from dimension 3 to dimension 2, using the upper bound on the volumes of the U_i's, which ultimately comes from the collapsing hypothesis.

One concludes the argument by simply quoting Gromov's vanishing theorem (Theorem 13.2.6).

The argument sketched above makes plausible the idea that it should be possible to detect the geometry of a 3-manifold (i.e., graph vs. hyperbolic) from collapsing via the simplicial volume. However, it is not sufficient to prove Theorem 13.1.3, since we do not know (yet) that the manifold M has a geometric decomposition. The trick to overcome this problem is explained in the next section. The strategy will consist in transforming the original problem into one about irreducible manifolds with nonempty boundary, where Thurston's geometrisation theorem can be used, since these manifolds are Haken (cf. Theorem 1.1.5).

13.3 Sketch of proof of Theorem 13.1.3

13.3.1 The collapsing case

For simplicity, we first do the sketch under the extra assumption that the sequence collapses in the sense of Perelman, i.e., the case treated in the paper [BBB$^+$10]. Assumption (2) is then vacuous, and as we will see, Assumption (1) is not needed.

Let M be a 3-manifold satisfying the topological hypotheses of Theorem 13.1.3 and let g_n be a sequence of Riemannian metrics on M which collapses in the sense of Perelman.

We begin by covering M by local models as explained in the previous section. The next step, Proposition 14.3.1, consists in finding a local model U such that all connected components of $M \setminus U$ are Haken. This requirement is equivalent to irreducibility of each connected component of $M \setminus U$, because each connected component of $M \setminus U$ has nonempty boundary. Since M is irreducible, it suffices to show that U is not contained in a 3-ball. This is in particular the case if U is *homotopically nontrivial*, i.e., the homomorphism $\pi_1(U) \to \pi_1(M)$ has nontrivial image.

The proof of the existence of a homotopically nontrivial local model U is done by contradiction: assuming that all local models are homotopically trivial, we construct a covering of M of dimension less than or equal to 2 by homotopically trivial open sets (Assertion 14.3.2). By a result of J. C. Gómez-Larrañaga and F. González-Acuña [GLGA92], corresponding to Corollary A.4.2 here, a closed irreducible 3-manifold admitting such a covering must have trivial fundamental group. This is where we use the hypothesis that M is not simply connected.

The last step (Subsection 14.3.3) is again a covering argument, but done relatively to some fixed homotopically nontrivial local model U. It shows that any manifold obtained by Dehn filling on $Y := M \setminus U$ has a covering of dimension less than or equal to 2 by virtually abelian open sets, and therefore has vanishing simplicial volume. We conclude using Proposition 14.0.1, which states that if Y is a Haken manifold with boundary a collection of tori and such that the simplicial volume of every Dehn filling on Y vanishes, then Y is a graph manifold.[1]

[1] As already mentioned in [BP01], [BLP05], Proposition 14.0.1 is a consequence of the geometrisation

13.3.2 The general case

A few arguments are needed in order to carry over the previous arguments to the general case: the thick part may now be non-empty, and needs to be taken care of. As already mentioned, we will assume that M contains no incompressible tori, and show that M is a graph manifold.

Let (M, g_n) satisfy the hypotheses of Theorem 13.1.3. In the first step (Section 14.1), we cover the thick part by submanifolds H_n^i approximating compact cores of limiting hyperbolic manifolds. This is where Hypothesis (1) gets used, via a Margulis constant, to control the number of such 'quasi-hyperbolic' pieces. Their boundaries consist of tori, which by assumption are compressible.

We then cover the thin part by local models (Section 14.2) as explained before. The bulk of the proof is in Section 14.3: we consider the covering of M by approximately hyperbolic submanifolds and local models of the thin part and perform two covering arguments: the first one (Assertion 14.3.2) proves the existence of a local model homotopically nontrivial M; the second one (Assertion 14.3.15) is done relatively to this homotopically nontrivial subset and proves that M is a graph manifold.

13.4 Comments

Theorem 13.1.3 is taken from the unpublished manuscript [BBB$^+$07], which was posted on arXiv in June 2007. A shortened version tackling only the collapsing case appeared as [BBB$^+$10]. The idea behind the notion of weak collapsing is to combine the concept of collapsing Riemannian manifolds with that of *volume comparison*, which can be traced back at least to early work of Thurston and Gromov. Our direct sources of inspiration are, besides Perelman's collapsing theorem [Per03b], Theorem 7.4, the argument for incompressibility of tori outlined in [Per03b], §8, and the observation (which we learned from M. Gromov, and can also be found in an earlier version of the Kleiner–Lott notes) that the scalar curvature can be used in place of Perelman's λ-invariant. For a detailed discussion of this, and the relation with the σ-constant, see [KL08], [And05].

There is an abundant literature on collapsing in Riemannian geometry. In particular, there are by now several texts on Perelman's collapsing theorem or some variant of it: an article by Shioya and Yamaguchi [SY05], the above-mentioned paper by the authors of the present monograph [BBB$^+$10], an arXiv preprint by Morgan and Tian [MT08], and two more recent preprints by Cao and Ge [CG09], and Kleiner and Lott [KL10].

We shall not attempt a detailed comparison between all versions of the theorem, in particular various minor technical differences such as the restriction $\rho \leqslant 1$ in the definition of the thin part (hence flat manifolds are not collapsed), or the number of

of Haken manifolds, additivity of the simplicial volume mentioned above, and Thurston's hyperbolic Dehn filling theorem.

derivatives involved in the curvature control hypothesis. It is important to note that the primary role of such a theorem is to be a step in a proof of the Geometrisation Conjecture. Viewed *a posteriori*, i.e., after the Geometrisation Conjecture has been established, the result(s) become easier to prove, and some of the hypotheses appear to be unnatural and/or superfluous. More interesting is to see which techniques are used or avoided in the proof of the collapsing theorem, and in the deduction of the Geometrisation Conjecture from it.

Perelman's statement (Theorem 7.4 of [Per03b]), allows the manifold to have a nonempty boundary, imposing an *ad hoc* hypothesis on the geometry near the boundary. This is essential if one wishes to prove the Geometrisation Conjecture without using Thurston's hyperbolisation theorem for Haken manifolds (Theorem 1.1.5).

Shioya–Yamaguchi [SY05] and Cao–Ge [CG09] do not have the hypothesis of curvature control in the sense of Perelman. In counterpart, they use deep results on Alexandrov spaces, such as Perelman's stability theorem [Per91] (see also the paper by V. Kapovitch [Kap07]).

The Morgan–Tian's proof is closest in spirit to Perelman's and their statement is nearly identical to [Per03b], Theorem 7.4. Thanks to the curvature control hypothesis, they only need elementary facts about Alexandrov spaces. In our treatment, Alexandrov spaces do not appear at all, although it should be clear to the expert that they are lurking behind the scenes.

Unlike all other authors, we always assume that the manifold is irreducible. This hypothesis is consistent with our choice of starting the proof of the Geometrisation Conjecture by quoting Kneser's decomposition theorem (Theorem 1.1.3). This simplifies the exposition quite a bit; however, we do not feel that mathematically it makes such a great difference, because Kneser's theorem is an 'elementary' result in 3-manifold topology, not significantly more difficult than Waldhausen's classification of graph manifolds, which gets (at least implicitly) used in the proof anyway. By contrast, Thurston's hyperbolisation theorem is a deep result, whose proof requires completely different techniques than those used by Perelman's (e.g. Teichmüller theory and Kleinian group theory). Hence it is natural to ask for a proof of the Geometrisation Conjecture which does not use this theorem. This is Perelman's original viewpoint which is developed in [KL08], [MT08] or [CG09].

In Theorem 13.1.3, we assume that the manifold is not simply-connected. This restriction is needed for our proof, but of course *a posteriori* unnecessary, since the only closed, simply-connected 3-manifold is S^3, which is a graph manifold. Logically, this is not a problem, since the Poincaré Conjecture follows from the independent arguments expounded in Chapter 1 and Parts I and II. However, it would be nice to extend the proof of Theorem 13.1.3 given in the next chapter to cover the simply-connected case.

We believe weak collapsing to be interesting in its own right. In particular, it leads to the refined invariant $V_0'(M)$ discussed in Chapter 15, which may have other applications. It also permits us to embed the argument for incompressibility of the boundary tori in the thick part into a general result about sequences of metrics, independent of the Ricci flow part of the proof. Using Hamilton's minimal surface arguments would not make this possible, since his computation seems to be special to Ricci flow.

To sum up, there are several results and techniques which can be used to good effect in proving the Geometrisation Conjecture, but can also be avoided: Kneser's decomposition theorem, the Jaco–Shalen–Johannson splitting theorem, Thurston's hyperbolisation theorem, minimal surface theory, Alexandrov space theory... so far, all authors (including us) purposefully avoid one or more of those, to the expense of making some or other part of the proof more complicated or convoluted. If on the contrary one is willing to use all of them, one gets a more streamlined and conceptually appealing proof. Here is a brief sketch of such a proof: first use 3-manifold theory to reduce geometrisation to the irreducible, atoroidal case (Theorem 1.1.6); construct Ricci flow with $(r(\cdot), \delta(\cdot), \kappa(\cdot))$-bubbling-off. When $\pi_1(M)$ is finite, use Colding–Minicozzi to show that M is spherical. When $\pi_1(M)$ is infinite, prove a simplified thin-thick decomposition theorem (Theorem 1.3.4 without the condition of controlled curvature in the sense of Perelman). Then consider the sequence $g_n := (4n)^{-1} g(n)$. If this sequence collapses, show that M is a graph manifold. Otherwise there is at least one hyperbolic limit. Following Hamilton [Ham99], prove persistence of a hyperbolic limit H. If H were noncompact, then by Hamilton's minimal surface argument, M would contain an incompressible torus, contradicting the hypothesis that M is atoroidal. Hence H is closed, and M is hyperbolic.

Chapter 14
Proof of the weak collapsing theorem

In this chapter we prove Theorem 13.1.3. Firstly, in Section 14.1 we investigate the structure of the thick part. In Section 14.2 we construct local models for the thin part. The core of the proof consists of the two covering arguments of Section 14.3.

Throughout this chapter, M is a 3-manifold satisfying the topological hypotheses of Theorem 13.1.3, i.e., it is closed, orientable, irreducible and not simply connected. We let g_n be a sequence of Riemannian metrics satisfying Hypotheses (1)–(3) of the same theorem. ***We further assume that M is atoroidal***. Our goal is to show that M is a graph manifold. As already mentioned in the previous chapter, this implies that M is in fact Seifert fibred.

For general facts and notation about 3-manifolds, the reader is referred to Appendix A.

We shall conclude the proof by applying the following consequence of Thurston's hyperbolic Dehn filling theorem (for the proof, see. [BLP05], Proposition 10.17, or [BMP03], Proposition 9.36):

Proposition 14.0.1. *Let Y be a Haken 3-manifold whose boundary is a union of tori. Assume that any manifold obtained from Y by Dehn filling has vanishing simplicial volume. Then Y is a graph manifold.*

This proposition will be applied to the complement of a well-chosen submanifold $\mathcal{V}_0 \subset M$, which is homeomorphic to $S^1 \times D^2$, $T^2 \times I$, or $K^2 \tilde{\times} I$. In particular $\mathcal{V}_0$ is a circle bundle, so once we have proved that its complement is a graph manifold, it follows that M itself is a graph manifold, which is the required conclusion.

For the sake of simplicity, we use the notation $M_n := (M, g_n)$. It is implicit that any quantity depending on a point $x \in M_n$ is computed with respect to the metric g_n, and thus depends also on n. For $\varepsilon > 0$, we denote by $M_n^-(\varepsilon)$ the ε-thin part of M_n (i.e., the set of ε-thin points) and by $M_n^+(\varepsilon)$ its complement.

14.1 Structure of the thick part

Proposition 14.1.1. *Up to taking a subsequence of M_n, there exists a finite (possibly empty) collection of pointed hyperbolic manifolds $(H^1, *^1), \dots, (H^m, *^m)$ and for every $1 \leqslant i \leqslant m$ a sequence $x_n^i \in M_n$ satisfying the following:*

(i) $\lim_{n\to\infty} (M_n, x_n^i) = (H^i, *^i)$ *in the $\mathcal{C}^2$ topology.*

(ii) *For all sufficiently small $\varepsilon > 0$, there exist $n_0(\varepsilon)$ and $C(\varepsilon)$ such that for all $n \geqslant n_0(\varepsilon)$ one has $M_n^+(\varepsilon) \subset \bigcup_i B(x_n^i, C(\varepsilon))$.*

Proof. By assumption, the sequence $\operatorname{vol}(M_n)$ is bounded above. Let $\mu_0 > 0$ be a universal number such that any hyperbolic manifold has volume at least μ_0.

If for all $\varepsilon > 0$ we have $M_n^+(\varepsilon) = \emptyset$ for n large enough, then Proposition 14.1.1 is vacuously true, with $m = 0$. Otherwise, we use Hypothesis (2) of Theorem 13.1.3: up to taking a subsequence of M_n, there exists $\varepsilon_1 > 0$ and a sequence of points $x_n^1 \in M_n^+(\varepsilon_1)$ such that (M_n, x_n^1) converges to a pointed hyperbolic manifold $(H^1, *^1)$.

If for all $\varepsilon > 0$ there exists $C(\varepsilon)$ such that, for n large enough, $M_n^+(\varepsilon)$ is included in $B(x_n^1, C(\varepsilon))$, then we are done. Otherwise there exists $\varepsilon_2 > 0$ and a sequence $x_n^2 \in M_n^+(\varepsilon_2)$ such that $d(x_n^1, x_n^2) \to \infty$. Again Hypothesis (2) of Theorem 13.1.3 ensures that, after taking a subsequence, the sequence (M_n, x_n^2) converges to a pointed hyperbolic manifold $(H^2, *^2)$. We iterate this construction.

Note that for each i, and for n sufficiently large, M_n contains a submanifold $\mathcal{C}^2$-close to some compact core of H_i and whose volume is greater than or equal to $\mu_0/2$. Moreover, for n fixed and large, these submanifolds are pairwise disjoint. Since the volume of the manifolds M^n is uniformly bounded above this construction has to stop. Condition (ii) of the conclusion of Proposition 14.1.1 is then satisfied for $0 < \varepsilon < \min\{\varepsilon_1, \dots, \varepsilon_m\}$. □

Remark 14.1.2. By Proposition 14.1.1, up to subsequence one can choose sequences $\varepsilon_n \to 0$ and $r_n \to \infty$ such that the ball $B(x_n^i, r_n)$ is arbitrarily close to a metric ball $B(*^i, r_n) \subset H^i$, for $i = 1, \dots, m$, and every point of $M_n \setminus \bigcup_i B(x_n^i, r_n)$ is ε_n-thin.

Let us fix a sequence of positive real numbers $\varepsilon_n \to 0$. Let $H^1, \dots, H^m$ be hyperbolic limits given by Proposition 14.1.1. For each i we choose a compact core $\bar{H}^i$ for H^i and for each n a submanifold $\bar{H}_n^i \subset M_n$ and an approximation $\phi_n^i \colon \bar{H}_n^i \to \bar{H}^i$. Up to renumbering, one can assume that for all n we have $M_n \setminus \bigcup \bar{H}_n^i \subset M_n^-(\varepsilon_n)$, and that the $\bar{H}_n^i$'s are disjoint.

The hypothesis in Theorem 13.1.3 that the volume of each hyperbolic limit H^i is less than V_0 implies that for n sufficiently large no component $\bar{H}_n^i$ is homeomorphic to the exterior of a link in M. Also, since we assume that M is atoroidal, all boundary components of each $\bar{H}_n^i$ are compressible. By classical 3-manifold arguments based on the loop theorem and recalled in Appendix A, this implies that each $\bar{H}_n^i$ is contained in a 3-ball or a solid torus. Hence the thick part is 'topologically inessential', which explains why we can hope to show that M is a graph manifold.

More precisely, Lemma A.3.3 implies that for all sufficiently large n and each $i \in \{1, \dots, m\}$, there exists a connected submanifold $Y_n^i \subset M_n$ containing $\bar{H}_n^i$, whose boundary is a boundary component of $\bar{H}_n^i$, and which is a solid torus, or contained in a 3-ball and homeomorphic to the exterior of a knot in S^3.

In the sequel we shall pass to a subsequence, so that this holds for all n.

Proposition 14.1.3. *Either M is a lens space or there exists for each n a (possibly empty) submanifold $W_n \subset M_n$ such that the following holds:*

(i) $\bigcup \bar{H}_n^i \subset W_n$.

(ii) *Each connected component of W_n is a solid torus, or contained in a 3-ball embedded in M_n and homeomorphic to the exterior of a knot in S^3.*

(iii) *The boundary of each connected component of W_n is a connected component of $\bigcup_i \partial \bar{H}_n^i$.*

Proof. We take W_n to be the union of the Y_n^i defined above. This readily implies assertion (i). Note also that those manifolds have disjoint boundaries.

If there exist indices i, j such that $M_n = Y_n^i \cup Y_n^j$, then by Lemma A.4.3, M is a lens space. Suppose this is not the case. Then for all $i \neq j$, the submanifolds Y_n^i and Y_n^j are disjoint or one contains the other. Thus each connected component of W_n is homeomorphic to one of the Y_n^i's. This yields assertions (ii) and (iii) of Proposition 14.1.3. □

14.2 Local structure of the thin part

Let us choose a sequence $\varepsilon_n \to 0$ (see Remark 14.1.2). For all $x \in M_n^-(\varepsilon_n)$, we choose a radius $0 < \rho(x) \leqslant 1$, such that on the ball $B(x, \rho(x))$ the sectional curvature is $\geqslant -\rho^{-2}(x)$ and the volume of this ball is $< \varepsilon_n \rho^3(x)$.

Let δ be a positive real number. Recall that a diffeomorphism $f : X \to Y$ is $(1+\delta)$-*bi-Lipschitz* if both f and f^{-1} are $(1+\delta)$-Lipschitz. In this chapter, we say that two Riemannian manifolds X, Y are δ-*close* if there exists a $(1+\delta)$-bi-Lipschitz diffeomorphism between them. A $(1+\delta)$-*bi-Lipschitz embedding* is a map $f : X \to Y$ which is a $(1+\delta)$-bi-Lipschitz diffeomorphism onto its image.

In the following proposition we use the Cheeger–Gromoll soul theorem (Theorem B.2.1).

Proposition 14.2.1. *For all $D > 1$ there exists $n_0(D)$ such that if $n > n_0(D)$, then we have the following alternative. Either*

(a) *M_n is $\frac{1}{D}$-close to some closed Riemannian 3-manifold of nonnegative sectional curvature.*

Or

(b) *For all $x \in M_n^-(\varepsilon_n)$ there exists a radius $\nu(x) \in (0, \rho(x))$ and a complete open Riemannian 3-manifold X_x, with nonnegative sectional curvature and soul S_x, such that the following properties are satisfied:*

(1) *$B(x, \nu(x))$ is $\frac{1}{D}$-close to a metric ball in X_x,*

(2) *there is a $(1+\frac{1}{D})$-bi-Lipschitz embedding $f_x \colon B(x, \nu(x)) \to X_x$ such that*

$$\max\{d(f(x), S_x), \operatorname{diam} S_x\} \leqslant \tfrac{\nu(x)}{D},$$

(3) *$\operatorname{vol} B(x, \nu(x)) \leqslant \frac{1}{D}\nu^3(x)$.*

Remark 14.2.2. Since $\nu(x) < \rho(x)$, the sectional curvature on $B(x, \nu(x))$ is greater than or equal to $-\frac{1}{\rho^2(x)}$, which is in turn bounded below by $-\frac{1}{\nu^2(x)}$.

Remark 14.2.3. By Theorem B.2.5, if Case (a) occurs, then M is a graph manifold. In case (b) we can apply Corollary B.2.3. Since M is orientable and irreducible, some cases are excluded. In fact the only possibilities for the soul are a point, a circle, a 2-sphere, a 2-torus or a Klein bottle. Thus the ball $B(x, \nu(x))$ is homeomorphic to $\mathbb{R}^3$, $S^1 \times \mathbb{R}^2$, $S^2 \times \mathbb{R}$, $T^2 \times \mathbb{R}$ or to the twisted $\mathbb{R}$-bundle on the Klein bottle.

Before starting the proof of this proposition, we prove the following lemma and its consequence.

Lemma 14.2.4. *There exists a small universal constant $C > 0$ such that for all $\varepsilon > 0$, for all $x \in M_n$, and for all $r > 0$, if the ball $B(x, r)$ has volume $\geqslant \varepsilon r^3$ and sectional curvature $\geqslant -r^{-2}$, then for all $y \in B(x, \frac{1}{3}r)$ and all $0 < r' < \frac{2}{3}r$, the ball $B(y, r')$ has volume $\geqslant C\varepsilon(r')^3$ and sectional curvature $\geqslant -(r')^{-2}$.*

We use the function $v_{-\kappa^2}(r)$ to denote the volume of the ball of radius r in the 3-dimensional hyperbolic space with curvature $-\kappa^2$. Notice that $v_{-\kappa^2}(r) = \kappa^{-3} v_{-1}(\kappa r)$.

Proof. The lower bound on the curvature is a consequence of the monotonicity of the function $-r^{-2}$ with respect to r. In order to estimate from below the normalised volume we apply Bishop–Gromov's inequality (Theorem B.1.1) twice. First, we apply it to the ball around y, increasing the radius r' to $\frac{2}{3}r$:

$$\operatorname{vol} B(y, r') \geqslant \operatorname{vol} B(y, \tfrac{2}{3}r) \frac{v_{-r^{-2}}(r')}{v_{-r^{-2}}(\frac{2}{3}r)}.$$

Since $v_{-r^{-2}}(r') = r^3 v_{-1}(\frac{r'}{r}) \geqslant r^3 (\frac{r'}{r})^3 C_1$ for $C_1 > 0$ uniform, $v_{-r^{-2}}(\frac{2}{3}r) = r^3 v_{-1}(\frac{2}{3})$, and $B(x, \frac{1}{3}r) \subset B(y, \frac{2}{3}r)$, we get

$$\operatorname{vol} B(y, r') \geqslant \operatorname{vol} B(x, \tfrac{1}{3}r) \big(\tfrac{r'}{r}\big)^3 C_2.$$

Applying the Bishop–Gromov inequality again, we obtain

$$\operatorname{vol} B(x, \tfrac{1}{3}r) \geqslant \operatorname{vol} B(x, r) \frac{v_{-r^{-2}}(\frac{1}{3}r)}{v_{-r^{-2}}(r)} \geqslant r^3 \varepsilon \frac{v_{-1}(\frac{1}{3})}{v_{-1}(1)} = r^3 \varepsilon\, C_3.$$

Hence $\operatorname{vol} B(y, r') \geqslant (r')^3\, \varepsilon\, C_4$. □

The next corollary is similar to the condition of controlled curvature in the sense of Perelman. The differences are that the conclusion is valid at each point of some metric ball, not only the centre, and the constants may have changed. This fact will be useful for technical purposes.

Corollary 14.2.5. *For all $\varepsilon > 0$ there exist $\bar{r}'(\varepsilon) > 0$, $K'_0(\varepsilon)$, $K'_1(\varepsilon)$ such that for n large enough, if $0 < r \leqslant \bar{r}'(\varepsilon)$, $x \in M_n$ and the ball $B(x,r)$ has volume $\geqslant \varepsilon\, r^3$ and sectional curvatures $\geqslant -r^{-2}$ then, for all $y \in B(x, \frac{1}{3}r)$, $|\operatorname{Rm}(y)| < K'_0\, r^{-2}$ and $|\nabla \operatorname{Rm}(y)| < K'_1\, r^{-3}$.*

Proof. It suffices to apply Lemma 14.2.4, setting $\bar{r}'(\varepsilon) = \bar{r}(C\varepsilon)$, $K'_0(\varepsilon) = K_0(C\varepsilon)$ and $K'_1(\varepsilon) = K_1(C\varepsilon)$, so that we can apply the controlled curvature condition on $y \in B(x, \frac{1}{3}r)$. □

Proof of Proposition 14.2.1. Suppose that there exists $D_0 > 1$ and, after re-indexing, a sequence $x_n \in M_n$ such that neither of the conclusions of Proposition 14.2.1 holds with $D = D_0$.

Set $\varepsilon_0 := \frac{v_0(1)}{v_{-1}(1)} \frac{1}{D_0}$. We shall rescale the metrics using the following radii:

Definition 14.2.6. For $x \in M_n$, define

$$\operatorname{rad}(x) = \inf\{r > 0 \mid \operatorname{vol} B(x,r)/r^3 \leqslant \varepsilon_0\}.$$

Notice that $\operatorname{rad}(x) < \infty$ because M_n has finite volume, and that $\operatorname{rad}(x) > 0$, because $\operatorname{vol} B(x,r)/r^3 \to \frac{4}{3}\pi$ when $r \to 0$. In the following lemma, we gather some properties which will be useful for the proof:

Lemma 14.2.7. (i) *For n large enough and $x \in M_n^-(\varepsilon_n)$, one has* $0 < \operatorname{rad}(x) < \rho(x)$.

(ii) *For $x \in M_n^-(\varepsilon_n)$, one has*

$$\frac{\operatorname{vol} B(x, \operatorname{rad}(x))}{\operatorname{rad}(x)^3} = \varepsilon_0.$$

(iii) *For $L > 1$, there exists $n_0(L)$ such that for $n > n_0(L)$ and for $x \in M_n^-(\varepsilon_n)$ we have*

$$L \operatorname{rad}(x) \leqslant \rho(x).$$

In particular $\lim_{n\to\infty} \operatorname{rad}(x_n) = \lim_{n\to\infty} \frac{\operatorname{rad}(x_n)}{\rho(x_n)} = 0$.

Proof of Lemma 14.2.7. Property (i) holds as long as $\varepsilon_n < \varepsilon_0$, since rad is defined as an infimum.

Assertion (ii) is proved by continuity.

Let us prove (iii): for $L > 1$, we choose $n_0(L)$ so that $\varepsilon_n < \frac{\varepsilon_0}{L^3}$ for $n > n_0(L)$. Then, for $x \in M_n^-(\varepsilon_n)$ and $n > n_0(L)$,

$$\frac{\operatorname{vol} B(x, \frac{\rho(x)}{L})}{(\frac{\rho(x)}{L})^3} \leqslant L^3 \frac{\operatorname{vol} B(x, \rho(x))}{\rho(x)^3} \leqslant L^3 \varepsilon_n \leqslant \varepsilon_0.$$

This implies that $\rho(x)/L$ is greater than or equal to $\operatorname{rad}(x)$. □

Remark 14.2.8. For n large enough, from Lemma 14.2.7 (iii), we have $\operatorname{rad}(x_n) < \bar{r}'(\varepsilon_0)$, where $\bar{r}'(\varepsilon_0)$ is the parameter from Corollary 14.2.5.

Corollary 14.2.9. *There exists a constant $C > 0$ such that every sequence $x_n \in M_n^-(\varepsilon_n)$ satisfies*

$$\frac{\operatorname{inj}(x_n)}{\operatorname{rad}(x_n)} \geqslant C$$

for n large enough.

Proof. Since $\operatorname{rad}(x_n) < \rho(x_n)$, the sectional curvatures on $B(x_n, \operatorname{rad}(x_n))$ are bounded from below by $-\frac{1}{\rho(x_n)^2} > -\frac{1}{\operatorname{rad}(x_n)^2}$. Moreover, since $\operatorname{rad}(x_n) < \bar{r}'(\varepsilon_0)$, Corollary 14.2.5 implies that the sectional curvatures on $B(x_n, \frac{\operatorname{rad}(x_n)}{3})$ are bounded from above by $K_0'(\varepsilon_0)/\operatorname{rad}(x_n)^2$. This rescaled ball

$$\frac{1}{\operatorname{rad}(x_n)} B(x_n, \tfrac{1}{3}\operatorname{rad}(x_n))$$

has volume $\geqslant \varepsilon_0/27$ (because $\operatorname{vol} B(x_n, \frac{1}{3}\operatorname{rad}(x_n)) \geqslant \varepsilon_0(\operatorname{rad}(x_n)/3)^3$ by definition of $\operatorname{rad}(x_n)$), radius $1/3$ and curvatures with absolute value $\leqslant K_0'(\varepsilon_0)$. Then the local injectivity radius estimate in Theorem B.1.2 gives a uniform lower bound on the injectivity radius at x_n with this rescaled metric. This proves Corollary 14.2.9. □

Having proved Lemma 14.2.7 and its corollary, we proceed with the proof of Proposition 14.2.1. Let us consider the rescaled manifold $\bar{M}_n = \frac{1}{\operatorname{rad}(x_n)} M_n$. We look for a limit of the sequence $(\bar{M}_n, \bar{x}_n)$, where $\bar{x}_n$ is the image of x_n. The ball $B(\bar{x}_n, \frac{\rho(x_n)}{\operatorname{rad}(x_n)}) \subset \bar{M}_n$ has sectional curvature bounded below by $-\left(\frac{\operatorname{rad}(x_n)}{\rho(x_n)}\right)^2$, which goes to 0 when $n \to \infty$, as follows from assertion (iii) of Lemma 14.2.7.

Given $L > 1$, the ball $B(\bar{x}_n, 3L)$ is obtained by rescaling the ball $B(x_n, 3L\operatorname{rad}(x_n))$. Since, for large n, $3L\operatorname{rad}(x_n) < \rho(x_n)$, the sectional curvature on $B(x_n, 3L\operatorname{rad}(x_n))$ is $\geqslant -\frac{1}{\rho(x_n)^2} \geqslant -\frac{1}{(3L\operatorname{rad}(x_n))^2}$. Moreover, by Lemma 14.2.7 (ii), we have

$$\frac{\operatorname{vol} B(x_n, 3L\operatorname{rad}(x_n))}{(3L\operatorname{rad}(x_n))^3} \geqslant \frac{\operatorname{vol} B(x_n, \operatorname{rad}(x_n))}{\operatorname{rad}(x_n)^3}\frac{1}{(3L)^3} \geqslant \frac{\varepsilon_0}{(3L)^3}.$$

By applying Corollary 14.2.5 for n sufficiently large so that we have $3L\operatorname{rad}(x_n) \leqslant \bar{r}'(\frac{\varepsilon_0}{(3L)^3})$, one gets that the curvature is locally controlled in the sense of Perelman at each point of the ball $B(x_n, L\operatorname{rad}(x_n))$. Therefore the curvature and its first derivative can be bounded above on any ball $B(\bar{x}_n, L) \subset \bar{M}_n$ with a given radius $L > 1$.

Since the injectivity radius of the basepoint $\bar{x}_n$ is bounded below along the sequence, this upper bound on the curvature allows us to use Gromov's compactness theorem. By Theorem B.1.4 of the appendix, one obtains pointed $\mathcal{C}^{1,\alpha}$-convergence. Using the bound on $|\nabla \operatorname{Rm}|$, this convergence is improved to $\mathcal{C}^2$ (see [Ham95a], Theorem 2.3, or [Fuk90], Theorem 5.10). It follows that the pointed sequence $(\bar{M}_n, \bar{x}_n)$ subconverges in the $\mathcal{C}^2$-topology towards a 3-dimensional smooth manifold $(\bar{X}_\infty, x_\infty)$, with a complete Riemannian metric of class $\mathcal{C}^2$ with nonnegative sectional curvature. This limit manifold cannot be closed, because that would contradict the assumption that the conclusion of Proposition 14.2.1 does not hold.

Hence $\bar{X}_\infty$ is open. By pointed convergence, assertion (b) (1) of Proposition 14.2.1 holds. Let $\bar{S}$ be a soul of $\bar{X}_\infty$. Let us set

$$\nu(x_n) := L\,\mathrm{rad}(x_n), \quad \text{where } L \geqslant 2\,\mathrm{diam}(\bar{S} \cup \{x_\infty\})\,D_0.$$

For n large (to be specified later) we set

$$X_{x_n} := \mathrm{rad}(x_n)\bar{X}_\infty \quad \text{and} \quad S_{x_n} = \mathrm{rad}(x_n)\bar{S}.$$

We then have

$$\mathrm{diam}(S_{x_n}) = \mathrm{rad}(x_n)\,\mathrm{diam}(\bar{S}) < \nu(x_n)/D_0.$$

Suppose that $\bar{f}_n \colon B(\bar{x}_n, L) \to (\bar{X}_\infty, x_\infty)$ is a $(1+\delta_n)$-bi-Lipschitz embedding such that $d(\bar{f}_n(\bar{x}_n), x_\infty) < \delta_n$, where δ_n is a sequence going to 0. After rescaling, $f_n \colon B(x_n, L\,\mathrm{rad}(x_n)) \to X_{x_n}$ is also a $(1+\delta_n)$-bi-Lipschitz embedding. We deduce:

$$\begin{aligned}
d(f_n(x_n), S_{x_n}) &= \mathrm{rad}(x_n) d(\bar{f}_n(\bar{x}_n), \bar{S}) \\
&\leqslant \mathrm{rad}(x_n)(d(\bar{f}_n(\bar{x}_n), x_\infty) + d(x_\infty, \bar{S})) \\
&\leqslant \mathrm{rad}(x_n)\delta_n + \frac{\nu(x_n)}{2D_0} \leqslant \frac{\nu(x_n)}{D_0}.
\end{aligned}$$

This proves assertion (2) of Proposition 14.2.1 (b).

Using the facts that $\nu(x_n) = L\,\mathrm{rad}(x_n) < \rho(x_n)$, the curvature on $B(x_n, \nu(x_n))$ is $\geqslant -1/\nu(x_n)^2$, $L > 1$ and the Bishop–Gromov inequality, we get

$$\begin{aligned}
\frac{\mathrm{vol}\,B(x_n, \nu(x_n))}{\upsilon_{-\frac{1}{\nu^2(x_n)}}(\nu(x_n))} &\leqslant \frac{\mathrm{vol}\,B(x_n, \mathrm{rad}(x_n))}{\upsilon_{-\frac{1}{\nu^2(x_n)}}(\mathrm{rad}(x_n))} \\
&= \varepsilon_0 \frac{\mathrm{rad}(x_n)^3}{\upsilon_{-\frac{1}{\nu^2(x_n)}}(\mathrm{rad}(x_n))} \\
&= \varepsilon_0 \left(\frac{\mathrm{rad}(x_n)}{\nu(x_n)}\right)^3 \frac{1}{\upsilon_{-1}(\frac{\mathrm{rad}(x_n)}{\nu(x_n)})} \\
&= \varepsilon_0 \frac{1}{L^3} \frac{1}{\upsilon_{-1}(\frac{1}{L})}.
\end{aligned}$$

Since $\upsilon_{-1}(\frac{1}{L}) \geqslant \upsilon_0(\frac{1}{L}) = \upsilon_0(1)\frac{1}{L^3}$, we find that

$$\mathrm{vol}\,B(x_n, \nu(x_n)) \leqslant \varepsilon_0 \frac{1}{L^3} \frac{\upsilon_{-1}(1)}{\upsilon_{-1}(\frac{1}{L})} \nu^3(x_n) \leqslant \varepsilon_0 \frac{\upsilon_{-1}(1)}{\upsilon_0(1)} \nu^3(x_n) = \frac{1}{D_0} \nu^3(x_n),$$

where the last equality comes from the definition of ε_0, see p. 192.

This provides the contradiction required to conclude the proof of Proposition 14.2.1. □

14.3 Constructions of coverings

14.3.1 Embedding thick pieces in solid tori

Let us start by making some reductions.

If case (a) of Proposition 14.2.1 occurs for arbitrarily large D, then M is a closed, orientable, irreducible 3-manifold admitting a metric of nonnegative sectional curvature. By [Ham82], [Ham86], M is spherical or Euclidean, hence a graph manifold. Therefore we may assume that all local models X_x produced by Proposition 14.2.1 are noncompact.

For the same reasons, since lens spaces are graph manifolds, we can also assume that M is not homeomorphic to a lens space, and in particular does not contain a projective plane (cf. the beginning of Section 1.2.2).

Consider the submanifolds W_n given by Proposition 14.1.3. Assume that there exists a component X of some W_n which is not a solid torus. From Proposition 14.1.3(ii), X is a knot exterior and contained in a 3-ball $B \subset M_n$. By Lemma A.3.1, it is possible to replace X by a solid torus Y without changing the global topology. Let us denote by M_n' the manifold thus obtained. We can endow M_n' with a Riemannian metric g_n', equal to g_n away from Y and such that an arbitrarily large collar neighbourhood of ∂Y in Y is isometric to a collar neighbourhood ∂X in X, because the end of the interior of X is close to a hyperbolic cusp. When n is large, this neighbourhood is thus almost isometric to a long piece of a hyperbolic cusp, and this geometric property will be sufficient for our covering arguments.

Repeating this construction for each connected component of W_n which is not a solid torus, we obtain a Riemannian manifold (M_n'', g_n'') together with a submanifold W_n'' satisfying the following properties:

(i) M_n'' is homeomorphic to M_n.

(ii) $M_n'' \setminus W_n''$ is equal to $M_n \setminus W_n$ and the metrics g_n and g_n'' coincide on this set.

(iii) $M_n'' \setminus W_n'' = M_n \setminus W_n$ is ε_n-thin.

(iv) When n goes to infinity, there exists a collar neighbourhood of $\partial W_n''$ in W_n'' of arbitrarily large diameter isometric to the corresponding neighbourhood in W_n.

(v) Each component of W_n'' is a solid torus.

For simplicity, we use the notation M_n, g_n, W_n instead of M_n'', g_n'', W_n''. This amounts to assuming in the conclusion of Proposition 14.1.3 that all components of W_n are solid tori.

14.3.2 Existence of a homotopically nontrivial open set

We say that a path-connected subset $U \subset M$ is *homotopically trivial* (in M) if the image of the homomorphism $\pi_1(U) \to \pi_1(M)$ is trivial. More generally, we say that the subset $U \subset M$ is homotopically trivial if all its path-connected components have this property.

We recall that the *dimension* of a finite covering $\{U_i\}_i$ of M is the dimension of its nerve, hence the dimension plus one equals the maximal number of U_i's containing any given point.

Proposition 14.3.1. *There exists $D_0 > 1$ such that for all $D > D_0$, for every $n \geqslant n_0(D)$ (where $n_0(D)$ is given by Proposition* 14.2.1*), one of the following assertions is true:*

(a) *some connected component of W_n is not homotopically trivial, or*
(b) *there exists $x \in M_n \setminus W_n$ such that the image of $\pi_1(B(x, \nu(x))) \to \pi_1(M_n)$ is not homotopically trivial, where $\nu(x)$ is also given by Proposition* 14.2.1.

In the proof of Proposition 14.3.1 we argue by contradiction using Corollary A.4.2 and the fact that $\pi_1(M)$ is not trivial.

With the notation of Proposition 14.2.1, we may assume that for arbitrarily large D there exists $n \geqslant n_0(D)$ such that the image of $\pi_1(B(x, \nu(x))) \to \pi_1(M_n)$ is trivial for all $x \in M_n \setminus \mathrm{Int}(W_n)$ as well as for each connected component of W_n.

Then for all $x \in M_n \setminus \mathrm{Int}(W_n)$ we set

$$\mathrm{triv}(x) = \sup \left\{ r \;\middle|\; \begin{array}{l} \pi_1(B(x,r)) \to \pi_1(M_n) \text{ is trivial and} \\ B(x,r) \text{ is contained in } B(x',r') \text{ with} \\ \text{curvature } \geqslant -\frac{1}{(r')^2} \end{array} \right\}.$$

Notice that $\mathrm{triv}(x) \geqslant \nu(x)$, by Proposition 14.2.1 (cf. Remark 14.2.2).

The proof of Proposition 14.3.1 follows by contradiction with the following assertion.

Assertion 14.3.2. *There exists a covering of M_n by open sets $U_1, \dots, U_p$ such that the following is true:*

- *Each U_i is either contained in some $B(x_i, \mathrm{triv}(x_i))$ or in a subset that deformation retracts to a component of* $\mathrm{Int}(W_n)$. *In particular, U_i is homotopically trivial in M.*
- *The dimension of this covering is at most* 2.

Since M is irreducible and non-simply connected, this contradicts Corollary A.4.2.

To prove Assertion 14.3.2, for $x \in M_n \setminus W_n$ we set

$$r(x) := \min\left\{\tfrac{1}{11}\,\mathrm{triv}(x), 1\right\}.$$

Lemma 14.3.3. (1) $\frac{\nu(x)}{11} \leqslant r(x)$*, where $\nu(x) < \rho(x) \leqslant 1$ is given by Proposition* 14.2.1.

(2) *$B(x, 11\, r(x))$ is contained in some ball $B(x', r'(x))$ with curvature $\geqslant -\frac{1}{(r')^2}$ and satisfying $r(x) \leqslant \frac{1}{11}\,\mathrm{triv}(x) \leqslant \frac{2}{11} r'(x)$.*

Proof. Assertion (1) uses the assumption that $B(x, \nu(x))$ is homotopically trivial in M_n and the inequality $\nu(x) < \rho(x)$, because the curvature on $B(x, \rho(x))$ is $\geqslant -\rho(x)^{-2} \geqslant -\nu(x)^{-2}$.

To prove (2), we shall show that $B(x, \mathrm{triv}(x))$ is contained in some metric ball $B(x', r')$ with curvatures $\geqslant -\frac{1}{(r')^2}$ and satisfying $r' \geqslant \frac{1}{2}\mathrm{triv}(x)$. By definition, there exists a sequence of radii $r_k \nearrow \mathrm{triv}(x)$ satisfying that $\pi_1(B(x, r_k)) \to \pi_1(M_n)$ is trivial and that $B(x, r_k)$ is contained in some $B(x'_k, r'_k)$ with curvature $\geqslant -\frac{1}{(r'_k)^2}$. Since $\pi_1(M_n)$ is nontrivial, $M_n \not\subset B(x, r_k)$, therefore there is a point $y_k \in M_n$ such that $d(x, y_k) = r_k$. By applying the triangle inequality to x, y_k and x'_k, we get $r_k \leqslant 2r'_k$. Then the claim follows by taking a partial subsequence so that both x'_k and r'_k converge, since we are working in a fixed M_n, that has bounded diameter. □

Lemma 14.3.4. *Let $x, y \in M_n \setminus \mathrm{int}(W_n)$. If $B(x, r(x)) \cap B(y, r(y)) \neq \emptyset$, then*

(a) $3/4 \leq r(x)/r(y) \leqslant 4/3$;

(b) $B(x, r(x)) \subset B(y, 4r(y))$.

Proof. To prove (a), we may assume that $r(x) \leqslant r(y)$ and that $r(x) = \frac{1}{11}\mathrm{triv}(x) < 1$. Since $B(x, \mathrm{triv}(y) - r(x) - r(y)) \subset B(y, \mathrm{triv}(y))$, we get

$$\mathrm{triv}(x) \geqslant \mathrm{triv}(y) - r(x) - r(y),$$

hence

$$11\, r(x) = \mathrm{triv}(x) \geqslant 11r(y) - r(x) - r(y) \geqslant 9r(y).$$

Thus $1 \geqslant r(x)/r(y) \geqslant 9/11 \geqslant 3/4$, which proves assertion (a).

Assertion (b) follows because $2\,r(x) + r(y) \leqslant (\frac{8}{3} + 1)r(y) < 4r(y)$. □

For n large enough, we pick points $x_1, \dots, x_q \in \partial W_n$ in such a way that a tubular neighbourhood of each connected component of the boundary of W_n contains precisely one of the x_j's and that the balls $B(x_j, 1)$ are disjoint, have volume $\leqslant \frac{1}{D}$ and sectional curvature close to -1. Furthermore, we may assume that $B(x_j, 1)$ is contained in some submanifold W'_n which contains W_n and can be retracted by deformation onto it. In particular, W'_n and $B(x_j, 1)$ are homotopically trivial, since we have assumed that W_n is. This implies that $\mathrm{triv}(x_j)$ is close to 1.

Moreover, for n large enough, we may assume that $B(x_j, \frac{2}{3}r(x_j))$ contains an almost horospherical torus corresponding to a boundary component of W_n. We can even arrange for both components of $B(x_j, r(x_j)) \setminus B(x_j, \frac{2}{3}r(x_j))$ to also contain a parallel almost horospherical torus (by going far enough into the cusp). This allows us to retract W_n on the complement of $B(x_j, \frac{2}{3}r(x_j))$.

We complete the previous finite sequence to a sequence of points $x_1, x_2, \dots$ in $M_n \setminus \mathrm{Int}(W_n)$ such that the balls $B(x_1, \frac{1}{4}r(x_1))$, $B(x_2, \frac{1}{4}r(x_2)), \dots$ are pairwise disjoint. Such a sequence is necessarily finite, since M_n is compact, and Lemma 14.3.4 implies a positive local lower bound for the function $x \mapsto r(x)$. Let us choose a maximal finite sequence $x_1, \dots, x_p$ with this property.

Lemma 14.3.5. *The balls $B(x_1, \frac{2}{3}r(x_1)), \dots, B(x_p, \frac{2}{3}r(x_p))$ cover $M_n \setminus \mathrm{Int}(W_n)$.*

Proof. Let $x \in M_n \setminus \mathrm{Int}(W_n)$ be an arbitrary point. By maximality, there exists a point x_j such that $B(x, \frac{1}{4}r(x)) \cap B(x_j, \frac{1}{4}r(x_j)) \neq \emptyset$. From Lemma 14.3.4, we have $r(x) \leqslant \frac{4}{3}r(x_j)$ and $d(x, x_j) \leqslant \frac{1}{4}(r(x)+r(x_j)) \leqslant \frac{7}{12}r(x_j)$, hence $x \in B(x_j, \frac{2}{3}r(x_j))$. □

Set $r_i := r(x_i)$. If $W_{n,1}, \dots, W_{n,q}$ are the components of W_n, so that the almost horospherical torus $\partial W_{n,i} \subset B(x_i, \frac{2}{3}r_i)$, for $i = 1, \dots, q$, we define

- $V_i := B(x_i, r_i) \cup W_{n,i}$, for $i = 1, \dots, q$;
- $V_i := B(x_i, r_i)$, for $i = q + 1, \dots, p$.

Furthermore, each connected component of $W_{n,i}$ can be retracted in order not to intersect $B(x_j, \frac{2}{3}r_j)$ when $j \neq i$.

Since $r_i \leqslant \frac{1}{11} \operatorname{triv}(x_i)$, the V_i's are homotopically trivial. The construction of the V_i's and Lemma 14.3.5 imply the following:

Lemma 14.3.6. *The open sets $V_1, \dots, V_p$ cover M_n.* □

Let K be the nerve of the covering $\{V_i\}$. The following lemma shows that the dimension of K is bounded above by a uniform constant.

Lemma 14.3.7. *There exists a universal upper bound N on the number of open sets V_i which intersect a given V_k.*

Proof. Assume that $V_i \cap V_k \neq \emptyset$. Then $B(x_i, r_i) \cap B(x_k, r_k) \neq \emptyset$ and $B(x_i, r_i) \subset B(x_k, 2r_i + r_k) \subseteq B(x_k, 4r_k)$. On the other hand, for all $i_1 \neq i_2$ such that V_{i_1} and V_{i_2} intersect V_k one has $d(x_{i_1}, x_{i_2}) \geqslant \frac{1}{4}(r_{i_1} + r_{i_2}) \geqslant \frac{3}{8}r_k$. Thus $B(x_{i_1}, \frac{3}{16}r_k) \cap B(x_{i_2}, \frac{3}{16}r_k) = \emptyset$ and $B(x_i, \frac{3}{16}r_k) \subset B(x_k, 4r_k)$. This motivates the following inequalities:

$$\frac{\operatorname{vol} B(x_k, 4r_k)}{\operatorname{vol} B(x_i, \frac{3}{16}r_k)} \leqslant \frac{\operatorname{vol} B(x_i, 8r_k)}{\operatorname{vol} B(x_i, \frac{3}{16}r_k)} \leqslant \frac{\operatorname{vol} B(x_i, 11r_i)}{\operatorname{vol} B(x_i, \frac{r_i}{8})}.$$

Since $B(x_i, 11r_i)$ is contained in some $B(x', r')$ with curvatures $\geq -\frac{1}{(r')^2}$, by the Bishop–Gromov inequality this quotient is bounded from above by

$$\frac{v_{-\frac{1}{(r')^2}}(11r_i)}{v_{-\frac{1}{(r')^2}}(\frac{r_i}{8})} = \frac{v_{-1}(\frac{11r_i}{r'})}{v_{-1}(\frac{r_i}{8r'})} \leqslant N.$$

The existence of a uniform N uses $r' \geqslant \frac{11}{2}r_i$ (Lemma 14.3.3). □

Let $\Delta^{p-1} \subset \mathbb{R}^p$ denote the standard unit $(p-1)$-simplex. Using $\frac{4}{r_i}$-Lipschitz test functions ϕ_i supported on the V_i's, we obtain a Lipschitz map:

$$f := \frac{1}{\sum_i \phi_i}(\phi_1, \dots, \phi_p) \colon M_n \to \Delta^{p-1} \subset \mathbb{R}^p.$$

Remark that the coordinate functions of f are a partition of unity subordinated to (V_i).

We view K as a subcomplex of Δ^{p-1}, so that the range of f is contained in K. Crucial for the following arguments is that the dimension of K is bounded from above by the universal number N from Lemma 14.3.7.

We first show that if the ϕ_i's are chosen properly, then the local Lipschitz constant of f can be estimated from above:

Lemma 14.3.8. *For an adequate choice of the ϕ_i's, there is a uniform constant $L > 0$ such that for each k, the restriction $f|_{V_k}$ is $\frac{L}{r_k}$-Lipschitz.*

Proof. Let $\tau\colon [0, +\infty) \to [0, 1]$ be an auxiliary function with Lipschitz constant bounded by 4, which vanishes in a neighbourhood of 0 and satisfies $\tau|_{[\frac{1}{3}, +\infty]} \equiv 1$. Let us define $\phi_k := \tau(\frac{1}{r_k} d(\partial V_k, \cdot))$ on V_k and let us extend it trivially on M_n. Then ϕ_k is $\frac{4}{r_k}$-Lipschitz.

Let $x \in V_k$. The functions ϕ_i have Lipschitz constant $\leqslant \frac{4}{3} \cdot \frac{4}{r_k}$ on V_k, and all ϕ_i vanish at x except at most $N + 1$ of them. Since the functions

$$(y_0, \dots, y_N) \mapsto \frac{y_k}{\sum_{i=0}^{N} y_i}$$

are Lipschitz on

$$\big\{y \in \mathbb{R}^{N+1} \mid y_0 \geqslant 0, \dots, y_N \geqslant 0 \text{ and } \textstyle\sum_{i=0}^{N} y_i \geqslant 1\big\},$$

and each $x \in M_n$ belongs to some V_k with $d(x, \partial V_k) \geqslant \frac{r_k}{3}$ by construction of the V_k and Lemma 14.3.5, the conclusion follows. □

We shall now inductively deform f by homotopy into the 3-skeleton $K^{(3)}$, while keeping the local Lipschitz constant under control.

We recall that the *open star* of a vertex of K is the union of the interiors of all simplices whose closure contains the given vertex.

Lemma 14.3.9. *For all $d \geqslant 4$ and $L > 0$ there exists $L' = L'(d, L) > 0$ such that the following assertion holds.*

Let $g\colon M_n \to K^{(d)}$ be a map that is $\frac{L}{r_k}$-Lipschitz on V_k and such that the pull-back of the open star of the vertex $v_{V_k} \in K^{(0)}$ is contained in V_k. Then g is homotopic relative to $K^{(d-1)}$ to a map $\tilde{g}\colon M_n \to K^{(d-1)}$ which is $\frac{L'}{r_k}$-Lipschitz on V_k and such that the pull-back of the open star of v_{V_k} is still contained in V_k.

Proof. It suffices to find a constant $\theta = \theta(d, L) > 0$ such that each d-simplex $\sigma \subset K$ contains a point z whose distance to $\partial\sigma$ and to the image of g is $\geqslant \theta$. In order to push g into the $(d-1)$-skeleton, we compose it on σ with the radial projection from z. This increases the Lipschitz constant by a multiplicative factor bounded from above by a function of $\theta(d, L)$, and decreases the inverse image of the open stars of the vertices.

If θ does not satisfy the required property for some d-simplex σ, then one can tile an open subset of $\sigma \setminus (\theta -$ neighbourhood of $\partial\sigma)$ with at least $C'(d) \cdot \frac{1}{\theta^d}$ cubes of length 2θ, and each cube contains a point of $\mathrm{Im}(g)$ in its interior. By choosing one point of $\mathrm{Im}(g)$ inside each cube whose tiling coordinates are even, we find a subset of cardinality at least $C(d) \cdot \frac{1}{\theta^d}$ of points in $\mathrm{Im}(g) \cap \mathrm{Int}(\sigma)$ whose pairwise distances are $\geqslant \theta$. Let $A \subset M_n$ be a set containing exactly one point of the inverse image of each of these points. By hypothesis, $A \subset V_k$ for some k corresponding to a vertex of σ; we may assume that $A \cap W_n = \emptyset$, hence $A \subset B(x_k, r_k)$ for the same k. As g is $\frac{L}{r_k}$-Lipschitz on V_k, the distance between any two distinct points in A is bounded from below by $\frac{1}{L} r_k \theta$. In order to bound the cardinal of A, we use the following inequality for $y \in A$:

$$\frac{\operatorname{vol} B(x_k, r_k)}{\operatorname{vol} B(y, \frac{r_k\theta}{2L})} \leqslant \frac{\operatorname{vol} B(y, 2r_k)}{\operatorname{vol} B(y, \frac{r_k\theta}{2L})} \leqslant \frac{v_{-\frac{1}{(r')^2}}(2r_k)}{v_{-\frac{1}{(r')^2}}(\frac{r_k\theta}{2L})} = \frac{v_{-1}(2\frac{r_k}{r'})}{v_{-1}(\frac{\theta}{2L}\frac{r_k}{r'})} \leqslant C\Big(\frac{L}{\theta}\Big)^3.$$

Since $r_k \leqslant \frac{2}{11} r'$, such a uniform C exists. Thus the cardinal of A is at most $C(L/\theta)^3$. Note that in order to apply Bishop–Gromov, we used the fact that $B(x_k, 11r_k)$ is contained in some metric ball of radius r' with curvatures $\geqslant -1/(r')^2$. The inequality $C(d) \cdot \frac{1}{\theta^d} \leqslant C \cdot (\frac{L}{\theta})^3$ yields a positive lower bound $\theta_0(d, L)$ for θ. Consequently, any $\theta < \theta_0$ has the desired property. □

Lemma 14.3.10. *There exists a universal constant C such that the following holds. Let $D > 1$ and $n > n_0(D)$ be as in Proposition* 14.2.1. *Then*

$$\operatorname{vol} B(x_i, r_i) \leqslant C \frac{1}{D} r_i^3 \quad \textit{for all } i.$$

Proof. We know that, by Proposition 14.2.1, $\operatorname{vol} B(x_i, \nu(x_i)) \leqslant \frac{1}{D}\nu(x_i)^3$. Moreover, $r_i \geqslant \frac{\nu(x_i)}{11}$, and $B(x_i, r_i)$ is contained in some metric ball $B(x', r')$ with curvatures $\geqslant -\frac{1}{r'^2}$. As $r' \geqslant \frac{11}{2} r_i > r_i$, the curvatures on $B(x_i, r_i)$ are at least $-\frac{1}{r_i^2}$. The Bishop–Gromov inequality gives

$$\frac{\operatorname{vol} B\big(x_i, \frac{\nu(x_i)}{11}\big)}{v_{-\frac{1}{r_i^2}}(\frac{\nu(x_i)}{11})} \geqslant \frac{\operatorname{vol} B(x_i, r_i)}{v_{-\frac{1}{r_i^2}}(r_i)}.$$

Equivalently,

$$\operatorname{vol} B\big(x_i, \tfrac{\nu(x_i)}{11}\big) \geqslant \frac{\operatorname{vol} B(x_i, r_i)}{v_{-1}(1)} v_{-1}\Big(\frac{\nu(x_i)}{11\, r_i}\Big) \geqslant \operatorname{vol} B(x_i, r_i) \frac{1}{C}\Big(\frac{\nu(x_i)}{r_i}\Big)^3,$$

for some uniform $C > 0$. Hence, using Proposition 14.2.1 (b) (3), we get

$$\operatorname{vol} B(x_i, r_i) \leqslant C\Big(\frac{r_i}{\nu(x_i)}\Big)^3 \operatorname{vol} B\big(x_i, \tfrac{\nu(x_i)}{11}\big)) \leqslant C r_i^3 \frac{1}{D}.$$

□

Finally we push f into the 2-skeleton.

Lemma 14.3.11. *For a suitable choice of $D > 1$, there exists a map $f^{(2)}: M_n \to K^{(2)}$ such that the following holds:*

(i) *$f^{(2)}$ is homotopic to f relative to $K^{(2)}$.*

(ii) *The inverse image of the open star of each vertex $v_{V_k} \in K^{(0)}$ is contained in V_k.*

Proof. The inverse image by f of the open star of the vertex $v_{V_k} \in K^{(0)}$ is contained in V_k. Using Lemma 14.3.9 several times, we find a map $f^{(3)}: M_n \to K^{(3)}$ homotopic to f and a universal constant $\hat{L}$ such that $(f^{(3)})^{-1}(\text{star}(v_{V_k})) \subset V_k$ and $f^{(3)}_{|V_k}$ is $\frac{\hat{L}}{r_k}$-Lipschitz.

Sublemma 14.3.12. *Let σ be a 3-simplex of K. Then σ is not contained in the image of $f^{(3)}$.*

Proof. Arguing by contradiction, we consider a 3-simplex σ contained in the image of $f^{(3)}$. The inverse image of $\text{Int}(\sigma)$ by $f^{(3)}$ is a subset of the intersection of those V_j's such that v_{V_j} is a vertex of σ. Let V_k be one of them. Since $\text{vol}(f^{(3)}(V_k)) \leqslant \text{vol}(f^{(3)}(B(x_k, r_k)))$, Lemma 14.3.10 yields

$$\text{vol}(\text{Im}(f^{(3)}) \cap \sigma) \leqslant \text{vol}(f^{(3)}(V_k)) \leqslant \left(\frac{\hat{L}}{r_k}\right)^3 \text{vol}\, B(x_k, r_k) \leqslant C\hat{L}^3 \frac{1}{D}$$

with uniform constants C and $\hat{L}$. Thus we have $\text{vol}(\text{Im}(f^{(3)}) \cap \sigma) < \text{vol}(\sigma)$ for sufficiently large D. □

Having proved Sublemma 14.3.12, we conclude the proof of Lemma 14.3.11 by simply choosing in each 3-simplex of K a point which is not in the image of $f^{(3)}$, and compose $f^{(3)}$ with central projections from those points. Note that here no metric estimate is required in the conclusion. □

The inverse images of the open stars of the vertices v_k satisfy $(f^{(2)})^{-1}(\text{star}(v_{V_k})) \subset V_k$, and therefore $(f^{(2)})^{-1}(\text{star}(v_{V_k}))$ is homotopically trivial in M. This proves Assertion 14.3.2 and ends the proof by contradiction of Proposition 14.3.1. □

14.3.3 End of the proof: covering by virtually abelian subsets

The following is a consequence of Proposition 14.3.1, where the constants D_0 and $n_0(D)$ are provided by Propositions 14.3.1 and 14.2.1 respectively.

Corollary 14.3.13. *There exists $D_0 > 0$ such that if $D > D_0$ and $n \geqslant n_0(D)$, then there exists a compact submanifold $\mathcal{V}_0 \subset M_n$ with the following properties:*

(i) *$\mathcal{V}_0$ is either a connected component of W_n or $\frac{1}{D}$-close to a tubular neighbourhood of a soul of the local model of some point $x_0 \in M_n$.*

(ii) *$\mathcal{V}_0$ is a solid torus, a thickened torus or the twisted I-bundle on the Klein bottle.*

(iii) *$\mathcal{V}_0$ is homotopically non-trivial in M_n.*

Proof. We recall that each connected component of W_n is a solid torus. If one of them is homotopically non-trivial, then we choose it to be $\mathcal{V}_0$. Otherwise, by Proposition 14.3.1, there exists a point $x_0 \in M_n$ such that $B(x_0, \nu(x_0))$ is homotopically non-trivial; Remark 14.2.3 shows that $B(x_0, \nu(x_0))$ is necessarily a solid torus, a thickened torus or a twisted I-bundle over the Klein bottle. Indeed, the soul S_0 of the local model X_0 can be neither a point nor a 2-sphere, otherwise $B(x_0, \nu(x_0))$ would be homeomorphic to B^3 or $S^2 \times I$, which have trivial fundamental group. Let $f_{x_0}\colon B(x_0, \nu_0) \to X_0$ denote the $(1 + \frac{1}{D})$-bi-Lipschitz embedding provided by Proposition 14.2.1. We take $\mathcal{V}_0 = f_{x_0}^{-1}(\overline{\mathcal{N}_\delta(S_0)})$, the inverse image of the closed tubular neighbourhood of radius δ of S_0, for some $0 < \delta < \nu_0/D$. □

As $\mathcal{V}_0$ is not contained in any 3-ball, each connected component Y of its complement is irreducible, hence a *Haken manifold* whose boundary is a union of tori (possibly compressible). In particular, Y admits a geometric decomposition. As remarked at the beginning of this chapter, in order to prove that M_n is a graph manifold, it is sufficient to show that each connected component of $M_n \setminus \operatorname{Int}(\mathcal{V}_0)$ is a graph manifold. By Proposition 14.0.1 and Remark 13.2.5, all we have to do is prove the following proposition:

Proposition 14.3.14. *For n large enough, one can choose the submanifold $\mathcal{V}_0$ as above such that every Dehn filling on each connected component Y of $M_n \setminus \operatorname{Int}(\mathcal{V}_0)$ has vanishing simplicial volume.*

We choose the set $\mathcal{V}_0$ as follows:

- If some component of W_n is homotopically non-trivial, then we choose it as $\mathcal{V}_0$.
- If all components of W_n are homotopically trivial, then there exists a point $x \in M_n$ such that $B(x, \nu(x))$ is homotopically non-trivial. We choose $x_0 \in M_n$ such that

$$\nu_0 = \nu(x_0) \geqslant \frac{1}{2} \sup_{x \in M_n} \{\nu(x) \mid \pi_1(B(x, \nu(x))) \to \pi_1(M_n) \text{ is non-trivial}\}.$$

Let S_0 be the soul of the local model X_0 of $B(x_0, \nu_0)$, provided by Proposition 14.2.1. If $f_{x_0}\colon B(x_0, \nu_0) \to X_0$ is a $(1 + \frac{1}{D})$-bi-Lipschitz embedding, we choose

$$\mathcal{V}_0 = f_{x_0}^{-1}(\overline{\mathcal{N}_\delta(S_0)})$$

where $\overline{\mathcal{N}_\delta(S_0)}$ denotes the closed tubular neighbourhood of radius δ of S_0, with $0 < \delta < \frac{\nu_0}{D}$, where $D > D_0$ is given by Corollary 14.3.13.
After possibly shrinking W_n, one has $\mathcal{V}_0 \cap W_n = \emptyset$, as $\nu_0 \leqslant 1$.

We say that a subset $U \subset M_n$ is *virtually abelian relative to* $\mathcal{V}_0$ if the image in $\pi_1(M_n \setminus \mathcal{V}_0)$ of the fundamental group of each connected component of $U \cap (M_n \setminus \mathcal{V}_0)$ is virtually abelian.

We set

$$\mathrm{ab}(x) = \sup \left\{ r \;\middle|\; \begin{array}{l} B(x,r) \text{ is virtually abelian relative to } \mathcal{V}_0 \\ \text{and } B(x,r) \text{ is contained in a ball } B(x',r') \\ \text{with curvature } \geqslant -\frac{1}{(r')^2} \end{array} \right\}$$

and

$$r(x) = \min\left\{\frac{1}{11}\,\mathrm{ab}(x), 1\right\}.$$

Since we may assume that M_n is not virtually abelian relative to $\mathcal{V}_0$, for every $x \in M_n$, $M_n \not\subset B(x, \mathrm{ab}(x))$. Therefore, by the same argument as Lemma 14.3.3, $r(x) < \frac{2}{11}r'$, where $B(x',r')$ contains $B(x,\mathrm{ab}(x))$ and has curvature $\geqslant -\frac{1}{(r')^2}$.

We are now led to prove the following assertion:

Assertion 14.3.15. *With this choice of $\mathcal{V}_0$, for n large enough, M_n can be covered by a finite collection of open sets U_i such that the following holds:*

- *Each U_i is either contained in a component of W_n or a metric ball $B(x_i, r(x_i))$ for some $x_i \in M_n^-(\varepsilon_n)$. In particular, U_i is virtually abelian relative to $\mathcal{V}_0$.*
- *The dimension of this covering is not greater than 2, and it is zero on $\mathcal{V}_0$ (equivalently $\mathcal{V}_0$ only intersects one open set of the covering).*

Let us first show why this assertion implies Proposition 14.3.14. This implication is a direct consequence of Gromov's vanishing theorem.

Proof. The covering described in the assertion naturally induces a covering on every closed and orientable manifold $\widehat{Y}$ obtained by gluing solid tori to ∂Y. By the second point of Assertion 14.3.15, it is a 2-dimensional covering of $\widehat{Y}$ by open sets which are virtually abelian and thus amenable in $\widehat{Y}$. Hence by Theorem 13.2.6, $\widehat{Y}$ has zero simplicial volume. □

We now prove Assertion 14.3.15. The argument for the construction of a 2-dimensional covering by abelian open sets is similar to the one used in the proof of Assertion 14.3.2, replacing everywhere the triviality radius triv by the virtual abelianity radius ab. The construction of the covering takes care of $\mathcal{V}_0$, in particular we require Lemma 14.3.16. The direct analogues of Lemmas 14.3.4 to 14.3.11 apply here, with three small fixes. Firstly, the proof of Lemma 14.3.3 (2) needs to be adapted in this setting (Lemma 14.3.17). Secondly, Lemma 14.3.3 (1) must be replaced by a similar statement with different constants (Lemma 14.3.18). Finally, the analogue of Lemma 14.3.5 also needs further discussion (Lemma 14.3.19).

We now make the details precise.

If $\mathcal{V}_0$ is a connected component of W_n, then for n large enough we choose points $x_0, x_1, \dots, x_q \in \partial W_n$, with $x_0 \in \partial\mathcal{V}_0$, in the following way:

- Every boundary component of W_n contains exactly one of the x_j's.
- The balls $B(x_j, 1)$ are pairwise disjoint.
- Every $B(x_j, 1)$ has normalised volume $\leqslant \frac{1}{D}$ and sectional curvature close to -1.
- Every $B(x_j, 1)$ is contained in a thickened torus (which implies that this ball is abelian).

Furthermore, going sufficiently far in the cusp and taking n large enough, one can assume that $B\left(x_j, \frac{1}{9}r(x_j)\right)$ contains an almost horospherical torus corresponding to a boundary component of W_n. In this case the proof previously done applies without any change, since the dimension of the original covering and all those obtained by shrinking is zero on W_n (or on a set obtained by shrinking W_n).

From now on we shall assume that *all connected components of W_n are homotopically trivial.* We then choose $x_0 \in \mathcal{V}_0 \subset M_n \setminus \mathrm{Int}(W_n)$ as above and points $x_1, \dots, x_q \in \partial W_n$ as before.

We complete the sequence $x_0, x_1, \dots, x_q$ to a maximal finite sequence

$$x_0, x_1, x_2, \dots, x_p$$

in $M_n \setminus \mathrm{Int}(W_n)$ such that the balls $B\left(x_i, \frac{1}{4}r(x_i)\right)$ are disjoint.

We set $r_i = r(x_i)$, and, if $W_{n,1}, \dots W_{n,q}$ are the connected components of W_n, then we set

- $V_0 := B(x_0, r_0)$,
- $V_i := B(x_i, r_i) \cup W_{n,i}$ for $i = 1, \dots, q$,
- $V_i := B(x_i, r_i) \setminus \mathcal{V}_0$ for $i = q+1, \dots, p$.

After possibly shrinking W_n, we have $\mathcal{V}_0 \cap B(x_i, r_i) = \emptyset$ for $i = 1, \dots, q$, since $r_i \leqslant 1$ and $\mathcal{V}_0 \cap W_n = \emptyset$. It follows that $\mathcal{V}_0 \cap V_i = \emptyset$ for $i \neq 0$.

We first need the following result about $\mathcal{V}_0$.

Lemma 14.3.16. *If n is large enough, then*

$$\mathcal{V}_0 \subset B\left(x_0, \frac{4\nu(x_0)}{D}\right) \subset B\left(x_0, \frac{r_0}{9}\right) \subset V_0.$$

Proof. The first inclusion uses the properties of Proposition 14.2.1: $\mathrm{diam}(S_0) < \frac{\nu_0}{D}$, $d(f_{x_0}(x_0), S_0) < \frac{\nu_0}{D}$, the construction $\mathcal{V}_0 = f_{x_0}^{-1}(\overline{\mathcal{N}_\delta(S_0)})$ with $\delta < \frac{\nu_0}{D}$ and the fact that f_{x_0} is a $(1+\frac{1}{D})$-bi-Lipschitz embedding. The second inclusion follows from $r_0 \geqslant \frac{1}{11}\nu(x_0)$ by taking $D \geqslant 4 \cdot 9 \cdot 11$. □

Then we have the proof of the following analogue of Lemma 14.3.3 (2).

Lemma 14.3.17. *We can assume that, for every $x \in M_n \setminus \mathrm{Int}(W_n)$, $B(x, 11\, r(x))$ is contained in some ball $B(x', r'(x))$ with curvature $\geqslant -\frac{1}{(r')^2}$ and satisfying $r(x) \leqslant \frac{1}{11}\,\mathrm{ab}(x) \leqslant \frac{2}{11} r'(x)$.*

Proof. If M_n is virtually abelian relative to $\mathcal{V}_0$, then Assertion 14.3.15 holds true by taking a single open subset, the whole M_n. Hence we may assume that M_n is not virtually abelian relative to $\mathcal{V}_0$, thus for every $x \in M_n$, $M_n \not\subset B(x, \mathrm{ab}(x))$. Therefore, by the same argument as in the proof of Lemma 14.3.3 applies. □

The next lemma is the analogue of Lemma 14.3.3 (1), with a new constant c that just entails a change of constant in the analogue of Lemma 14.3.10.

Lemma 14.3.18. *There exists a universal $c > 0$ such that, if n is sufficiently large, then $r_i \geqslant c\,\nu(x_i)$ for all i.*

Proof. One has $r_0 \geqslant \frac{1}{11}\nu(x_0)$ by construction. For all $i > 0$, if $B(x_i, \nu(x_i)) \cap \mathcal{V}_0 = \emptyset$, then $\mathrm{ab}(x_i) \geqslant \nu(x_i)$ so $r_i \geqslant \frac{1}{11}\nu(x_i)$. Hence we assume $B(x_i, \nu(x_i)) \cap \mathcal{V}_0 \neq \emptyset$, and we claim that $d(x_i, \mathcal{V}_0) > c'\,\nu(x_i)$ for a uniform $c' > 0$.

We also assume from now on that $r_i = \frac{1}{11}\,\mathrm{ab}(x_i) < 1$ (otherwise ($r_i = 1 \geqslant \rho(x_i) > \nu(x_i)$ and we are done). Since $\mathcal{V}_0 \subset B(x_0, \frac{1}{9}r_0)$:

$$\begin{aligned} r_i &\geqslant \frac{1}{11} d(x_i, \mathcal{V}_0) \geqslant \frac{1}{11}\left(d(x_i, x_0) - \frac{1}{9}r_0\right) \\ &\geqslant \frac{\frac{1}{4}r_0 - \frac{1}{9}r_0}{11} > \frac{r_0}{88} \geqslant \frac{1}{1000}\nu(x_0). \end{aligned} \tag{14.1}$$

We distinguish two cases, according to whether $\mathcal{V}_0$ is contained in $B(x_i, \nu(x_i))$ or not.

If $\mathcal{V}_0 \subset B(x_i, \nu(x_i))$, then the image of $\pi_1(B(x_i, \nu(x_i))) \to \pi_1(M_n)$ cannot be trivial, since the image of $\pi_1(\mathcal{V}_0) \to \pi_1(M_n)$ is not. In addition,

$$\nu(x_i) \leqslant 2\,\nu(x_0), \tag{14.2}$$

by the choice of x_0 and $\nu(x_0)$. Equations (14.1) and (14.2) give $r_i > \nu(x_i)/2000$.

If $\mathcal{V}_0 \not\subset B(x_i, \nu(x_i))$, then since $\mathcal{V}_0 \cap B(x_i, \nu(x_i)) \neq \emptyset$ and $\mathcal{V}_0 \subset B\left(x_0, \frac{r_0}{9}\right)$, we have

$$11r_i = \mathrm{ab}(x_i) \geqslant d(x_i, x_0) - r_0/9, \quad \nu(x_i) \leqslant d(x_i, x_0) + r_0/9.$$

Since $d(x_i, x_0) \geqslant \frac{1}{4}r_0$, this yields $\frac{r_i}{\nu(x_i)} \geqslant \frac{1}{11} \cdot \frac{1/4 - 1/9}{1/4 + 1/9} \geqslant \frac{1}{30}$. □

Finally the analogues of Lemmas 14.3.4 to 14.3.11 apply with no changes, except for Lemma 14.3.5 in the context of the new boundary created by $\mathcal{V}_0$, and it is used in the control of the Lipschitz constant of the characteristic map, Lemma 14.3.8.

Lemma 14.3.19. *Each $x \in M_n$ belongs to some V_k such that $d(x, \partial V_k) \geqslant \frac{1}{3}r_k$.*

Proof of Lemma 14.3.19. If $x \in B(x_0, \frac{2}{3}r_0)$ we may choose $k = 0$. Let us then assume that $x \notin B\left(x_0, \frac{2}{3}r_0\right)$. There exists k such that $x \in B\left(x_k, \frac{2}{3}r_k\right)$, by the analogue of Lemma 14.3.5. If $B(x_k, r_k)$ and V_0 are disjoint, then $B(x_k, r_k) \cap \mathcal{V}_0 = \emptyset$ and we are done. Hence we assume that $B(x_k, r_k) \cap V_0 \neq \emptyset$. By Lemma 14.3.16 and the analogue of Lemma 14.3.4, one has:

$$d(x, \mathcal{V}_0) \geqslant d(x, x_0) - \frac{1}{9}r_0 \geqslant \frac{2}{3}r_0 - \frac{1}{9}r_0 \geqslant \frac{3}{4} \cdot \frac{5}{9}r_k > \frac{1}{3}r_k.$$

This implies that $d(x, \partial V_k) \geqslant \frac{1}{3}r_k$. □

Chapter 15
A rough classification of 3-manifolds

In this final chapter, which is not part of our proof of the Geometrisation Conjecture, we prove the refined Theorem 1.3.12 using an invariant $V_0'(M)$ which is closely related to $V_0(M)$.

For a closed 3-manifold M, let $V_0'(M)$ denote the minimum of the volumes of all hyperbolic submanifolds $H \subset M$ having the property that either H is the complement of a link in M or ∂H has at least one component which is incompressible in M.

By definition, we always have $0 < V_0'(M) \leqslant V_0(M)$. If M is orientable and irreducible, then it admits a metric of positive scalar curvature if and only if it is spherical. If this is not the case (which amounts to saying that $\pi_1(M)$ is infinite), then the invariant $\bar{V}(M)$ can be defined as the infimum of $\hat{V}$ taken on all Riemannian metrics (see (1.5), p. 14) and may be zero.

Here is the main result of this chapter. It characterises hyperbolic manifolds and graph manifolds among irreducible 3-manifolds with infinite fundamental group.

Theorem 15.1.1. *Let M be a closed, orientable, irreducible 3-manifold with infinite fundamental group. Then, one of the following occurs:*

(i) $0 < \bar{V}(M) = V_0'(M) = V_0(M)$ *if M is hyperbolic.*

(ii) $0 < V_0'(M) \leqslant \bar{V}(M) < V_0(M)$ *if M contains an incompressible torus and has some hyperbolic pieces in the JSJ decomposition.*

(iii) $0 = \bar{V}(M) < V_0'(M) \leqslant V_0(M)$ *if M is a graph manifold.*

Before giving the proof, we note the following variant of Theorem 13.1.3, where V_0 is replaced by V_0', and the conclusion is changed to 'M is a graph manifold':

Theorem 15.1.2. *Let M be a closed, orientable, non-simply connected, irreducible 3-manifold. Let g_n be a sequence of Riemannian metrics satisfying the following conditions:*

(1) *The sequence* $\mathrm{vol}(g_n)$ *is bounded.*

(2) *For all $\varepsilon > 0$, if $x_n \in M$ is a sequence such that for all n, x_n is in the ε-thick part of (M, g_n), then (M, g_n, x_n) subconverges in the $\mathcal{C}^2$ topology to a pointed hyperbolic manifold with volume strictly less than $V_0'(M)$.*

(3) *The sequence (M, g_n) has locally controlled curvature in the sense of Perelman.*

Then M is a graph manifold.

Proof. Let $H^1, \dots, H^m$ be hyperbolic limits given by Proposition 14.1.1 and let $\varepsilon_n \to 0$ be a sequence chosen as in the remark after Proposition 14.1.1, or in the beginning of Section 14.2 to describe the local structure of the thin part. As in the proof of

Theorem 13.1.3, for each i we fix a compact core $\bar{H}^i$ of H^i and for each n a submanifold $\bar{H}^i_n$ and an approximation $\phi^i_n \colon \bar{H}^i_n \to \bar{H}^i$. The fact that the volume of each hyperbolic manifold H^i is less than $V_0'(M)$ implies the following result:

Lemma 15.1.3. *Up to taking a subsequence of M_n, for all $i \in \{1, \dots, m\}$ each connected component of $\partial \bar{H}^i_n$ is compressible in M for all n.*

Proof. Indeed, if the conclusion of Lemma 15.1.3 does not hold, then up to extracting a subsequence we may assume that there exists an integer $i_0 \in \{1, \dots, m\}$ such that $\partial \bar{H}^{i_0}_n$ contains an incompressible torus for all n. From the definition of $V_0'(M)$ this would contradict the inequality $\mathrm{vol}(H^{i_0}) < V_0'(M)$. □

From this lemma on, the proof of Theorem 15.1.2 is identical to the proof of Theorem 13.1.3. □

Remark 15.1.4. In fact, Theorem 15.1.2 is also true if M is simply-connected, since this implies that M is diffeomorphic to S^3, which is a graph manifold.

Proof of Theorem 15.1.1. (i) If M is hyperbolic, then M is atoroidal, hence $V_0'(M) = V_0(M)$. We already saw that $\bar{V}(M) = V_0(M)$ in this case, so there is nothing to prove.

(iii) If M is a graph manifold, then by Cheeger–Gromov [CG86], one can construct Riemannian metrics on M with sectional curvature pinched between -1 and 1 whose volume is arbitrarily small. In particular, $\bar{V}(M) = 0$.

(ii) Let us suppose that M is neither a hyperbolic manifold nor a graph manifold. We already know that $\bar{V}(M) < V_0(M)$, because M is not hyperbolic. We show that $V_0'(M) \leqslant \bar{V}(M)$ by contradiction. If $V_0'(M) > \bar{V}(M)$ then M admits a Riemannian metric g_0 such that $\hat{V}(g_0) < V_0'(M)$. Using Corollary 1.3.7 with initial condition g_0, we find a sequence g_n satisfying the hypotheses of Theorem 15.1.2. This theorem implies that M is a graph manifold, contrary to the hypothesis. □

Appendix A
3-manifold topology

General references for this appendix are [Hem76], [Jac80] and [Hat00].

A.1 General notation

For $n \geqslant 1$, we let S^n (resp. T^n, resp. B^n, resp. $\mathbb{R}P^n$) denote the n-dimensional sphere (resp. torus, resp. ball, resp. real projective space). By abuse of language, we say that S is *a* (*smoothly embedded*) *2-sphere* in a 3-manifold M if S is a smooth submanifold of M which is diffeomorphic to S^2. Similar terminology is used for 2-tori, 3-balls, $\mathbb{R}P^2$'s etc.

We let I denote the real interval $[0, 1]$ (which for a topologist is the same as B^1). Consequently, $S^2 \times I$ and $T^2 \times I$ are compact, orientable 3-manifolds (with nonempty boundary), sometimes called the *thickened sphere* and the *thickened torus* respectively. We let K^2 denote the Klein bottle. If S is a surface, then we let $S \tilde{\times} I$ denote the twisted I-bundle over S. This manifold is orientable if and only if S is nonorientable; thus we shall be interested only in $\mathbb{R}P^2 \tilde{\times} I$ and $K^2 \tilde{\times} I$.

We let $\mathrm{Int}(M)$ denote the interior of a manifold M with boundary.

A.2 Alexander's theorem and consequences

Let M be a 3-manifold. Two closed, embedded surfaces $F_1, F_2 \subset M$ are *parallel* if they cobound a submanifold diffeomorphic to $F_1 \times I$.

Theorem A.2.1 (Alexander, see e.g. [Hat00]). *Let S be a smoothly embedded 2-sphere in B^3. Then S bounds a 3-ball B. Furthermore, the complement of* $\mathrm{Int}\, B$ *is diffeomorphic to $S^2 \times I$.*

Corollary A.2.2. *Let S be a smoothly embedded 2-sphere in $S^2 \times I$.*

(i) *If S is null-homologous, then it bounds a 3-ball.*
(ii) *Otherwise, it is isotopic to $S^2 \times \{0\}$.*

Proof. Capping-off $S^2 \times \{0\}$ by a 3-ball B, we obtain a 3-manifold B' which is also diffeomorphic to B^3. By Theorem A.2.1, S bounds a 3-ball $B'' \subset B'$. Then either $B'' \subset S^2 \times I$, or $B \subset B''$ and S cobounds with $S^2 \times \{0\}$ a submanifold X such that $B'' = X \cup B$.

If S is null-homologous in $S^2 \times I$ the former case must happen. This proves the first part of Corollary A.2.2. If not, then the latter case must happen. By Theorem A.2.1 again, X is diffeomorphic to $S^2 \times I$. This proves the second assertion. □

A similar argument gives:

Corollary A.2.3. *Let S_1, S_2 be two smoothly embedded 2-spheres in $S^2 \times I$. Assume that S_1, S_2 are isotopic to each other, disjoint, and not null-homologous. Then they are parallel.*

A.3 Submanifolds with compressible boundary

Let X be an orientable, irreducible 3-manifold and T be a compressible torus embedded in X. The loop theorem shows the existence of a *compression disc* D for T, that is, a disc D embedded in M such that $D \cap T = \partial D$ and the curve ∂D is not null homotopic in T. By cutting open T along an open small regular neighbourhood of D and gluing two parallel copies of D along the boundary curves, one constructs an embedded 2-sphere S in X. We say that S is obtained by *compressing T along D*.

Since X is assumed to be irreducible, S bounds a 3-ball B. There are two possible situations depending on whether B contains T or not. The following lemma collects some standard results that we shall need.

Lemma A.3.1. *Let X be an orientable, irreducible 3-manifold and T be a compressible torus embedded in X. Let D be a compression disc for T, S be a sphere obtained by compressing T along D, and B a ball bounded by S. Then the following holds:*

(i) *$X \setminus T$ has two connected components U, V, and D is contained in the closure of one of them, say U.*

(ii) *If B does not contain T, then B is contained in $\bar{U}$, and $\bar{U}$ is a solid torus.*

(iii) *If B contains T, then B contains V, and $\bar{V}$ is homeomorphic to the exterior of a knot in S^3. In this case, there exists a homeomorphism f from the boundary of $S^1 \times D^2$ into T such that the manifold obtained by gluing $S^1 \times D^2$ to $\bar{U}$ along f is homeomorphic to X.*

Remark A.3.2. If T is a component of ∂X and T is a compressible torus, the same argument shows that X is a solid torus.

To understand the situation in Lemma A.3.1 (iii), one has to think of the connected sum of any manifold with a solid torus $S^1 \times D^2$, which is glued to any knot exterior in S^3.

Lemma A.3.3. *Let X be a closed, orientable, irreducible 3-manifold. Let $\bar{H} \subset X$ be a connected, compact, orientable, irreducible submanifold of X whose boundary is a collection of compressible tori. If $\bar{H}$ is not homeomorphic to the exterior of a (possibly* *empty) link in X, then $\bar{H}$ is included in a connected submanifold Y whose boundary is one of the tori of $\partial\bar{H}$ and which satisfies one of the following properties:*

(i) *Y is a solid torus, or*

(ii) *Y is homeomorphic to the exterior of a knot in S^3 and contained in a ball $B \subset X$.*

Proof. By hypothesis the boundary of $\bar{H}$ is not empty. We denote by $T_1, \dots, T_m$ the components of $\partial\bar{H}$. If one of them bounds a solid torus containing $\bar{H}$, then we can choose this solid torus as Y. Henceforth we assume that this is not the case.

Each T_j being compressible, it separates and thus bounds a submanifold V_j not containing $\bar{H}$. Up to renumbering the boundary components of $\bar{H}$, we may assume that $V_1, \dots, V_k$ are disjoint solid tori, but $V_{k+1}, \dots, V_m$ are not solid tori. At least one of the V_j's is not a solid torus, otherwise $\bar{H}$ would be homeomorphic to the exterior of a link in X.

For the same reason, at least one V_j, for some $j > k$, is not contained in the 3-ball bounded by the sphere obtained after compressing T_j. Otherwise each of the $V_{k+1}, \dots, V_m$ is homeomorphic to the exterior of a knot in S^3, by Lemma A.3.1, and one could then replace each V_j, $k+1 \leqslant j \leqslant m$ by a solid torus without changing the topological type of X. Hence $\bar{H}$ would be homeomorphic to the exterior of a link in X.

Pick a V_j, for $j > k$, which is not contained in a ball. Then compressing surgery on the torus $T_j = \partial V_j$ yields a sphere S bounding a ball B in X, which contains $\bar{H}$ by the choice of V_j. This shows that conclusion (ii) is satisfied with $Y = X \setminus \operatorname{int} V_j$. □

A.4 Covering 3-manifolds by abelian subsets

In [GLGA92] J. C. Gómez-Larrañaga and F. González-Acuña have computed the 1-dimensional Lusternik–Schnirelmann category of a closed 3-manifold. One step of their proof gives the following proposition (cf. [GLGA92], Proof of Proposition 2.1):

Proposition A.4.1. *Let X be a closed, connected 3-manifold. If X has a covering of dimension 2 by open subsets which are homotopically trivial in X, then there is a connected 2-dimensional complex K and a continuous map $f\colon X \to K$ such that the induced homomorphism $f_\star\colon \pi_1(X) \to \pi_1(K)$ is an isomorphism.* □

Standard homological arguments show the following, cf. [GLGA92], §3:

Corollary A.4.2. *Let X be a closed, connected, orientable, irreducible 3-manifold. If X has a covering of dimension 2 by open subsets which are homotopically trivial in X, then X is simply connected.*

Proof. Following Proposition A.4.1, let $f\colon X \to K$ be a continuous map from X to a connected 2-dimensional complex K, such that the induced homomorphism

$f_\star\colon \pi_1(X) \to \pi_1(K)$ is an isomorphism. Let Z be a $K(\pi_1(X),1)$ space. Let $\phi\colon X \to Z$ be a map from X to Z realising the identity homomorphism on $\pi_1(X)$ and let $\psi\colon K \to Z$ be the map from K to Z realising the isomorphism $f_\star^{-1}\colon \pi_1(K) \to \pi_1(X)$. Then ϕ is homotopic to $\psi \circ f$ and the induced homomorphism $\phi_*\colon H_3(X;\mathbb{Z}) \to H_3(Z;\mathbb{Z})$ factors through $\psi_*\colon H_3(K;\mathbb{Z}) \to H_3(Z;\mathbb{Z})$. Since $H_3(K;\mathbb{Z}) = \{0\}$, the homomorphism ϕ_* must be trivial.

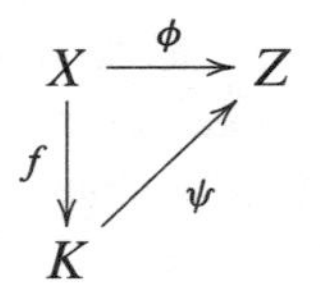

If $\pi_1(X)$ is infinite, then X is aspherical and $\phi_*\colon H_3(X;\mathbb{Z}) \to H_3(Z;\mathbb{Z})$ is an isomorphism. Therefore $\pi_1(X)$ is finite.

If $\pi_1(X)$ is finite of order $d > 1$, then let $\widetilde{X}$ be the universal covering of X. The covering map $p\colon \widetilde{X} \to X$ induces an isomorphism between the homotopy groups $\pi_k(\widetilde{X})$ and $\pi_k(X)$ for $k \geqslant 2$. Since $\pi_2(X) = \{0\}$, $\pi_2(\widetilde{X}) = \{0\}$, and by the Hurewicz theorem, the canonical homomorphism $\pi_3(\widetilde{X}) \to H_3(\widetilde{X};\mathbb{Z}) = \mathbb{Z}$ is an isomorphism. It follows that the canonical map $\pi_3(X) = \mathbb{Z} \to H_3(X;\mathbb{Z}) = \mathbb{Z}$ is the multiplication by the degree $d > 1$ of the covering $p\colon \widetilde{X} \to X$. It is well known that one can construct a $K(\pi_1(X),1)$ space Z by adding a 4-cell to kill the generator of $\pi_3(X) = \mathbb{Z}$, and adding further cells of dimension $\geqslant 5$ to kill the higher homotopy groups. Then the inclusion $\phi\colon X \to Z$ induces the identity on $\pi_1(X)$ and a surjection $\phi_*\colon H_3(X;\mathbb{Z}) = \mathbb{Z} \to H_3(Z;\mathbb{Z}) = \mathbb{Z}/d\mathbb{Z}$. Therefore, since ϕ_* vanishes, X must be simply connected. □

Finally, we need the following fact:

Lemma A.4.3. *Let X be a closed, orientable, irreducible* 3*-manifold. Assume that X is the union of two submanifolds X_1, X_2 which are solid tori or* 3*-balls. Then X is a lens space.*

Proof. When X_1 and X_2 are both solid tori, this is proved in [GLGAH04]. Otherwise X is covered by three 3-balls and thus homeomorphic to the 3-sphere S^3 by [HM69] (see also [GLGAH07]). □

Appendix B
Comparison geometry

Let us recall the fundamental comparison theorem below due to M. Gromov. It is used extensively in Perelman's work as well as in this book.

B.1 Comparison and compactness theorems

Theorem B.1.1 (Bishop–Gromov inequality [Gro99]). *Let (M, g) be a Riemannian manifold satisfying $\mathrm{Ric}_g \geqslant (n-1)kg$, where $k \in \mathbb{R}$. Let $v_k^n(r)$ denote the volume of an r-ball in the simply connected n-space of constant sectional curvature equal to k. Then for all $0 < r \leqslant R$, for all $x \in M$,*

$$\frac{\mathrm{vol}_g\, B(x,r)}{\mathrm{vol}_g\, B(x,R)} \geqslant \frac{v_k^n(r)}{v_k^n(R)}.$$

From [CGT82], Theorem 4.3, one has the following local version of Cheeger's injectivity radius estimate [Che70]:

Theorem B.1.2. *For any $K, v, D > 0$ and any integer $n \geqslant 2$, there exists $C = C(K, v, D, n) > 0$ with the following property. Suppose that (M, g) is a Riemannian n-manifold, and let $B(x, D) \subset M$ be a geodesic ball of radius D compactly embedded in M. Suppose also that $|\mathrm{Rm}| \leqslant K$ on $B(x, D)$ and that $\mathrm{vol}\, B(x, D) \geqslant v$. Then the injectivity radius at x of (M, g) satisfies $\mathrm{inj}(M, g, x) \geqslant C$.*

Proof. Bounding the injectivity radius at x is equivalent to bounding two quantities: on the one hand, the distance l_1 between x and the closest conjugate point (if any); on the other hand, the length l_2 of the shortest geodesic loop based at x.

By the Rauch comparison theorem, l_1 is at least $\min(\pi/\sqrt{K}, D)$. Note that this only uses the upper bound on the curvature along geodesic segments starting at x. To estimate l_2 from below, we use Cheeger–Gromov–Taylor [CGT82]. Indeed, Theorem 4.3 from [CGT82] applies here because it only requires a bound of the curvature on the ball. More precisely, taking $r = \min(\pi/\sqrt{K}, D)$ (and $r_0 = s = r/4$ in the notation of [CGT82]) by [CGT82], Inequality 4.22, we have

$$l_2 \geqslant \frac{r/8}{1 + v_{-K}(r/2)/\,\mathrm{vol}\, B(x, r/4)}.$$

In addition, by the Bishop–Gromov inequality applied to the balls of radius $r/4$ and D centred at x,

$$\mathrm{vol}\, B(x, r/4) \geqslant \frac{v_{-K}(r/4)}{v_{-K}(D)} v.$$

The combination of both inequalities gives the required lower bound for l_2. See [BBB$^+$10], Appendix A, for more details. □

We shall now state Gromov's compactness theorem. Before we need to define the notion of Lipschitz convergence.

Definition B.1.3. Let M_1 and M_2 be two metric spaces we define

$$d_L(M_1, M_2) = \inf\{|\ln(\mathrm{dil}(f))| + |\ln(\mathrm{dil}(f^{-1}))|\},$$

where the infimum is taken over the bi-Lipschitz homeomorphisms f between M_1 and M_2 (possibly $d_L(M_1, M_2) = +\infty$). Now, a sequence (M_k, x_k) of pointed metric spaces is said to converge in pointed Lipschitz topology towards (M_∞, x_∞) if for all $R > 0$, $d_L(B(x_\infty, R), B(x_k, R))$ goes to zero as k goes to infinity.

We next state the main compactness theorem, whose proof follows from [Gro99], [Pet87], [GW88].

Theorem B.1.4. *Let $0 < \alpha < 1$. Let (M_k, g_k, x_k) be a sequence of pointed smooth n-dimensional complete Riemannian manifolds such that*

(a) *for all $R > 0$, there exists $\Lambda(R) > 0$ such that the sectional curvature is bounded in absolute value by $\Lambda(R)$ on $B(x_k, R)$ for all k,*

(b) *there exists $V > 0$ such that* $\mathrm{vol}\, B(x_k, 1) \geq V$ *for all k.*

Then, there exists a subsequence which converges in the pointed Lipschitz topology to a smooth n-dimensional complete Riemannian manifold $(M_\infty, g_\infty, x_\infty)$ where the Riemannian metric g_∞ is of class $\mathcal{C}^{1,\alpha}$.

Remarks B.1.5. (i) From [GW88], the convergence takes place in the $\mathcal{C}^{1,\alpha}$ norm. More precisely, there exist diffeomorphisms f_k from $B(x_\infty, R)$ to $B(x_k, R)$ such that $(f_k)^* g_k$ converges to g_∞ in the $\mathcal{C}^{1,\alpha}$ norm (cf. [GW88], p. 139–140).

(ii) If the manifolds M_k have uniformly bounded diameter, then M_∞ is compact and diffeomorphic to M_k for k large enough.

(iii) If assumption (a) is weakened to "For some $R > 0$, there exists $\Lambda > 0$ such that...", one obtains sub-convergence of $(B(x_k, R), g_k, x_k)$ to a (possibly non complete) smooth n-dimensional Riemannian manifold $(M_\infty, g_\infty, x_\infty)$.

B.2 Manifolds with nonnegative curvature

Next we describe the Cheeger–Gromoll soul theorem and some of its consequences. Let M be a Riemannian n-manifold. A subset $S \subset M$ is *totally convex* if every (possibly nonminimising) geodesic segment M with endpoints in S is contained in S. This is stronger than being totally geodesic. We say that S is a *soul* of M if S is a closed, totally convex submanifold of M.

Theorem B.2.1 ([CG72]). *Every Riemannian manifold M of nonnegative sectional curvature has a soul S.*

If M is closed, then it turns out that $S = M$. In the open case, Theorem B.2.1 has strong topological consequences thanks to the following lemma:

Lemma B.2.2 ([CG72]). *Let M be an open Riemannian manifold of nonnegative sectional curvature, and let $S \subset M$ be a soul. Then M is diffeomorphic to the normal bundle of S in M. In addition, for every $r > 0$, the tubular neighbourhood*

$$\mathcal{N}_r(S) = \{x \in M \mid d(x, S) < r\}$$

is diffeomorphic to the disc subbundle of the normal bundle.

This lemma is proved by constructing a vector field that averages all directions between a point and S. Since S is totally convex, this average is nonzero and the vector field is well defined, it is in fact an almost gradient vector field for minus the distance function to S. The integral curves of this field correspond to the fibres of the normal bundle.

Since the soul S is itself a closed manifold of nonnegative curvature, we easily classify the diffeomorphism type of low-dimensional manifolds with nonnegative curvature, [CG72]:

Corollary B.2.3 (Classification of open 3-manifolds of nonnegative curvature). *Let M be an open, orientable 3-manifold M of nonnegative sectional curvature. Then exactly one of the following conclusions holds:*

(i) *If $\dim S = 0$, then S is a point, and $M \cong \mathbb{R}^3$.*
(ii) *If $\dim S = 1$, then S is a circle, and $M \cong S^1 \times \mathbb{R}^2$.*
(iii) *If $\dim S = 2$, then S is diffeomorphic to S^2, T^2, $\mathbb{R}P^2$, or K^2. Thus M is diffeomorphic to $S^2 \times \mathbb{R}$, $T^2 \times \mathbb{R}$, $\mathbb{R}P^2 \tilde{\times} \mathbb{R}$, or $K^2 \tilde{\times} \mathbb{R}$, respectively.*

Remark B.2.4. All these manifolds are irreducible except $S^2 \times \mathbb{R}$ and $\mathbb{R}P^2 \tilde{\times} \mathbb{R}$. This follows from Alexander's theorem (Theorem A.2.1) and the elementary fact that if M is a 3-manifold whose universal cover is irreducible, then M itself is irreducible.

In the closed 3-dimensional case, the classification has been obtained by R. Hamilton using Ricci flow [Ham82], [Ham86]:

Theorem B.2.5 (Classification of closed 3-manifolds of nonnegative curvature). *Every closed, orientable 3-manifold of nonnegative sectional curvature is spherical, Euclidean, or diffeomorphic to $S^2 \times S^1$ or $\mathbb{R}P^3 \# \mathbb{R}P^3$. In particular, any such manifold is a graph manifold.*

Next we discuss the consequences of the classification for the fundamental group. We say that a group is *virtually abelian* if it contains an abelian subgroup of finite index.

Corollary B.2.6. *Let M be an orientable 3-manifold of nonnegative sectional curvature. Then $\pi_1(M)$ is virtually abelian.*

Proof. If M is noncompact, then we apply Corollary B.2.3. By elementary computations, $\pi_1(M)$ is isomorphic to the trivial group, $\mathbb{Z}$, $\mathbb{Z}/2\mathbb{Z}$, $\mathbb{Z}^2$ or the Klein bottle group. The latter has a subgroup of index 2 isomorphic to $\mathbb{Z}^2$.

If M is compact, apply Theorem B.2.5. Thus $\pi_1(M)$ is finite, virtually $\mathbb{Z}^3$, or virtually $\mathbb{Z}$. □

Remark B.2.7. This corollary is in fact also true if M is nonorientable, by passing to the orientation cover.

Appendix C
Ricci flow

C.1 Existence and basic properties

Theorem C.1.1 ([Ham82], [Shi89b], [CK04], [CZ06c]). *Let (M, g_0) be a complete Riemannian manifold such that $|\mathrm{Rm}| \leqslant \Lambda$ for some $\Lambda < +\infty$. Then there exists a unique complete Ricci flow $g(\cdot)$ of bounded curvature defined on $M \times [0, (16\Lambda)^{-1}]$, such that $g(0) = g_0$. Moreover, we have $|\mathrm{Rm}| \leqslant 2\Lambda$ everywhere.*

Proposition C.1.2. *Let $(M, \{g(t)\}_{t \in [0,b)})$ be an n-dimensional Ricci flow. If the interval of definition $[0, b)$ is maximal, then $\sup_{(x,t)} |\mathrm{Rm}(x,t)| = +\infty$.*

C.2 Consequences of the maximum principle

We recall some basic formulas (cf. [CK04]):

Proposition C.2.1. *For a Ricci flow, the quantities $R(x,t)$ and $V(t)$ satisfy*

(i) $\dfrac{dV}{dt} = -\displaystyle\int_M R\, dV,$

(ii) $\dfrac{dR}{dt} = \Delta R + 2|\operatorname{Ric}^0|^2 + \dfrac{2}{3} R^2.$

Corollary C.2.2.

$$\frac{dV}{dt} \leqslant -R_{\min} V, \tag{C.1}$$

$$\frac{dR_{\min}}{dt} \geqslant \frac{2}{3} R_{\min}^2, \tag{C.2}$$

$$\frac{d\hat{R}}{dt} \geqslant \frac{2}{3} \hat{R} V^{-1} \int_M (R_{\min} - R)\, dV. \tag{C.3}$$

Proof.

$$\begin{aligned} \frac{d}{dt} V(t) &= \frac{d}{dt} \int_M dv_{g(t)} \\ &= \int \frac{1}{2} \operatorname{tr}_g\left(\tfrac{dg}{dt}\right) dv_{g(t)} \\ &= \int -R\, dv \\ &\leqslant -R_{\min} V(t). \end{aligned}$$

Equation (C.2) follows from Proposition C.2.1 (ii) and the maximum principle.

$$\begin{aligned}
\frac{d}{dt}\hat{R} &= R'_{\min} V^{\frac{2}{3}} + \frac{2}{3} R_{\min} \int -R \, dv V^{-\frac{1}{3}} \\
&\geqslant \frac{2}{3} R^2_{\min} V^{\frac{2}{3}} + \frac{2}{3} R_{\min} \int -R \, dv V^{-\frac{1}{3}} \\
&= \frac{2}{3} R_{\min} V^{\frac{2}{3}} \Big(R_{\min} V^{-1} \int_M dv + \int -R \, dv V^{-1} \Big) \\
&= \frac{2}{3} R_{\min} V^{\frac{2}{3}} V^{-1} \Big(\int_M (R_{\min} - R) \, dv \Big) \\
&= \frac{2}{3} \hat{R}(g) V^{-1} \Big(\int_M (R_{\min} - R) \, dv \Big)
\end{aligned}$$

□

C.3 Compactness

A *pointed evolving metric* is a triple $(M, \{g(t)\}_{t\in I}, (x_0, t_0))$ where M is a manifold, $g(\cdot)$ is an evolving metric on M, and (x_0, t_0) belongs to $M \times I$. We say that a sequence of pointed evolving metrics $(M_k, \{g_k(t)\}_{t\in I}, (x_k, t_0))$ *converges smoothly* to a pointed evolving metric $(M_\infty, \{g_\infty(t)\}_{t\in I}, (x_\infty, t_0))$ if there exists an exhaustion of M by open sets U_k, such that $x \in U_k$ for all k, and smooth maps $\psi_k : U_k \to M_k$, diffeomorphic onto their images, sending x to x_k, such that $\psi_k^* g_k(\cdot) - g(\cdot)$ and all its derivatives converge to zero uniformly on compact subsets of $M \times I$.

Theorem C.3.1 (Hamilton's compactness). *Let $(M_k, \{g_k(t)\}_{t\in(a,b]}, (x_k, t_0))$ be a sequence of complete pointed Ricci flows of the same dimension. Assume that*

(i) *for all $\rho > 0$, $\sup_{k\in\mathbb{N}} \sup_{B(x_k,t_0,\rho)\times(a,b]} |\mathrm{Rm}| < +\infty$, and*
(ii) $\inf_{k\in\mathbb{N}} \mathrm{inj}(M_k, g_k(t_0), x_k) > 0$.

Then $(M_k, \{g_k(t)\}_{t\in(a,b]}, (x_k, t_0))$ subconverges smoothly to a complete Ricci flow of the same dimension, defined on $(a, b]$.

Remark C.3.2. If $g(t)$ is defined on $[a, b]$, one can take $t_0 = a$ if one has also uniform bounds on the derivatives of the curvature operator at time t_0, that is: for any $\rho > 0$, for any integer p, $\sup_{k\in\mathbb{N}} \sup_{B(x_k,t_0,\rho)\times\{t_0\}} |\nabla^p \mathrm{Rm}| < +\infty$.

An inspection of the proof of Theorem C.3.1 shows that the following natural extension holds. The definition of a partial Ricci flow is given in Section 2.2.

Theorem C.3.3 (Local compactness for flows). *Let $(U_k, \{g_k(t)\}_{t\in(a,0]}, (x_k, 0))$ be a sequence of pointed Ricci flows of the same dimension. Suppose that for some $\rho_0 \in (0, +\infty]$, all the balls $B(x_k, 0, \rho)$ of radius $\rho < \rho_0$ are relatively compact in U_k and that the following holds:*

(i) *For any* $\rho \in (0, \rho_0)$*, there exists* $\Lambda(\rho) < +\infty$ *and* $\tau(\rho) > 0$ *such that* $|\mathrm{Rm}| < \Lambda(\rho)$ *on all* $P(x_k, 0, \rho, -\tau(\rho))$.

(ii)
$$\inf_{k \in \mathbb{N}} \mathrm{inj}(U_k, g_k(0), x_k) > 0.$$

Then there is a Riemannian ball $B(x_\infty, \rho_0)$ *of the same dimension such that the pointed sequence* $(B(x_k, 0, \rho_0), g_k(\cdot), x_k)$ *subconverges smoothly to a partial Ricci flow* $g_\infty(\cdot)$ *defined on* $\bigcup_{\rho<\rho_0}(B(x_\infty, \rho) \times (-\tau(\rho), 0])$*. Moreover, if* $\rho_0 = +\infty$ *then for any* $t \in [\sup_\rho -\tau(\rho), 0]$*,* $g_\infty(t)$ *is complete.*

C.4 Harnack inequalities for the Ricci Flow

We state first the matrix Harnack inequality. Consider the following tensors P and M associated to the curvature:

$$P_{abc} = \nabla_a R_{bc} - \nabla_b R_{ac},$$
$$M_{ab} = \Delta R_{ab} - \frac{1}{2}\nabla_a \nabla_b R + R_{abcd} R_{cd} - R_{ac} R_{bc}.$$

For a 2-form U and 1-form W, Hamilton defines:

$$Z(U, W) = M_{ab} W_a W_b + 2P_{abc} U_{ab} W_c + R_{abcd} U_{ab} U_{cd}.$$

Theorem C.4.1 (Hamilton's matrix Harnack inequality [Ham93]). *Suppose we have a Ricci flow on* $t \in [0, T]$ *such that every time slice is complete with bounded curvature. If* $\mathrm{Rm} \geqslant 0$ *then, for every 2-form* U *and 1-form* W,

$$Z(U, W) \geqslant 0.$$

Given two vector fields X, Y, taking $U = X^\flat \wedge Y^\flat$ and $W = Y^\flat$, and tracing over Y, with the same hypothesis as in Theorem C.4.1, we obtain Hamilton's trace Harnack inequality:

$$\frac{\partial R}{\partial t} + \frac{1}{t} R + 2g(t)(\nabla R, X) + 2\,\mathrm{Ric}(X, X) \geqslant 0.$$

In particular for $X = 0$ and reparametrizing the time, we get:

Corollary C.4.2 ([Ham93]). *Suppose we have a Ricci flow for* $t \in [t_0, T]$ *such that every time slice is complete with bounded curvature. If* $\mathrm{Rm} \geqslant 0$*, then*

$$\frac{\partial R}{\partial t} + \frac{1}{t - t_0} R \geqslant 0.$$

If in addition the flow is ancient ($t_0 = -\infty$)*, then*

$$\frac{\partial R}{\partial t} \geqslant 0.$$

C.5 Ricci Flow on cones

The following lemma is used in Section 4 for the proof of Theorem 6.1.1. It rests on Hamilton's strong maximum principle (cf. [KL08], Theorem 41.2, [MT07], Proposition 4.22):

Lemma C.5.1. *Let (C, g_0, p) be a Riemannian cone with base point p, such that the curvature operator of g_0 is nonnegative and nonzero. Let $U \subset C \setminus p$ be an open subset and $g(\cdot)$ be a Ricci flow on $U \times [-\tau, 0]$, for some $\tau > 0$, with final time slice $g(0)$ equal to g_0. Then $g(\cdot)$ cannot have nonnegative curvature operator.*

Appendix D
Alexandrov spaces

We recall a few facts about Alexandrov spaces that are used in this book. General references are [BBI01], [BGP92].

Definition D.1.2 ([BGP92], 2.3, [BBI01], 10.1.1). A locally complete length space M is called an *Alexandrov space of nonnegative curvature* if in some neighbourhood U_p of any point $p \in M$ the following condition is satisfied:

For any four (distinct) points (a, b, c, d) in U_p we have the inequality

$$\tilde{\measuredangle}(bac) + \tilde{\measuredangle}(bad) + \tilde{\measuredangle}(cad) \leqslant 2\pi.$$

Remarks D.1.3. 1) Here $\tilde{\measuredangle}$ denotes the comparison angle in $\mathbb{R}^2$. For a precise definition see the cited references.

2) For spaces in which locally, any two points are joined by a geodesic, the condition can be replaced by: For any triangle $\Delta(pqr)$ with vertices in U_p and any $s \in [qr]$, one has $|ps| \geq |\tilde{p}\tilde{s}|$, where $\tilde{s}$ is a point on $[\tilde{q}\tilde{r}]$, side of the comparison triangle $\tilde{\Delta}(pqr)$ such that $|qs| = |\tilde{q}\tilde{s}|$ and $|rs| = |\tilde{r}\tilde{s}|$. We will say that $\Delta(pqr)$ is *thicker* than $\tilde{\Delta}(pqr)$. This definition applies in Section 6.1.2.

We recall that given a metric space (X, d) and $x \in X$, one calls a tangent cone of X at x, a pointed Gromov–Hausdorff limit, if it exists, of $(X, \lambda_i d, x)$, where λ_i is a sequence of real numbers going to $+\infty$.

In Section 6.1.2 we consider finite-dimensional Alexandrov spaces of nonnegative curvature, for which we have:

Theorem D.1.4 ([BBI01], p. 391). *Let X be a finite-dimensional Alexandrov space of nonnegative curvature and $x \in X$. Then the tangent cone of X at x exists and is isometric to the cone on the space of directions at x.*

Remark D.1.5. In fact, Theorem D.1.4 is stated in [BBI01] for *complete* spaces, but as observed in the beginning of [BBI01], Chapter 10, local results hold for *locally complete* spaces. Hence this theorem applies in our situation.

Lemma D.1.6 (Sub-ball lemma). *For all $w > 0$, there exists $\theta_0 = \theta_0(w) > 0$, such that if $B(x, 1)$ is a ball with volume at least w, which is relatively compact in a manifold without boundary whose sectional curvature is bounded below by -1, then, there exists a $B(y, \theta_0) \subset B(x, 1)$, such that every sub-ball $B(z, r) \subset B(y, \theta_0)$ of any radius $r > 0$ has volume bounded below by $(1 - \varepsilon_0)$ times the volume of a Euclidean ball of the same radius.*

In such a context, we will say that $B(z, r)$ has $(1 - \varepsilon_0)$-*almost Euclidean volume*.

Appendix E
A sufficient condition for hyperbolicity

Below we give a proof of Proposition 1.3.10 which we recall here.

Proposition E.1.7. *Let M be a closed and orientable 3-manifold. Suppose that the inequality $\overline{V}(M) \geqslant V_0(M)$ holds. Then equality holds, M is hyperbolic, and the hyperbolic metric realises $\overline{V}(M)$.*

Proof. Let H_0 be a hyperbolic manifold homeomorphic to the complement of a link L_0 in M and whose volume realises $V_0(M)$. To prove Proposition E.1.7, it is sufficient to show that L_0 is empty. Let us assume that it is not true and prove that M carries a metric g_ε such that $\mathrm{vol}(g_\varepsilon) < V_0(M)$ and $R_{\min}(g_\varepsilon) \geqslant -6$. This can be done by a direct construction as in [And02]. We give here a different argument relying on Thurston's hyperbolic Dehn filling theorem.

If $L_0 \neq \emptyset$, then we consider the orbifold $\mathcal{O}$ with underlying space M, singular locus L_0 local group $\mathbb{Z}/n\mathbb{Z}$ with $n > 1$ sufficiently large so that the orbifold carries a hyperbolic structure, by the hyperbolic Dehn filling theorem [Thu80] (cf. [BP01], Appendix B). We then desingularise the conical metric on M corresponding to the orbifold structure, in a tubular neighbourhood of L_0:

Lemma E.1.8 (Salgueiro [Sal09]). *For each $\varepsilon > 0$ there exists a Riemannian metric g_ε on M with sectional curvature bounded below by -1 and such that* $\mathrm{vol}(M, g_\varepsilon) < (1+\varepsilon)\,\mathrm{vol}(\mathcal{O})$.

For completeness we give the proof of this lemma, following [Sal09], Chapter 3.

Proof. Let g be the hyperbolic cone metric on M induced by the hyperbolic orbifold $\mathcal{O}$. Let $\mathcal{N} \subset \mathcal{O}$ be a tubular neighbourhood of radius $r_0 > 0$ around the singular locus L_0. In $\mathcal{N}$ the local expression of the singular metric g in cylindrical coordinates (r, θ, h) is:

$$ds^2 = dr^2 + \left(\frac{1}{n}\sinh(r)\right)^2 d\theta^2 + \cosh^2(r)dh^2,$$

where $r \in (0, r_0)$ is the distance to L_0, h is the linear coordinate of the projection to L_0, and $\theta \in (0, 2\pi)$ is the rescaled angle parameter.

The deformation depends only on the parameter r and consists in replacing $\mathcal{N}$ with the metric g by a smaller cylinder $\mathcal{N}'$ with smooth metric g' of the form

$$ds^2 = dr^2 + \phi^2(r)d\theta^2 + \psi^2(r)dh^2,$$

where for some $\delta = \delta(\varepsilon) > 0$ sufficiently small the functions

$$\phi, \psi \colon [0, r_0 - \delta] \to [0, +\infty)$$

are smooth and satisfy the following properties:

(1) In a neighbourhood of 0, $\phi(r) = r$ and $\psi(r)$ is constant.

(2) In a neighbourhood of $r_0 - \delta$, $\phi(r) = \frac{1}{n}\sinh(r+\delta)$ and $\psi(r) = \cosh(r+\delta)$.

(3) $\forall r \in (0, r_0 - \delta)$, $\frac{\phi''(r)}{\phi(r)} \leqslant 1 + \varepsilon$, $\frac{\psi''(r)}{\psi(r)} \leqslant 1 + \varepsilon$ and $\frac{\phi'(r)\psi'(r)}{\phi(r)\psi(r)} \leqslant 1 + \varepsilon$.

The new metric is Euclidean near L_0 (hence non-singular) by (1), it matches the previous one away from $\mathcal{N}$ by (2) and has sectional curvature $\geqslant -1 - \varepsilon$ by (3).

We first deal with the construction of ψ, satisfying

- on $[0, \eta]$, ψ is constant, where $\eta^2 = \delta$,
- on $[\eta, r_0 - \delta]$, $\psi \geqslant \cosh(r+\delta)$, $\psi' \leqslant \sinh(r+\delta)$, $\psi'' \leqslant \cosh(r+\delta)(1+\varepsilon)$.

We start with a function $f_\delta = \psi'$ which vanishes in $[0, \eta]$, $f_\delta = \sinh(r+\delta)$ on a neighbourhood of $r_0 - \delta$, and satisfies $f_\delta \leqslant \sinh(r+\delta)$, $f_\delta' \leqslant \cosh(r+\delta)(1+\varepsilon)$. This f_δ can be achieved by a suitable deformation of $f_0(r) = \sinh(r)$, because $f_0'(r) < \cosh(r)(1+\varepsilon)$. Then ψ is just the primitive of f that equals $\cosh(r+\delta)$ at $r = r_0 - \delta$.

Next we deal with the construction of ϕ. Let $r_1 = r_1(\delta) > 0$ be the smallest positive real number satisfying $r_1 = \frac{1}{n}\sinh(r_1 + \delta)$. Consider the piecewise smooth function

$$r \mapsto \begin{cases} r & \text{on } [0, r_1], \\ \frac{1}{n}\sinh(r+\delta) & \text{on } [r_1, r_0 - \delta]. \end{cases}$$

Then, since $r_1 \sim \frac{1}{n-1}\delta \ll \eta$, ϕ can be chosen as a smooth modification of this function satisfying

- on $[0, r_1]$, $\phi \geqslant r/(1+\varepsilon)$, $\phi'' \leqslant 0$,
- on $[r_1, \eta]$, $\phi \geqslant \frac{1}{n}\sinh(r+\delta)\frac{1}{1+\varepsilon}$, $\phi'' \leqslant \frac{1}{n}\sinh(r+\delta)$,
- on $[\eta, r_0 - \delta]$, $\phi = \frac{1}{n}\sinh(r+\delta)$.

As $\delta \to 0$, $\mathrm{vol}(M, g') \to \mathrm{vol}(\mathcal{O})$, since $\phi \to \frac{1}{n}\sinh(r)$ and $\psi \to \cosh^2(r)$. Hence, given $\varepsilon > 0$, for a choice of δ sufficiently small, one obtains a smooth Riemannian metric g' on M with sectional curvature $\geqslant -1 - \varepsilon$ and volume $\mathrm{vol}(M, g') \leqslant (1+\varepsilon)\,\mathrm{vol}(\mathcal{O})$. Then the rescaled metric $g_\varepsilon = \sqrt{1+\varepsilon}\, g'$ on M has sectional curvature $\geqslant -1$ and volume $\mathrm{vol}(M, g_\varepsilon) \leqslant (1+\varepsilon)^{\frac{5}{2}}\,\mathrm{vol}(\mathcal{O})$. □

As $\mathrm{vol}(\mathcal{O}) < \mathrm{vol}(H_0)$ [Thu80], [Hod86], [NZ85], for $\varepsilon > 0$ sufficiently small we obtain a Riemannian metric on M such that $\mathrm{vol}(M, g_\varepsilon) < \mathrm{vol}(H_0)$ and $R_{\min}(g_\varepsilon) \geqslant -6$. In particular

$$\bar{V} \leqslant \mathrm{vol}(M, g_\varepsilon) < \mathrm{vol}(H_0) = V_0(M),$$

which contradicts the hypothesis. The link L_0 is thus empty and we have $M = H_0$. This proves Proposition E.1.7 □

Bibliography

[AK04] S. Angenent and D. Knopf, An example of neckpinching for Ricci flow on S^{n+1}. *Math. Res. Lett.* **11** (2004), 493–518. 6

[And02] M. T. Anderson, Scalar curvature and the existence of geometric structures on 3-manifolds, I. *J. Reine Angew. Math.* **553** (2002), 125–-182. 14, 21, 181, 222

[And04] M. T. Anderson, Geometrization of three-manifolds via the Ricci flow. *Notices Amer. Math. Soc.* **51** (2004), 184–193. vi

[And05] M. T. Anderson, Canonical metrics on 3-manifolds and 4-manifolds. *Asian J. Math.* **10** 2006, 127–163. 15, 185

[AST07] I. Agol, P. A. Storm, and W. P. Thurston, Lower bounds on volumes of hyperbolic Haken 3-manifolds. With an appendix by N. Dunfield. *J. Amer. Math. Soc.* **20** (2007), 1053–1077. 21

[Bam07] R. H. Bamler, Ricci flow with surgery. Diploma thesis, Ludwig-Maximilians University, München 2007.
http://www.math.princeton.edu/ rbamler/Ricci-flow-with-surgery.pdf 74

[BBB06] L. Bessières, G. Besson, and M. Boileau, La preuve de la conjecture de Poincaré d'après Perelman. In *L'héritage scientifique de Poincaré*, Échelles, Belin, Paris 2006, 263–277. vi

[BBB+07] L. Bessières, G. Besson, M. Boileau, S. Maillot, and J. Porti, Weak collapsing and geometrisation of aspherical 3-manifolds. Preprint 2007. arXiv:0706.2065 v, 14, 185

[BBB+10] L. Bessières, G. Besson, M. Boileau, S. Maillot, and J. Porti, Collapsing irreducible 3-manifolds with nontrivial fundamental group. *Invent. Math.* **179** (2010), 435–460. v, 184, 185, 214

[BBI01] D. Burago, Y. Burago, and S. Ivanov, *A course in metric geometry*. Grad. Stud. Math. 33, Amer. Math. Soc. Providence, RI, 2001. 66, 221

[BBM09] L. Bessières, G. Besson, and S. Maillot, Ricci flow on open 3-manifolds and positive scalar curvature. Preprint 2010. arXiv:1001.1458 21

[Bes00] L. Bessières, Sur le volume minimal des variétés ouvertes. *Ann. Inst. Fourier (Grenoble)* **50** (2000), 965–980. 13

[Bes05] L. Bessières, Conjecture de Poincaré: la preuve de R. Hamilton et G. Perelman. *Gaz. Math.* **106** (2005), 7–35. vi

[Bes06a] L. Bessières, Poincaré conjecture and Ricci flow. An outline of the work of R. Hamilton and G. Perelman. *EMS Newsletter* **59** (2006), 11–15. vi

[Bes06b] G. Besson, Preuve de la conjecture de Poincaré en déformant la métrique par la courbure de Ricci (d'après Perelman). In *Séminaire Bourbaki 2004/2005*, Exp. no. 947, Astérisque 307, Soc. Math. France, Paris, France 2006, 309–347. vi

[Bes07] G. Besson, The geometrization conjecture after R. Hamilton and G. Perelman. *Rend. Semin. Mat. Univ. Politec. Torino* **65** (2007), 397–411. vi

[Bes87] A. L. Besse, *Einstein manifolds*. Ergeb. Math. Grenzgeb. 10, Springer-Verlag, Berlin 1987. 78, 83

[BGP92] Yu. Burago, M. Gromov, and G. Perel′man, A. D. Aleksandrov spaces with curvatures bounded below. *Uspekhi Mat. Nauk* **47** (1992), no. 2, 3–51; English transl. *Russian Math. Surveys* **47** (1992), no. 2, 1–58. 221

[BGS85] W. Ballmann, M. Gromov, and V. Schroeder, *Manifolds of nonpositive curvature*. Progr. Math. 61, Birkhäuser, Boston 1985. 105

[BLP05] M. Boileau, B. Leeb, and J. Porti, Geometrization of 3-dimensional orbifolds. *Ann. of Math.* (2) **162** (2005), 195–290. v, 184, 188

[BM91] M. Bestvina and G. Mess, The boundary of negatively curved groups. *J. Amer. Math. Soc.* **4** (1991), 469–481. 20

[BMP03] M. Boileau, S. Maillot, and J. Porti, *Three-dimensional orbifolds and their geometric structures*. *Panor. Synthèses* 15, Soc. Math. France, Paris 2003. 4, 15, 188

[Bow04] B. H. Bowditch, Planar groups and the Seifert conjecture. *J. Reine Angew. Math.* **576** (2004), 11–62. 19

[BP01] M. Boileau and J. Porti, Geometrization of 3-orbifolds of cyclic type. *Astérisque* **272** (2001). 184, 222

[Bro82] K. S. Brown, *Cohomology of groups*. Grad. Texts in Math. 87, Springer-Verlag, New York 1982. 19

[CCG+07] B. Chow, S.-C. Chu, D. Glickenstein, C. Guenther, J. Isenberg, T. Ivey, D. Knopf, P. Lu, F. Luo, and L. Ni, *The Ricci flow: techniques and applications*. Part I, Math Surveys Monogr. 135, Amer. Math. Soc., Providence, RI, 2007. 129

[CG72] J. Cheeger and D. Gromoll, On the structure of complete manifolds of nonnegative curvature. *Ann. of Math.* (2) **96** (1972), 413–443. 215

[CG86] J. Cheeger and M. Gromov, Collapsing Riemannian manifolds while keeping their curvature bounded. I. *J. Differential Geometry* **23** (1986), 309–346. 179, 208

[CG90] J. Cheeger and M. Gromov, Collapsing Riemannian manifolds while keeping their curvature bounded. II. *J. Differential Geometry* **32** (1990), 269–298. 179

[CG09] J. Cao and J. Ge, A proof of Perelman's collapsing theorem for 3-manifolds. Preprint 2009. arXiv:0908.3229 v, 21, 180, 185, 186

[CGT82] J. Cheeger, M. Gromov, and M. Taylor, Finite propagation speed, kernel estimates for functions of the Laplace operator, and the geometry of complete Riemannian manifolds. *J. Differential Geom.* **17** (1982), 15–53. 213

[Che70] J. Cheeger, Finiteness theorems for Riemannian manifolds. *Amer. J. Math.* **92** (1970), 61–74. 213

[CK04] B. Chow and D. Knopf, *The Ricci flow: an introduction*. Math. Surveys Monogr. 110, Amer. Math. Soc., Providence, RI, 2004. v, 217

[CL06] A. V. Chernavsky and V. P. Leksine, Unrecognizability of manifolds. *Ann. Pure Appl. Logic* **141** (2006), 325–335. 17

[CLN06] B. Chow, P. Lu, and L. Ni, *Hamilton's Ricci flow*. Grad. Stud. Math. 77, Amer. Math. Soc., Providence, RI, 2006. 47

[CM05] T. H. Colding and W. P. Minicozzi, II, Estimates for the extinction time for the Ricci flow on certain 3-manifolds and a question of Perelman. *J. Amer. Math. Soc.* **18** (2005), 561–569. 9, 10

[CM07] T. H. Colding and W. P. Minicozzi, II, Width and finite extinction time of Ricci flow. *Geom. Topol.* **12** (2008), 2537–2586. v, 10

[CM09] F. C. Marques, Deforming three-manifolds with positive scalar curvature. Preprint 2009. arXiv:0907.2444 21

[CZ06a] H.-D. Cao and X.-P. Zhu, A complete proof of the Poincaré and geometrization conjectures – application of the Hamilton-Perelman theory of the Ricci flow. *Asian J. Math.* **10** (2006), 165–492. v

[CZ06b] B.-L. Chen and X.-P. Zhu, Ricci flow with surgery on four-manifolds with positive isotropic curvature. *J. Differential Geom.* **74** (2006), 177–264. 41

[CZ06c] B.-L. Chen and X.-P. Zhu, Uniqueness of the Ricci flow on complete noncompact manifolds. *J. Differential Geom.* **74** (2006), 119–154. 42, 93, 217

[Dav98] M. W. Davis, The cohomology of a Coxeter group with group ring coefficients. *Duke Math. J.* **91** (1998), 297–314. 19

[DL09] J. Dinkelbach and B. Leeb, Equivariant Ricci flow with surgery and applications to finite group actions on geometric 3-manifolds. *Geom. Topol.* **13** (2009), 1129–1173. 21, 36

[DS00] M. J. Dunwoody and E. L. Swenson, The algebraic torus theorem. *Invent. Math.* **140** (2000), 605–637. 20

[ECH+92] D. B. A. Epstein, J. W. Cannon, D. F. Holt, S. V. F. Levy, M. S. Paterson, and W. P. Thurston, *Word processing in groups*. Jones and Bartlett Publishers, Boston 1992. 18

[Eck87] B. Eckmann, Poincaré duality groups of dimension two are surface groups. In *Combinatorial group theory and topology* (Alta, Utah, 1984), Ann. of Math. Stud. 111, Princeton University Press, Princeton, NJ, 1987, 35–51. 19

[Fuk90] K. Fukaya, Hausdorff convergence of Riemannian manifolds and its applications. In *Recent topics in differential and analytic geometry*, Adv. Stud. Pure Math. 18, Academic Press, Boston, MA 1990, 143–238. 193

[GLGA92] J. C. Gómez-Larrañaga and F. J. González-Acuña, Lusternik-Schnirelmann category of 3-manifolds. *Topology* **31** (1992), 791–800. 184, 211

[GLGAH04] J. C. Gómez-Larrañaga, F. J. González-Acuña, and W. Heil, 3-manifolds that are covered by two open bundles. *Bol. Soc. Mat. Mexicana* (3) **10** (2004), 171–179. 212

[GLGAH07] J. C. Gómez-Larrañaga, F. J. González-Acuña, and W. Heil, A note on Hempel–McMillan coverings of 3-manifolds. *Topology Appl.* **154** (2007), 1363–1367. 212

[Gro81a] M. Gromov, Hyperbolic manifolds (according to Thurston and Jørgensen). In *Bourbaki Seminar, Vol. 1979/80*, Lecture Notes in Math. 842, Springer, Berlin 1981, 40–53. 13

[Gro81b] M. Gromov, *Structures métriques pour les variétés riemanniennes*. Textes Mathématiques 1, CEDIC-Fernand Nathan, Paris 1981.

[Gro82] M. Gromov, Volume and bounded cohomology. *Inst. Hautes Études Sci. Publ. Math.* **56** (1982), 5–99. 181, 182, 183

[Gro99] M. Gromov, *Metric structures for Riemannian and non-Riemannian spaces* (eds. J. LaFontaine and P. Pansu). Progr. Math. 152, Birkhäuser, Boston 1999. 213, 214

[GW88] R. E. Greene and H. Wu, Lipschitz convergence of Riemannian manifolds. *Pacific J. Math.* **131** (1988), 119–141. 214

[Hak61] W. Haken, Theorie der Normalflächen. *Acta Math.* **105** (1961), 245–375. 16

[Hak62] W. Haken, Über das Homöomorphieproblem der 3-Mannigfaltigkeiten. I. *Math. Z.* **80** (1962), 89–120. 16

[Ham82] R. S. Hamilton, Three-manifolds with positive Ricci curvature. *J. Differential Geom.* **17** (1982), 255–306. 5, 6, 195, 215, 217

[Ham86] R. S. Hamilton, Four-manifolds with positive curvature operator. *J. Differential Geom.* **24** (1986), 153–179. 5, 56, 195, 215

[Ham93] R. S. Hamilton, The Harnack estimate for the Ricci flow. *J. Differential Geom.* **37** (1993), 225–243. 5, 219

[Ham95a] R. S. Hamilton, A compactness property for solutions of the Ricci flow. *Amer. J. Math.* **117** (1995), 545–572. 93, 193

[Ham95b] R. S. Hamilton, The formation of singularities in the Ricci flow. In *Surveys in differential geometry* (Cambridge, MA, 1993), Vol. II, Internat. Press, Cambridge, MA, 1995, 7–136. 5, 6, 29

[Ham97] R. S. Hamilton, Four-manifolds with positive isotropic curvature. *Comm. Anal. Geom.* **1** (1997), 1–92. 4, 9, 30

[Ham99] R. S. Hamilton, Non-singular solutions of the Ricci flow on three-manifolds. *Comm. Anal. Geom.* **7** (1999), 695–729. 5, 6, 21, 44, 45, 88, 180, 181, 187

[Hat00] A. Hatcher, Notes on basic 3-manifold topology. Cornell University, Ithaca, NY, 2000. http://www.math.cornell.edu/ hatcher/3M/3Mdownloads.html 209

[Hem76] J. Hempel, *3-manifolds*. Ann. of Math. Stud. 86, Princeton University Press, Princeton 1976. 209

[Hem79] G. Hemion, On the classification of homeomorphisms of 2-manifolds and the classification of 3-manifolds. *Acta Math.* **142** (1979), 123–155. 16

[HH97] E. Hebey and M. Herzlich, Harmonic coordinates, harmonic radius and convergence of Riemannian manifolds. *Rend. Mat. Appl.* (7) **17** (1997), 569–605 (1998). 93

[Hil87] J. A. Hillman, Three-dimensional Poincaré duality groups which are extensions. *Math. Z.* **195** (1987), 89–92. 20

[Hil08] J. A. Hillman, Indecomposable PD_3-complexes. Preprint 2008. arXiv:808.1775 20

[HM69] J. Hempel and D. R. McMillan, Jr., Covering three-manifolds with open cells. *Fund. Math.* **64** (1969), 99–104. 212

[Hod86] C. D. Hodgson, Degeneration and regeneration of hyperbolic structures on three-manifolds. Ph.D. thesis, Princeton University, Princeton 1986. 223

[HS09] G. Huisken and C. Sinestrari, Mean curvature flow with surgeries of two-convex hypersurfaces. *Invent. Math.* **175** (2009), 137–221. 56

[Iva85] N. V. Ivanov, Foundations of the theory of bounded cohomology. *Zap. Nauchn. Sem. Leningrad. Otdel. Mat. Inst. Steklov.* (*LOMI*) **143** (1985), 69–109; English transl. *J. Soviet Math.* **37** (1987), 1090–1115. 181, 183

[Ive93] T. Ivey, Ricci solitons on compact three-manifolds. *Diff. Geom. Appl.* (1993), 301–307. 44, 45

[Jac80] W. Jaco, *Lectures on three-manifold topology*. CBMS Lecture Notes 43, Amer. Math. Soc., Providence, RI, 1980. 209

[JLR02] W. Jaco, D. Letscher, and J. H. Rubinstein, Algorithms for essential surfaces in 3-manifolds. In *Topology and geometry: commemorating SISTAG*, Contemp. Math. 314, Amer. Math. Soc., Providence, RI, 2002, 107–124. 16

[JO84] W. Jaco and U. Oertel, An algorithm to decide if a 3-manifold is a Haken manifold. *Topology* **23** (1984), 195–209. 16

[Joh79] K. Johannson, *Homotopy equivalences of* 3*-manifolds with boundary*. Lecture Notes in Math. 761, Springer, Berlin 1979. 3

[JR89] W. Jaco and J. H. Rubinstein, PL equivariant surgery and invariant decompositions of 3-manifolds. *Adv. Math.* **73** (1989), 149–191. 16

[JS79] W. H. Jaco and P. B. Shalen, *Seifert fibered spaces in* 3*-manifolds*. *Mem. Amer. Math. Soc.* **21** (1979), no. 220. 3

[JT95] W. Jaco and J. L. Tollefson, Algorithms for the complete decomposition of a closed 3-manifold. *Illinois J. Math.* **39** (1995), 358–406. 16

[Kap07] V. Kapovitch, Perelman's stability theorem. In *Surveys in differential geometry*, Vol. XI, Surv. Differ. Geom. 11, Internat. Press, Somerville, MA, 2007, 103–136. 21, 186

[KL08] B. Kleiner and J. Lott, Notes on Perelman's papers. *Geom. Topol.* **12** (2008), 2587–2855. v, 15, 30, 38, 39, 40, 41, 74, 91, 105, 107, 112, 121, 128, 129, 145, 175, 185, 186, 220

[KL10] B. Kleiner and J. Lott, Locally collapsed 3-manifolds. Preprint 2010. arXiv:1005.5106 v, 21, 180, 185

[Kne29] H. Kneser, Geschlossene Flächen in dreidimensionalen Mannigfaltigkeiten. *Jahresber. Deutsch. Math.-Verein.* **38** (1929), 248–260. 3

[Kue03] T. Kuessner, Efficient fundamental cycles of cusped hyperbolic manifolds. *Pacific J. Math.* **211** (2003), 283–313. 181, 182

[Li06] T. Li, An algorithm to find vertical tori in small Seifert fiber spaces. *Comment. Math. Helv.* **81** (2006), 727–753. 16

[Lot07] J. Lott, The work of Grigory Perelman. In *Proc. Internat. Congr. of Mathematicians* (Madrid, 2006), Vol. I, European Math. Soc. Publ. House, Zürich 2007, 66–76. vi

[LT05] P. Lu and G. Tian, Uniqueness of standard solutions in the work of Perelman. Preprint 2005. http://math.berkeley.edu/~lott/ricciflow/StanUniqWork2.pdf 41

[Mai08] S. Maillot, Some applications of Ricci flow to 3-manifolds. In *Actes de séminaire de théorie spectrale et géométrie. Année 2006–2007*, Sémin. Théor. Spectr. Géom. 25, Université Grenoble I, St. Martin d'Hères 2008, 121–148. http://tsg.cedram.org/cedram-bin/article/TSG_2006-2007_25_121_0.pdf vi, 10

[Mai09] S. Maillot, Ricci flow, scalar curvature and the Poincaré conjecture. In *Geometry, topology, quantum field theory and cosmology*, Hermann, Paris 2009, 97–116. vi, 10

[Mai10] S. Maillot, Variétés hyperboliques de petit volume (d'après D. Gabai, R. Meyerhoff, P. Milley, ...). In *Séminaire Bourbaki 2008–2009*, Exp. no. 1011, Astérisque 332, Soc. Math. France, Paris 2010. 13, 22

[Man02] J. Manning, Algorithmic detection and description of hyperbolic structures on closed 3-manifolds with solvable word problem. *Geom. Topol.* **6** (2002), 1–25. 17

[Mar60] A. A. Markov, Insolubility of the problem of homeomorphy. In *Proc. Internat. Congr. of Mathematicians* (Edinburgh, 1958), Cambridge University Press, Cambridge 1960, 300–306. 17

[Mat97] S. V. Matveev, On the recognition problem for Haken 3-manifolds. *Rend. Circ. Mat. Palermo* (2) *Suppl.* **49** (1997), 131–148. 16

[Mat03] S. Matveev, *Algorithmic topology and classification of* 3-*manifolds*. Algorithms Comput. Math. 9, Springer-Verlag, Berlin 2003. 16

[Mil62] J. Milnor, A unique decomposition theorem for 3-manifolds. *Amer. J. Math.* **84** (1962), 1–7. 3

[MM88] M. J. Micallef and J. D. Moore, Minimal two-spheres and the topology of manifolds with positive curvature on totally isotropic two-planes. *Ann. of Math.* (2) **127** (1988), 199–227. 10

[Mor05] J. W. Morgan, Recent progress on the Poincaré conjecture and the classification of 3-manifolds. *Bull. Amer. Math. Soc.* (*N.S.*) **42** (2005), 57–78. vi

[Mor07] J. W. Morgan, The Poincaré Conjecture. *Proc. Internat. Congr. of Mathematicians* (Madrid, 2006), Vol. I, European Math. Soc. Publ. House, Zürich 2007, 713–736. vi

[MT07] J. Morgan and G. Tian, *Ricci flow and the Poincaré conjecture*. Clay Math. Monogr. 3, Amer. Math. Soc., Providence, RI, 2007. v, 36, 41, 43, 45, 74, 92, 144, 220

[MT08] J. Morgan and G. Tian, Completion of the proof of the geometrization conjecture. Preprint 2008. arXiv:0809.4040 v, 21, 180, 185, 186

[Mum61] D. Mumford, The topology of normal singularities of an algebraic surface and a criterion for simplicity. *Inst. Hautes Études Sci. Publ. Math.* **9** (1961), 5–22. 179

[Mye82] R. Myers, Simple knots in compact, orientable 3-manifolds. *Trans. Amer. Math. Soc.* **273** (1982), 75–91. 13, 180

[NS97] W. D. Neumann and G. A. Swarup, Canonical decompositions of 3-manifolds. *Geom. Topol.* **1** (1997), 21–40. 3

[NZ85] W. D. Neumann and D. Zagier, Volumes of hyperbolic three-manifolds. *Topology* **24** (1985), 307–332. 223

[Orl72] P. Orlik, *Seifert manifolds*. Lecture Notes in Math. 291, Springer-Verlag, Berlin 1972. 16, 18

[Ota98] J.-P. Otal, Thurston's hyperbolization of Haken manifolds. In *Surveys in differential geometry* (Cambridge, MA, 1996), Vol. III, Internat. Press, Cambridge, MA, 1998, 77–194. 3

[Ota01] J.-P. Otal, *The hyperbolization theorem for fibered 3-manifolds*. SMF/AMS Texts Monogr. 7, Amer. Math. Soc., Providence, RI, 2001. 3

[OVZ67] P. Orlik, E. Vogt, and H. Zieschang, Zur Topologie gefaserter dreidimensionaler Mannigfaltigkeiten. *Topology* **6** (1967), 49–64. 17

[Per91] G. Perelman, Alexandrov spaces with curvatures bounded from below II. Preprint 1991. http://www.math.psu.edu/petrunin/papers/papers.html 21, 186

[Per02] G. Perelman, The entropy formula for the Ricci flow and its geometric applications. Preprint 2002. arXiv:math/0211159 v, 25, 37, 38, 39, 40, 45, 60, 74, 107, 111, 121, 126

[Per03a] G. Perelman, Finite extinction time for the solutions to the Ricci flow on certain three-manifolds. Preprint 2003. arXiv:math/0307245 v, 9

[Per03b] G. Perelman, Ricci flow with surgery on three-manifolds. Preprint 2003. arXiv:math/0303109 v, 12, 30, 39, 41, 45, 56, 74, 96, 107, 129, 132, 145, 175, 180, 185, 186

[Pet87] S. Peters, Convergence of Riemannian manifolds. *Compositio Math.* **62** (1987), 3–16. 214

[Pré05] J.-P. Préaux, Conjugacy problem in groups of non-oriented geometrizable 3-manifolds. Preprint 2005. arXiv:math/0512484 18

[Pré06] J.-P. Préaux, Conjugacy problem in groups of oriented geometrizable 3-manifolds. *Topology* **45** (2006), 171–208. 18

[Rub95] J. H. Rubinstein, An algorithm to recognize the 3-sphere. In *Proc. Internat. Congr. of Mathematicians* (Zürich, 1994), Vol. I, Birkhäuser, Basel 1995, 601–611. 16

[Rub97] J. H. Rubinstein, Polyhedral minimal surfaces, Heegaard splittings and decision problems for 3-dimensional manifolds. In *Geometric topology* (Athens, GA, 1993), AMS/IP Stud. Adv. Math. 2, Amer. Math. Soc., Providence, RI, 1997, 1–20. 16, 17

[Rub04] J. H. Rubinstein, An algorithm to recognise small Seifert fiber spaces. *Turkish J. Math.* **28** (2004), 75–87. 16, 17

[Sak96] T. Sakai, *Riemannian geometry*. Transl. Math. Monogr. 149, Amer. Math. Soc., Providence, RI, 1996. 77

[Sal09] A. Salgueiro, Orders and actions of branched coverings of hyperbolic links. *Topology Appl.* **156** (2009), no. 9, 1703–1710. 14, 222

[Sco83] P. Scott, There are no fake Seifert fibre spaces with infinite π_1. *Ann. of Math.* (2) **117** (1983), 35–70. 18

[Sei33] H. Seifert, Topologie dreidimensionaler gefaserter Räume. *Acta Math.* **60** (1933), 147–238. 2

[Sel95] Z. Sela, The isomorphism problem for hyperbolic groups I. *Ann. of Math.* (2) **141** (1995), 217–283. 15, 16, 17

[Shi89a] W.-X. Shi, Complete noncompact three-manifolds with nonnegative Ricci curvature. *J. Differential Geom.* **29** (1989), 353–360. 74

[Shi89b] W.-X. Shi, Ricci deformation of the metric on complete noncompact Riemannian manifolds. *J. Differential Geom.* **30** (1989), 303–394. 217

[Sma59] S. Smale, Diffeomorphisms of the 2-sphere. *Proc. Amer. Math. Soc.* **10** (1959), 621–626. 80

[Som81] T. Soma, The Gromov invariant of links. *Invent. Math.* **64** (1981), 445–454. 181, 182

[Sta62] J. R. Stallings, On the recursiveness of sets of presentations of 3-manifold groups. *Fund. Math.* **51** (1962), 191–194. 19

[SY05] T. Shioya and T. Yamaguchi, Volume collapsed three-manifolds with a lower curvature bound. *Math. Ann.* **333** (2005), 131–155. v, 21, 180, 185, 186

[Tao06] T. Tao, Perelman's proof of the Poincaré conjecture: a nonlinear PDE perspective. Preprint 2006. arXiv:math/0610903 vi

[Tho84] C. B. Thomas, Splitting theorems for certain PD^3-groups. *Math. Z.* **186** (1984), 201–209. 20

[Tho94] A. Thompson, Thin position and the recognition problem for S^3. *Math. Res. Lett.* **1** (1994), 613–630.

[Thu80] W. P. Thurston, *The geometry and topology of three-manifolds*. Lecture Notes, Princeton University, Princeton 1980.
http://www.msri.org/communications/books/gt3m 182, 222, 223

[TS31] W. Threlfall and H. Seifert, Topologische Untersuchung der Diskontinuitätsbereiche endlicher Bewegungsgruppen des dreidimensionalen sphärischen Raumes. *Math. Ann.* **104** (1931), 1–70. 2

[TS33] W. Threlfall and H. Seifert, Topologische Untersuchung der Diskontinuitätsbereiche endlicher Bewegungsgruppen des dreidimensionalen sphärischen Raumes (Schluß). *Math. Ann.* **107** (1933), 543–586. 2

[Wal67] F. Waldhausen, Eine Klasse von 3-dimensionalen Mannigfaltigkeiten. I. II. *Invent. Math.* 3 (1967), 308–333; ibid. **4** (1967), 87–117. 179

[Wal68a] F. Waldhausen, On irreducible 3-manifolds which are sufficiently large. *Ann. of Math.* (2) **87** (1968), 56–88. 18

[Wal68b] F. Waldhausen, The word problem in fundamental groups of sufficiently large irreducible 3-manifolds. *Ann. of Math. (2)* **88** (1968), 272–280. 18

[Wal04] C. T. C. Wall, Poincaré duality in dimension 3. In *Proceedings of the Casson Fest*, Geom. Topol. Monogr. 7, Geom. Topol. Publ., Coventry 2004, 1–26. 19

[Yam96] T. Yamaguchi, A convergence theorem in the geometry of Alexandrov spaces. In *Actes de la table ronde de géométrie différentielle* (Luminy, 1992), Sémin. Congr. 1, Soc. Math. France, Paris 1996, 601–642. 21

[Yan82] K. Yano, Gromov invariant and S^1-actions. *J. Fac. Sci. Univ. Tokyo Sect. IA Math.* **29** (1982), 493–501. 182

[Yau06] S.-T. Yau, Structure of three-manifolds – Poincaré and geometrization conjectures. Preprint 2006. arXiv:math/0607821 vi

[Ye08a] R. Ye, On the l-function and the reduced volume of Perelman I. *Trans. Amer. Math. Soc.* **360** (2008), 507–531. 112, 129

[Ye08b] R. Ye, On the l-function and the reduced volume of Perelman II. *Trans. Amer. Math. Soc.* **360** (2008), 533–544. 112, 129

Index